AF530785

AIR POLLUTION

AIR POLLUTION

Henry C. Perkins

Aerospace and Mechanical Engineering Department
University of Arizona

McGRAW-HILL BOOK COMPANY

New York St. Louis San Francisco Düsseldorf Johannesburg
Kuala Lumpur London Mexico Montreal New Delhi Panama
Paris São Paulo Singapore Sydney Tokyo Toronto

Library of Congress Cataloging in Publication Data

Perkins, Henry Crawford.
Air pollution.

Includes bibliographical references.
1. Air-Pollution. I. Title.
TD883.P45 628.5'3 73-22345
ISBN 0-07-049302-2

AIR POLLUTION

1 2 3 4 5 6 7 8 9 0 M A M M 7 9 8 7 6 5 4

This book was set in Century by Scripta Technica, Inc.
The editor was B. J. Clark; the designer was Scripta Technica, Inc.; and the production supervisor was Judi Frey.
The Maple Press, Inc., was printer and binder.

To
Jennifer, William the Poo, and
those to follow.
May the air be cleaner for them.

CONTENTS

PREFACE

This text was begun during a sabbatical year spent at Danmarks Tekniske Højskole in Copenhagen. The warm hospitality of the faculty of the Danish Technical University and most particularly the encouragement and support of Dr. Peder Worsøe-Schmidt were most important in producing this book. The laboratory for Køleteknik was an outstanding place for a year of work free of the usual interruptions which occur on one's home campus.

The text format results from an early decision to provide information in separate chapters on certain specific pollutants, notably particulates, SO_x, and NO_x, and yet to provide some information by also discussing specific sources of emission, for example the automobile. Additional chapters are included on the effects of pollutants, both local and global; on the history of federal legislative efforts in this area; and on necessary background in combustion to understand many of the control techniques for combustion-generated pollutants. A substantial portion of the book deals with meteorology and the dispersion of pollutants from point sources.

The book is written as an introductory *technical* work for students who have some background in chemistry, thermodynamics, and fluid mechanics. It has been tested with senior and masters-level students.

Since the manuscript went to the publisher, a number of important events have occurred in the area of air pollution control. Some of these changes have been included as footnotes to the book. Briefly these are: (a) the Environmental Protection Agency has revoked the national secondary annual average SO_2 standard; (b) an additional year has been granted for auto manufacturers to meet the CO, HC, and NO_x emission standards; (c) the NO_x ambient standards have been heavily criticized because of the questionable accuracy of the ambient level measurements and will, no doubt, be changed through future Congressional action; (d) the intermittent control technique has received EPA approval as a means of reducing ground-level concentrations during adverse meteorological conditions. This technique, which EPA calls the supplementary control system and which is referred to in Arizona as the closed-loop control system, enables industry to temporarily reduce production in order to reduce emissions.

Another very significant event occurred in June 1973, when the Supreme Court upheld a lower court ruling forbidding any substantial degradation of air which is already clean. Thus areas with air cleaner than the level set in the ambient air quality standards may not increase pollution to these higher concentrations. How this court ruling will be interpreted and enforced is not yet clear. The energy crisis is still another important variable in the efforts to achieve clean air. One of the effects of the energy shortage will undoubtedly be to permit limited burning of high sulfur fuel. Thus we may expect to see SO_2 levels rise in some of the urban centers as low sulfur fuel supplies are exhausted. The duration and amount of this increase is difficult to predict but as of this writing one must be pessimistic.

A number of people have aided in the efforts to complete this book. I would like to thank Inge-Lisa Munzert who typed the first draft—in English, Debbie Grimble and Pat Lou who typed the second, and Meta Anderson who did the final. The students at The University of Arizona and the Danish Technical University actively helped by contributing their comments and criticisms. Dr. Jost Wendt's comments were helpful on the chapter for combustion. My wife's encouragement and critical evaluation of style are greatly appreciated. Finally the contribution of the editors is to be commended; without their efforts I would never have known what a caput is . . . .

Henry C. Perkins

AIR POLLUTION

1

AN ENVIRONMENTAL PROBLEM

> What we want to stress is the indivisibility and complexity of the environment. For example the earth's atmosphere is so thoroughly mixed and so rapidly recycled through the biosphere that the next breath you inhale will contain atoms exhaled by Jesus at Gethsemane and Adolf Hitler at Munich. It will also contain atoms of radioactive strontium 90 and iodine 131 from atomic explosions and gases from the chimneys and exhaust pipes of the world.
>
> *The Oxygen Cycle, P. Cloud and A. Gibor,*

In recent years there has been great interest in the environment and many "new" words have become part of our vocabulary, words such as *ecology*, *environment*, *photochemical smog*, and *greenhouse effect*. Simultaneously we have been made aware of environmental problems caused by the high technology created to achieve the material comforts we demand, and to provide us with the military protection we think we need. Among these problems are the effects of air pollution. Air pollution causes increased respiratory illness in the old and the young, decreased visibility, damage to plants and animals, and has possibly catastrophic effects on a global scale. We shall examine some of the causes and effects of air pollution and some methods of control for various air pollutants.

We shall often be frustrated by inability to analyze completely a specific air pollution problem. For example, some scientists suggest the present slight cooling trend in the average global temperature is caused by the buildup of particulates in the atmosphere. Others say that, in fact, the particulates cause a warming effect. No one knows as yet the final answer to this question.

On the other hand we shall see that great progress has been made in understanding the formation of photochemical "Los Angeles-type" smog. Based on the many years of careful research which explained the Los Angeles phenomenon, engineers can now perhaps develop successful ways of coping with it.

Why at this time should we be concerned about air pollution? After all, man has lived centuries in atmospheres permeated by dust, methane from decomposing materials in bogs and swamps, and other hydrocarbon compounds emitted by trees in the forests. Society's concern exists now because urbanization and industrialization have brought together large concentrations of people in small areas. Concentrations of pollutants emitted from many of man's activities thus build up to levels sufficient to have adverse effects on plant, animal, and human health. Pollution concentrations on a global scale are reaching the point of potential serious consequences to the planet itself.

Much of our industrialization is the result of a standard of living which requires a car or two for every family. We use electric power which must be generated to meet a demand doubling every eight to ten years. And we generate immense quantities of solid wastes which are often incinerated or put onto open burning dumps.

Thus urbanization with most of the nation's people living on 1 percent of the land, the high standard of living demanded at a minimal cost in dollars without regard for the environment, and the expansion of industry to provide new products have all combined to cause an increase in concentrations of pollutants to the dangerous level.

MAN'S NEEDS

What are man's needs for air? Table 1.1 shows the amount of air that a man actually inhales per day. The body requires something like 50 lb of air a day to sustain its requirement for oxygen. That amount multiplied by the world population of about 4 billion means that about 200 billion lb of air are breathed every day by mankind.

TABLE 1.1 Man's biological air needs*

	liter/min	liter/day	lb/day	kg/day
Resting	7.4	10,600	26	12
Light work	28	40,400	98.5	45
Heavy work	43	62,000	152	69

*68.5 kg = 151-lb man.

We can estimate the quantity of food a man consumes per day at about 1.5 kg (about 3 lb). Thus man takes in about 15 to 20 times the amount of air as food. This explains why we require pollutant concentrations in air to be an order of magnitude or more lower than concentrations we allow in food. For air we desire annual average levels to be less than a few parts per hundred million (pphm) for pollutants such as NO_2, SO_2, and hydrocarbons; we allow 50 pphm (0.5 ppm) of mercury in fish and shellfish. We do not often regard air as a necessary requirement just because it is so available, but try getting along without it!

Consider the combustion of coal, which is largely carbon, as a further example of our use of air. We have the chemical reaction $C + O_2 \rightarrow CO_2$. Now let us take a 2,200-megawatt, multiple-unit plant, much like some now being built in the United States where 500 to 750 megawatt units are common. Such a plant will use about 21,000 tons per day of coal. To obtain an order of magnitude of the air required for the plant assume all of this coal is carbon. Since 1 lb-mole of C has 12 lb of C we will burn $(21{,}000 \times 2{,}000)/12 = 1{,}750 \times 2{,}000$ lb-mole of C. Each carbon mole requires a mole of oxygen at 32 lb so we consume for stoichiometric complete combustion $1{,}750 \times 2{,}000 \times 32 = 112 \times 10^6$ lb of O_2 per day. The nitrogen going through the plant will be at the rate of 3.76 mole of N_2 per mole of oxygen, at 28 lb per lb-mole. Or 366×10^6 lb of N_2 will go through the combustion chambers daily. We obviously consume immense quantities of oxygen in such combustion processes and give even larger quantities of nitrogen an opportunity to become oxides of nitrogen. Fortunately, only a very small percentage choose to do so.

Consider a car, or better yet, the 4×10^6 cars in Los Angeles County. For complete combustion about 15 lb of air per pound of fuel is required and the average United States car uses about 12 lb (2 gallons) of gas per day. Thus each car requires 180 lb of air per day. For 4 million cars this is $4 \times 10^6 \times 180 = 720{,}000{,}000$ lb of air per day! We are truly dependent on our supply of air.

AIR POLLUTION DEFINED

What is air pollution? The Engineers' Joint Council in "Air Pollution and Its Control" gives the following definition:

> Air pollution means the presence in the outdoor atmosphere of one or more contaminants, such as dust, fumes, gas, mist, odor, smoke, or vapor in quantities, of characteristics, and of duration, such as to be injurious to human, plant, or animal life or to property, or which unreasonably interferes with the comfortable enjoyment of life and property.

This is a broad definition and one referred to frequently in writing some of the legal statutes. Consider the definition of air pollution as quoted from the State of Arizona air pollution control regulations:

> 7-1-1.2; 3. Air pollution means the presence in the outdoor atmosphere of one or more air contaminants or combination thereof in such quantities and of such duration as are *or may tend to be* injurious to human, plant, or animal life, or property. [author's italics]
>
> 2. Air contaminants include smoke, vapors, charred paper, dust, soot, grime, carbon fumes, gases, mist, odors, particulate matter, radioactive materials, or noxious chemicals, or any other material in the outdoor atmosphere.

Note that the Arizona regulations require only that the contaminants "tend to be injurious." However, there is no aesthetic clause in these rules such as "unreasonably interferes with the comfortable enjoyment of life and property." Of course, whether the courts could determine what is an "unreasonable interference" is open to question. One man's business may be another man's "unreasonable interference" and the question of which is which evolves with time.

Other necessary definitions are given below, corresponding to those of the American Society for Testing and Materials in "Standards on Methods of Atmospheric Sampling and Analysis." We shall make use of these terms in the remainder of the text. *Particulate* is a general term meaning "existing in the form of minute separate particles" (AP-49, 1969), either solid or liquid. Particulate is used interchangeably with aerosol.

Aerosol: A dispersion of solid or liquid particles of microscopic size in gaseous media, such as smoke, fog, or mist.

Dust: A term loosely applied to solid particles predominantly larger than colloidal, and capable of temporary suspension in air or other gases. Dusts do not tend to flocculate except under electrostatic forces; they do not diffuse, but settle under the influence of gravity. Derivation from larger masses through the application of physical force is usually implied.

Droplet: A small liquid particle of such size and density as to fall under still conditions, but which may remain suspended under turbulent conditions.

Fly Ash: The finely divided particles of ash entrained in flue gases arising from the combustion of fuel. The particles of ash may contain incompletely burned fuel. The term has been applied principally to the gas-borne ash from boilers with spreader stoker, underfeed stoker, and pulverized fuel (coal) firing.

Fog: A loose term applied to visible aerosols in which the dispersed phase is liquid. Formation by condensation is usually implied. In meteorology, a dispersion of water or ice.

Fume: Properly, the solid particles generated by condensation from the gaseous state, generally after volatilization from melted substances, and often accompanied by a chemical reaction such as oxidation. Fumes flocculate and sometimes coalesce. The term is popularly used in reference to any or all types of contaminants, and in many laws or regulations, with the qualification that the contaminant have some unwanted action.

Gas: One of the three states of aggregation of matter, having neither independent shape nor volume, and tending to expand indefinitely.

Mist: A term loosely applied to low-concentration dispersions of liquid particles of large size. In meteorology, a light dispersion of water droplets of sufficient size to be falling.

Particle: A small, discrete mass of solid or liquid matter.

Smoke: Finely divided aerosol particles resulting from incomplete combustion. Consists mainly of carbon and other combustible material.

Soot: Agglomerations of particles of carbon impregnated with "tar," formed in the incomplete combustion of carbonaceous material.

Vapor: The gaseous phase of matter which normally exists in a liquid or solid state.

HISTORY

> Be it known to all within the sound of my voice, whosoever shall be found guilty of burning coal shall suffer the loss of his head.
>
> *King Edward I, ca. 1300*

Is air pollution a recent phenomenon, or have we only recently noted it? Some forms of pollution, such as photochemical smog, are certainly a new phenomenon. This form requires a concentration of sunshine and cars in a meteorologically stable region. The number of cars required has only lately been reached in urban centers. Smoke pollution, however, has been with man for centuries. During his reign, King Edward I of England prohibited the burning of coal in London during sessions of Parliament, owing to the smoke and odor produced. Because of extensive wood burning the English forests were soon reduced, and coal consumption increased in spite of royal disapproval. In 1661, John Evelyn wrote his "Fumifugium or the Inconvenience of Aer and Smoak of London Dissipated, together with some Remedies Humbly Proposed." A new edition was published in 1772, so one can see that the solutions proposed by Evelyn in the 1600s were not adopted. They involved moving industry from the city and planting greenbelts around the city, both of which plans are still used as "solutions." It is interesting to follow British air pollution history somewhat further. In 1819, a select committee was appointed to study the problem, in 1843 a second select committee was appointed, in 1845 a third, etc. Only after the 1952 smog disaster, which we shall be discussing at length later, did England finally take decisive measures.

When an air pollution problem exists, why isn't something done? We can turn to history for a little enlightenment here. Halliday (1961) quotes an anonymous writer in a British journal of 1889, *Builder*:

> He had pointed out that (a) smoke was a by-product of an activity which commanded the attention and the support of all financial interests in the country because it produced goods and profits, and therefore the willingness to give thought to the smoke aspect of the activity was very limited; (b) the damage done by smoke, though very considerable throughout the country, was not very clearly visible to the individual smoke producer, because it was widely spread, so that interest in the reduction of smoke was lacking; and (c) since the damage done by smoke was due to a very large number of small producers of smoke, a clear relationship between cause and effect was difficult to establish.

These words should sound familiar though written over 80 years ago. In short, smoking smokestacks meant jobs, and besides it was really the other fellow anyhow. Britain's air pollution problem continued to worsen. The Smoke Abatement League heard Dr. Des-Voeux talk about smoke-fog deaths in 1905 and 1911, and it is likely that the word, *smog*, was coined from his work.

Britain, however, never monopolized the air pollution scene for the United States had many cities plagued with problems of smoke and sulfur dioxide. Reports were filed for Chicago in 1910, St. Louis in 1907, and Pittsburgh in 1912. The United States has been spared a disaster of the magnitude of the London disasters only because meteorological conditions were not the same, and because oil and natural gas were substituted for coal as the fuel for space heating and electric power production in many large cities. The air pollution problem has been around for awhile. Only now are we aware and informed enough to be goaded into action. Table 1.2 shows dates of significance in this history.

PRESENT EMISSIONS

A sense of the magnitude of the present problem can be seen from Table 1.3, which lists the preliminary 1970 estimated United States emissions in *millions of tons* per year (Council on Environmental Quality (CEQ), 1972). Excluding CO, we find that over 116×10^6 tons of pollutants were emitted into the air over the United States. This is

TABLE 1.2 Significant dates in the history of air pollution

	Ca. 1300	1578	1661	1772
	Edward I proclamation prohibiting use of sea coal during sessions of Parliament	Q. Elizabeth I annoyed by coal smoke and complains to Parliament	John Evelyn's book	Second edition of Evelyn's book
	1819	1843	1845	1866
	1st English Select Committee	2d Select Committee	Parliament passes law requiring locomotives to consume own smoke!	1st paper on effects on health due to air pollution
	1875	1905-1911	1919	1930
	Cattle deaths at London shown due to air pollution episode	Dr. Des-Voeux uses term "smoke-fog" deaths, i.e., smog	1st paper on auto air pollution (CO effects)	Meuse Valley, Belgium, disaster
Los Angeles smog	1948	1952	1955	1956
	Donora, Pa., disaster	London disaster	1st U.S. federal legislation	British Clean Air Act
	1963	1966	1970	1975
	U.S. Clean Air Act	East Coast Thanksgiving episode	U.S. Amended Clean Air Act	U.S. primary standards to be reached

more than 1,000 lb per person each year. If we include CO, we roughly double these figures. Naturally these emissions are not uniformly dispersed. Concentrations are high in urban centers, or in rural areas with particular industrial complexes, such as oil refineries or smelters. While we have seen that there is a history of air pollution in industrial countries, it is the increasing concentrations from emissions in urban areas which give us urgent cause for concern. The increased emissions for the United States are noted in Table 1.4. This shows that, with the

TABLE 1.3 Estimated emissions of air pollutants by weight, nationwide, 1970 (preliminary data)

Source	CO	Particulates	SO_x	HC	NO_x
	In millions of tons per year				
Transportation	111.0	0.7	1.0	19.5	11.7
Fuel combustion in stationary sources	.8	6.8	26.5	.6	10.0
Industrial processes	11.4	13.1	6.0	5.5	.2
Solid waste disposal	7.2	1.4	.1	2.0	.4
Miscellaneous	16.8	3.4	.3	7.1	.4
Total	147.2	25.4	33.9	34.7	22.7
Percent change 1969–70	−4.5	−7.4	0	0	+4.5

Source: Environmental Protection Agency.

exception of particulates all of the major pollutants have increased significantly over the past 30 years.

Are all cities alike in their pollution problems? The answer is, of course, no. Different cities have different emitted pollutants and also different meteorological conditions. Figure 1.1, from Hafstad (1969), describes the pollution of several cities. Chicago suffers from an overabundance of SO_2, while Los Angeles complains of photochemical smog. Air quality thus depends on a number of variables including the local meteorological conditions and the emissions.

Weather is also an important variable. Pollutants affect weather and weather affects pollutants. Rain and snow will wash out the air and deposit the pollutants on soil and water. At the same time high concentrations of particulates will affect the likelihood of rain or snow. Temperature inversions put a lid over the atmosphere so that emissions

TABLE 1.4 Weight of emissions of air pollutants, 1940–1970 (tons $\times 10^6$)

Year	SO_x	CO	Particulates	HC	NO_x
1940	22	85	27	19	7
1950	24	103	26	26	10
1960	23	128	25	32	14
1968	31	150	26	35	21
1969	34	154	27	35	22
1970	34	147	25	35	23

Source: Environmental Protection Agency.

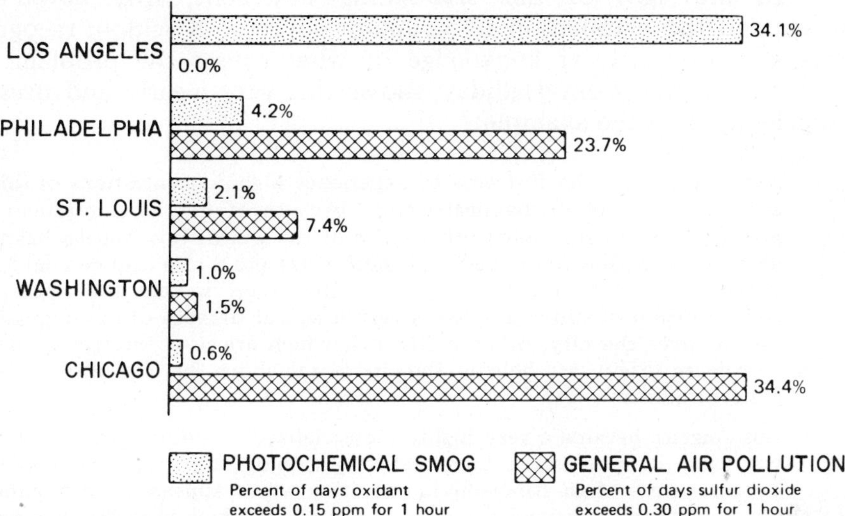

Fig. 1.1 Types of air pollution in various cities in 1965 and 1966. (*Hafstad, 1969.*)

occur in relatively small volumes with correspondingly high concentrations. Thus the East Coast of the United States had the Thanksgiving, 1966 "episode" of air pollution during a period of temperature inversions, with a warm air mass at a height of about 300 meters preventing vertical mixing.

Los Angeles has the combination of a very stable temperature inversion and strong solar input. As we shall see, the sun provides the necessary energy to initiate a complex set of chemical reactions among the pollutants emitted, primarily from cars. The result is the production of chemicals which make us cry, cause plant damage, and, on occasion, reduce visibility to practically nothing. The inversion prevents vertical mixing to dilute these concentrations, and light winds limit the horizontal introduction of fresh air. It is evident that the meteorology and climate of an area determine whether an air pollution problem will exist locally and, if so, what kind of problem it will be.

ATTEMPTS TO FIND SOLUTIONS

> The Public School System in Los Angeles County agreed that when the oxidant reached a concentration of 0.35 ppm, they would recommend that the school children not be allowed to engage in strenuous physical activity because of possible risk to health.
>
> *Atkisson and Gaines, p. 5*

In the light of this astounding statement, why haven't we demanded the elimination of pollutant sources? A seldom recognized reason is our lack of knowledge of what causes the problem. The following quote from Halliday shows this very clearly and uses Los Angeles again as the example:

> Los Angeles was the first area to experience high concentrations of this new pollutant (i.e., photochemical smog), because of two circumstances more marked there than in most other cities of the world. Los Angeles has a poor natural ventilation (on a meteorological scale) and it also uses no coal but has a very high consumption of petrol. The result is that the atmospheric concentration of sulfur dioxide is very low, but because of the stagnation of the air over the city, other pollutants, which are now thought to be of a petroleum origin, are held in the air-volume above the city until quite high concentrations result.
>
> Los Angeles became a very highly industrialized area during the recent world war, and its population increased very rapidly. By 1944 the inhabitants became aware of an atmospheric condition which appeared and disappeared, but sometimes remained for days at a time, in which visibility was reduced markedly by a light blue haze, and many persons suffered from sore throat, running nose and eyes, and varying degrees of headache. Air pollution officials from the eastern towns of the USA came to study the situation and decided that the cause was sulfur dioxide, which can produce some of the above symptoms, though usually not all of them. A programme for the reduction of sulfur dioxide emission by various industries, including the oil refineries, did not produce the expected reduction in the smog, as it was called, so further investigation was made and it was discovered that petrol vapour (hydrocarbons) in combination with perhaps nitrogen dioxide or some other reagent present in the air, and under the influence of the energy of sunlight, produced some organic compound which has still not been defined with certainty but which is thought to be a strong oxidizer. A substance which was made from hydrocarbons and nitrogen dioxide, irradiated by ultraviolet light under experimental conditions, was shown to possess the lachrymatory properties of Los Angeles smog, and so the smog control agency set out to prevent the escape of petrol vapours from storage tanks of the oil refineries. This activity still did not have the required effect, so further thought was given to the subject and it was realized that the very high number of two and a half million motor cars in Los Angeles used five million gallons of petrol a day and, because of the inefficiency of the automobile carburetor, emitted over a thousand tons of hydrocarbons to the atmosphere daily.

Thus action was taken which did not solve the problem. In the case of initial efforts to control hydrocarbon emissions from automobiles in the Los Angeles basin the smog was actually worsened. The oxides of nitrogen were increased by the control techniques used to reduce hydrocarbon emissions, notably increasing the air/fuel ratio to make the combustion mixture fuel lean.

In the case of factory-caused pollution, though the cause of the problem may be known, the technology to deal with it may be lacking. One cannot simply close down the offending plant. It is very unlikely that people will be willing to do without steel, copper, electricity, etc., in the future. As an example of the lack of available technology, consider the emission of SO_2 from coal-burning power plants. The reduction in SO_2 can be achieved by using a fuel which is either sulfur free, such as natural gas, or lower in sulfur, such as certain oil and coal sources, or by putting in SO_2 removal devices. What if no proven removal devices are commercially available, as in fact is the present case, and what if low sulfur fuels cost significantly more? The consumer pays either for the more expensive fuel (if it is available), or for the development of cleaning devices which may take several years to perfect. The point in either case is that technology is not always on the shelf ready to use.

A still further obstacle to a solution lies in our bureaucratic legal system. Until very recently, we had not set up effective agencies to deal with environmental problems, and it remains to be seen how successful the new Environmental Protection Agency (EPA) will be at the federal level in alleviating air pollution. In fact, the present laws leave the enforcement aspects to the states, and even the 1970 amendments to the Air Quality Act of 1967, while establishing a time table for setting up federal standards or goals (which were set up in 1971 on schedule), leaves the means of reaching these goals and the enforcement mechanisms to the states. (A more complete discussion of this aspect of air pollution is given in Esposito, 1970.) The weakness inherent in state enforcement is that state officials are more vulnerable to economic and political pressures from local industry than are the more distant federal officials. In favor of state enforcement is the necessity created for each state to enforce a cleanup appropriate to its particular air pollution problems.

We have deliberately left for last the economic and political reasons why air pollution control has not been achieved. It may be complained that some industry will not clean up, and suggested that the reason is economically and/or politically motivated. In fact, the engineers hired by the company may not have found a workable solution. Or their solution is not commercially available and will have to be developed, as is the case with the copper smelter emission of SO_2 in Arizona. The public has little appreciation of the time required for research and development, even with a sufficient input of money.

It is equally true that industry has all too often been led kicking and screaming into the environmental decade, as the following quotations illustrate:

> Our basic value assumption for air quality management is that some balance must be reached between improving the quality of the air and the cost of doing so.
>
> *G. Schwartz and G. Siegel* in *Atkisson and Gaines*

> We must conclude that it is impossible to signify a lower limit for any single pollutant which will be 'safe' for all persons.
>
> *Higgens and McCarrol, ibid.*

Thus the public is led to make an economic tradeoff between the cost of improving the air quality and an unknown on the other side of the equation. The "unknown" is the effect on health and welfare of not improving the air quality. The cost of cleaning up can be quantified, at least for the plant manager who must purchase equipment. Since the cost of not cleaning up is nebulous, it can be seen why most industries have remained environmentally unconcerned until recently.

It appears that only federal standards and federal enforcement can provide the economic environment in which no industry can gain advantage over another by failing to clean up. (Even this solution ignores the question of international competition where a dirty industry in country A may be able to undersell a clean one in country B.)

The past inability of most industries to see even a short time into the future and to be willing to invest in cleanup devices is now costing them both public relations' dollars and crash program dollars for "instant" cleanup. Haagen-Smit's article, citing the role of the automobile in producing photochemical smog, was originally published in June, 1952. It is difficult to be too sympathetic to the auto companies who must meet federal standards in 1976!

GLOBAL EFFECTS

Both the global effects of air pollution and the effects on plant and animal life will be considered in greater detail later in the text. Some introductory material, however, is appropriate here.

Consider the energy balance for the earth-atmosphere system as noted in Fig. 1.2. Of 100 units of incoming solar radiation, 35 units are reflected from the clouds, particulates and molecules, and the surface of the earth. Seventeen units are absorbed in the atmosphere: three by the ozone in the stratosphere, thus producing a warm region some 20 to 25 kilometers above the earth, two by CO_2 and particulates, and 12 by water vapor. The remaining 48 units are absorbed at the surface of the earth. The earth radiates back to space but this must also traverse the atmosphere. Some 114 units are radiated in the infrared at longer

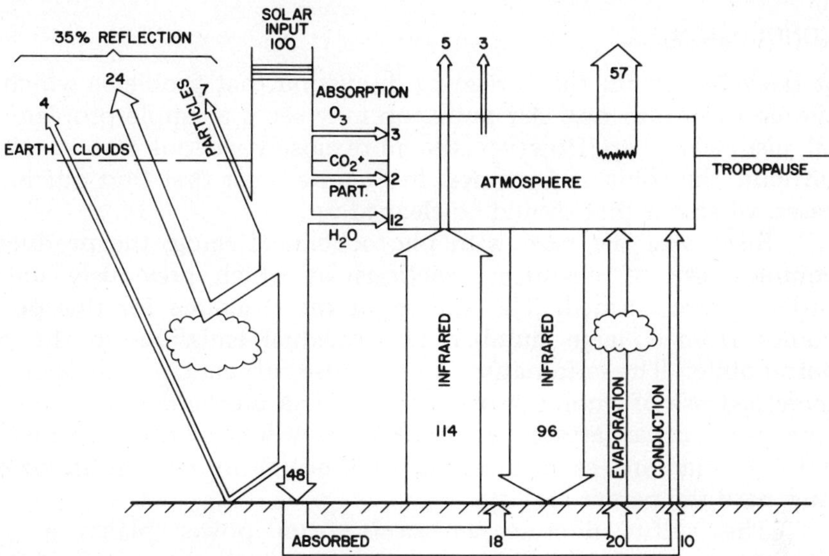

Fig. 1.2 Radiation-energy balance.

wavelengths than the solar input. Ninety-six units are, however, radiated back to the earth from the atmosphere so that 18 of our absorbed 48 go to make up the difference in the radiation-energy balance. This difference results because the average earth temperature is a little higher than the average atmospheric temperature. The remaining 30 units go back to the atmosphere through evaporation and conduction. Engineers will probably prefer to call this convection rather than conduction. Note that, of the 114 units radiated back from the earth, only about five go directly through the atmosphere; the rest are absorbed and re-emitted on the way out.

The crucial elements in this picture are the points at which man can disturb the balance, the so-called "leverage points." For example, by increasing the CO_2 in the atmosphere we increase the amount of radiation which is reradiated back to the earth, i.e., in the 96 units. By increasing the particulate load in the atmosphere we affect both the 96 units reradiated and the seven units of scattered incoming radiation. By increasing the energy dissipated as heat in mankind's industrial activities, we increase the earth-average surface temperature and affect the evaporation-condensation water cycle, the potential size of the polar ice caps, and therefore, the water level of the oceans. Man's ability to act at these points in this model means that small changes produced by emissions of pollutants may have large effects. We shall examine these effects in the next chapter.

CONCLUSION

It is evident from the foregoing discussion that problems which seem simple often are not. Air pollution may seem a simple problem—after all, just clean it up. However, the more closely we look into it, the more difficult the solution becomes. In fact we must first find out in many cases, what it is that should be cleaned up!

Such was the case with photochemical smog, the product of a complex set of chemical reactions in which previously unknown products are formed. The source of the reactants for this pollution comes from a large number of individual emissions in the private automobile. The alternative to the use of cars is not merely the increased use of public transit (if that is an alternative in many of our spread-out urban areas), but a whole new way of life. The love affair between man and car is a serious one, not easily broken up no matter how hard the parent may try.

The sulfur dioxide emissions from power plants are single concentrated sources and can be reduced by using alternative fuels. But our domestic supply of one such fuel, natural gas, is only about 12–13 years at the present rate of consumption. We can gasify coal but the resulting product is more expensive than burning the coal itself. Who pays and how much? Thus the simple air pollution problem becomes a complex one of energy supply and demand, human psychology, and man's life style, as well as "just" economics, politics, and technology.

In the following chapters we shall discuss sources of air pollution, the transport of the pollutants from the sources to the receptors and the resulting effects, control techniques for individual pollutants, and information on the present federal program for air quality and emission standards. While most of this material will pertain to the United States, the reader should be aware that air pollution problems exist throughout the industrialized world. Milan, Tokyo, Santiago, the Ruhr Valley, Mexico City, Tehran, and other cities and areas suffer from the problem of air pollution.

REFERENCES

"Air Quality Criteria for Particulate Matter," National Air Pollution Control Administration Publication AP-49, 1969.

Atkisson, A., and R. S. Gaines (eds.): "Development of Air Quality Standards," Merrill, Columbus, Ohio, 1970.

Council on Environmental Quality, Third Annual Report, 1972.

Esposito, J. C.: "Vanishing Air" (Nader report on air pollution), Grossman, New York, 1970.

Haagen-Smit, A. J.: Chemistry and Physiology of Los Angeles Smog, *Ind. Eng. Chem.*, vol. 44, pp. 1342-1346, 1952.

Hafstad, L. R.: Automobiles and Air Pollution, *Argonne Conference on Universities, National Laboratories, and Man's Environment, 1969.*

Halliday, E. C.: A Historical Review of Air Pollution, in "Air Pollution," World Health Organization, Geneva, 1961.

PROBLEMS

1.1 Calculate the total mass of air in the atmosphere if the mean world-surface pressure is 760 mm Hg (14.7 lb/in.^{2}). If the total world-particulate load is 10^7 tons, what is the concentration of particulate, assuming uniform dispersion? The height of the aerosol layer may be taken as 1 kilometer. For comparison, rural suspended particulate matter is about 15–30 μg/m^3.

1.2 Estimate the amount of oxygen existing in the atmosphere if air is 21 percent oxygen.

1.3 If a power plant produces 1,000 megawatts of electricity and is 35 percent efficient, how much coal does it require? How much waste heat is generated at the plant and how much oxygen is consumed? Assume that coal is 75 percent carbon and produces 13,000 Btu/lb.

1.4 The sun can be treated as a blackbody at 6,000°K and the energy received at the surface of our atmosphere from the sun is about 2 cal/cm^2-min. Calculate the energy intercepted by the earth atmosphere system per day in ergs, joules, and Btus. How much of this energy is actually absorbed at the earth's surface? (The distance between the sun and the earth is 93 million miles.)

1.5 If the average earth-surface temperature is 288°K with our present solar input, and we have an increase in energy to be radiated from the earth of 10 percent, how much must the temperature of the earth rise (neglecting increased water evaporation, etc.)?

1.6 The present generating capacity (1972) of the United States is about 400,000 megawatts. If the time for power consumption to double is about nine years, and each new plant is to produce 1,500 megawatts, then, by 1980, how many new sites will be needed? How much coal will be required per year for the new plants, assuming that half of them burn exclusively coal? How much SO_2 will be produced from these new plants if the coal averages 2 percent sulfur? Compare this figure with the 1968 emissions.

1.7 Find out the definition of air pollution in your location. Is this definition sufficient to cover the present problems? What agency has jurisdiction in matters of air pollution and what other duties are they also responsible for?

1.8 Plot the population growth and the registration figures for motor vehicles in your state (or country) for the past few decades. Now plot the ratio of cars per capita for the last few decades. What conclusions can you draw?

1.9 In 1945, an estimated 2.7 million tons of fuel, having an average combustible sulfur content of 1 percent, was burned in Copenhagen. Tests during that period measured a sulfate (SO_4) fallout of 1,750 mg/m^2 per month. At that time also the city was approximately a square of 12 km to a side. Calculate the sulfur fallout per year. What happened to the rest of the sulfur?

1.10 1945 measurements in greater Copenhagen of "total solid" fallout yielded a figure of 12.27 g/m^2 per month. If the city was a 12 km by 12 km square, how much particulate fell on the city in that year?

1.11 Denmark has about 1.4×10^6 cars and each one uses about 4 liters per day of "benzine." Carbon monoxide is known to be about 2-3 percent of the exhaust products on the average. How much CO is produced per year in Denmark? Pb is used in gasoline at about 0.5 g/l. How much lead is put into the atmosphere if three-quarters of that in the gasoline is exhausted? (It is estimated that, by the year 2000, there will be 2.5 times as many cars as in 1970.)

1.12 Look up the gross national product as a function of time and plot this for the last 25 years (i.e., since 1945). Electric power usage in the United States has been:

Year	Population	kwhr/capita per year
1900	76×10^6	49
1910	92	220
1920	106	540
1930	123	944
1940	132	1,379
1945	132	2,067
1950	151	2,582
1955	164	3,853
1960	180	4,718
1965	194	5,969
1970	205	8,100
(1980)	236	14,000
(2000)	285	31,000

Compare the GNP to the electric power use. What do these figures mean in terms of air pollution emissions?

1.13 Determine air pollution emissions for your state for particulate, SO_x, and CO. Compare the density of pollution for these three substances on the basis of kg/km^2 and kg/person. How do these densities compare to figures for the United States as a whole? Is this a valid way to compare?

1.14 One suggestion to clean up the air in the Los Angeles basin has been to pump out air to the East so that clean ocean air can come in from the West. Investigate the feasibility of this suggestion.

1.15 Consider a power plant operating with two 500-megawatt units and burning coal with an ash content of 10 percent. If the plant operates over a 35-year period, what is the size of the ash pile? Assume 10,500 Btu/lb m of coal.

1.16 Electric energy conservation: make an inventory of electric use in your home for a weekday and a weekend day. Estimate the waste use, e.g., lights on and no one in the room, T.V. on, etc. Would you include self-defrost refrigerators, dishwashers, etc., as waste use? What percent of your electric use is waste?

1.17 Coal produces 10,500 Btu/lb. A 1,000-megawatt power plant is to be built burning this coal. (*a*) If coal costs $6 per ton at the power plant, calculate the cost of this fuel per year, and for the 30-year life of the plant. Calculate the cost per 10^6 Btu. (*b*) If 10 percent of the coal is ash, how big is the ash pile after 30 years? (*c*) If 2 percent of the coal is S, how many pounds of SO_2 are produced per year?

2

GLOBAL EFFECTS OF AIR POLLUTION

> **Thus far a major disruption in the oceans and atmosphere is not apparent. We may, however, be looking at subtle changes in the environment such as the accumulation of lead and other toxic chemicals, an increase in carbon dioxide and organic carbon compounds, atmospheric dust, and acid rainfall which may have far reaching consequences in the years to come. The equilibrium existing between physical, chemical, geological, and biological forces is a subject for serious study on a global basis.**
>
> *F. D. Sisler, in Singer, 1970a*

THE ORIGINS OF THE ATMOSPHERE

To ask how it all began means to look back several billion years. What can be gleaned from geologic evidence as to how the atmosphere was formed?

There is agreement that the earth developed without an atmosphere. Evidence points to the fact that in time the atmosphere formed from gases evolving from the interior of the earth. In particular water vapor and carbon dioxide were released and subsequently formed both the earth's oceans and large quantities of limestone, i.e., calcium carbonate. Some oxygen was formed both through the decomposition of water vapor (through photodissociation with the subsequent loss of hydrogen to space), and from photosynthesis of the limited plant life then existing. Eventually enough oxygen had accumulated to begin formation of the ozone layer (ozone is O_3) in the upper atmosphere. This layer is crucial to life since it acts as an ultraviolet filter, absorbing much of the ultraviolet from the solar radiation before it reaches the surface of the earth.

Once a level of oxygen had developed that was sufficient to give an upper atmosphere ozone layer, life could expand rapidly both on land and in the surface water of the oceans. Photosynthesis then could increase the oxygen produced. The plant cycle is such that CO_2 is used by the plants and O_2 is released. However, when the plant dies the reverse mechanism occurs and oxidation of the plant removes the O_2 with a release of CO_2. At present, only about 1 part in 10^4 of the products of photosynthesis escapes decay and can provide us with the oxygen in our atmosphere (Johnson, in Singer, 1970*a*).

Nitrogen in the atmosphere appears to have evolved from the earth's interior and to have accumulated to its present level. The important point is to appreciate that the evolution of the atmosphere, which may seem quite simple, took place over some 3 to 5 billion years. Only in the last tens of million years has the world oxygen and nitrogen level been about what it is today. The large coal deposits that we are presently using for energy give evidence of possibly previous accelerated plant growth rates. If so, this indicates levels of CO_2, for photosynthesis, greater than those today. A fluctuating oxygen level may also have been the case some hundred million years ago (Cloud and Gibor, 1970).

What is clear is the complexity and number of multiple interactions abundant in the environment. Most of these are not well understood. Nature's secrets are not easily revealed and one is reminded of the story of the recent history of physics. Each new generation of physicists in its search for knowledge has opened a door at the end of a room and peered through to find another room full of new knowledge and understanding. Elated, they have swept the room clean, only to find another door leading to another new room, and so on and on.

In our present atmosphere, which we have seen has only recently evolved on a geologic time scale, the lowest region is called the *troposphere*. It extends from the earth's surface up through a region of generally decreasing temperature to the *tropopause*, which is the boundary between the troposphere and the stratosphere. The tropopause is a function of latitude and decreases in altitude with increasing latitude. At the midlatitudes it is located at an altitude of about 12 kilometers (7.5 miles).

Since the troposphere is the closest region to the surface of the earth it is the region of storms. Consequently, pollutants in this region can be washed out of the air by rain or snowfall. The *stratosphere* is a region, because of ozone absorption, of nearly constant, or of increasing temperature with increasing altitude, and extends upward to about 50 kilometers (31 miles). Pollutants entering this region tend to remain a long time since the nearly constant temperature distribution with height produces a very stable atmosphere, as we shall discuss in Chap. 7. Beyond the stratosphere is the *stratopause* which separates the stratosphere from the mesosphere. The change of temperature with height provides us with a good means to establish these atmospheric regions.

We shall examine in the remainder of this chapter the possible global effects of air pollution. To do so we must first consider some background about the temperature of the earth-atmosphere system. The earth is powered by the sun through the thermal radiation which we receive. Some 440 Btu/hr-ft^2 (2 cal/cm^2-min) are received outside of the earth's atmosphere and this value is known as the solar constant. After this interacts with the atmosphere, about 300 Btu/hr-ft^2 are available to reach the surface of the earth at the latitudes of the United States. These facts may be familiar to us, but what is not so generally known is the function of the atmosphere and the oceans to redistribute this incoming solar energy throughout the different latitudes (Oort, 1970). Any interference with these transfer media affects the distribution of energy. Furthermore, interference can, of course, affect the direct input by means such as increased scattering of the incoming radiation, or absorption of the outgoing radiation.

Of many of these possible interferences you have undoubtedly heard. We shall find, however, that there are now no *exact* answers as to the magnitude of these effects, and thus one cannot say with certainty that this or that will or will not happen. Figure 2.1 from Singer (1970*b*) shows the energy cycle for the earth. The air pollutants noted on the figure include heat, CO, SO_x, HC, NO_x, and particulates. The CO_2 release from the processes of combustion is also noted.

Most of our sources of air pollution are directly related to our need for energy. This need is expressed in electric power use, transportation requirements including the individual automobile, space heating, and waste disposal. Our material standard of living, therefore, is directly related to our use of energy. We shall see in this chapter some of the potential difficulties that this fact creates.

Since we seem able or willing to respond to problems only after they reach the crisis stage, our lack of knowledge in this area of global effects of air pollution is most unfortunate. The crisis stage for global effects of all forms of environmental pollution may well come too late to effect a change in a process which at that point may be irreversible. Since we cannot yet offer clear-cut answers to these problems, some may brush them off, hoping the forecasters of doom "may be wrong." And, in fact, the dire predictions of some will undoubtedly be wrong—one certainly hopes so. But let us consider whether some of the predictions have even a degree of validity.

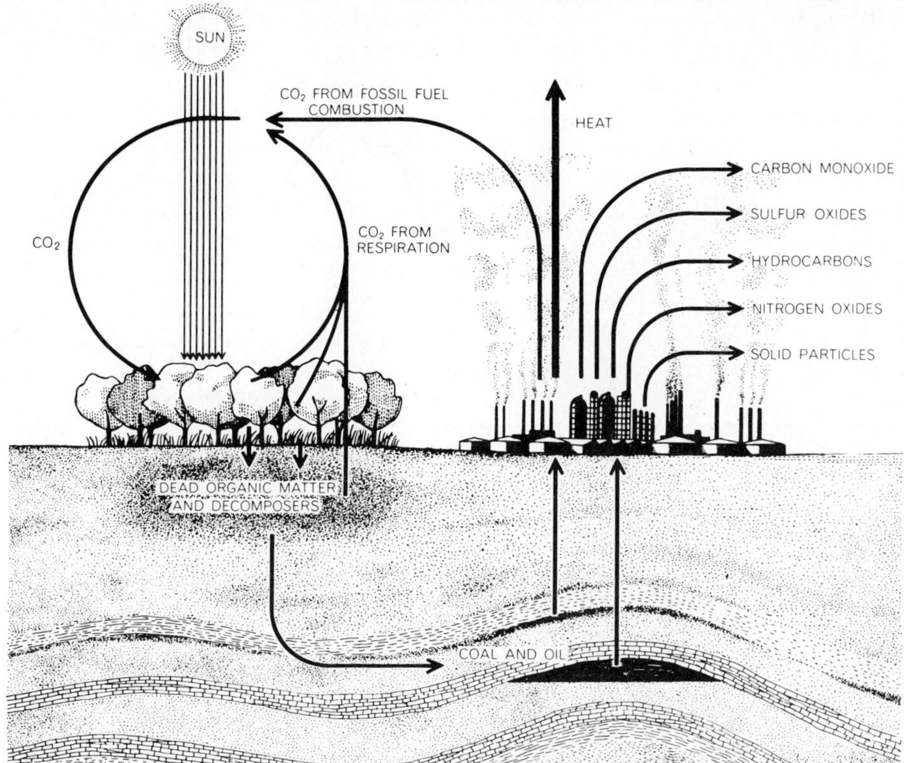

Fig. 2.1 Energy cycle involved in the combustion of fossil fuels begins with solar energy employed in photosynthesis millions of years ago. A small fraction of the plants is buried under conditions that prevent complete oxidation. The material undergoes chemical changes that transform it into coal, oil and other fuels. When they are burned to release their stored energy, only part of the energy goes into useful work. Much of the energy is returned to the atmosphere as heat, together with such byproducts of combustion as carbon dioxide and water vapor. Other emissions in fossil fuel combustion are listed at right in the relative order of their volume. (*Reprinted from Human Energy Production as a Process in the Biosphere by S. Fred Singer. Copyright © September 1970 by Scientific American, Inc. All rights reserved.*)

We can define three major global effects from air pollutants produced by both man and nature. These three are the so-called *greenhouse effect*, owing to CO_2, the effect of particulates on the earth-atmosphere heat balance, and the effect of climate change from man's massive use and dissipation of energy. We shall examine these in turn.

First, however, we need to consider recent trends in the average earth-surface temperature. What has world climate been in the twentieth century? Figure 2.2 presents the Mitchell's results (Singer,

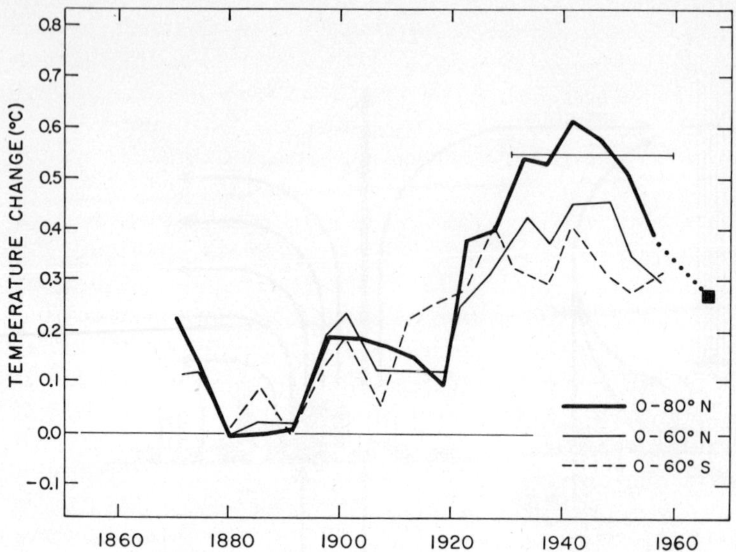

Fig. 2.2 Trends of hemispheric mean annual temperature for various latitude bands, 1870–1960. (*Mitchell, in Singer, 1970a.*)

1970*a*). These data show a net warming of about 0.6°C (1°F) from 1880 to the 1940 decade, and a net cooling since then of about 0.3°C (0.5°F). Is this a significant effect? As Mitchell points out, the magnitude of a 0.6°C fluctuation is about 10 percent of the 6°C change that distinguished the glacial from the interglacial conditions during the Pleistocene Ice Age. Sellers' (1969) calculations indicate that a decrease in the solar constant of 5 percent would be conclusively sufficient to start another ice age. This corresponds to an earth-surface temperature change of about 4°C.[1] He further indicates that a change of about 2 percent would increase the ice caps. This magnitude of change in the solar constant corresponds to a temperature change of only 2°C. Thus it is clear that we should give serious consideration to environmental effects that cause global changes on the order of 0.5°C.

What effects might bear consideration?

1. Does the CO_2 buildup cause a sufficient rise in the average global surface temperature to concern us?

[1] Remembering that thermal radiation is proportional to the fourth power of the absolute temperature, we can calculate the temperature change as follows, $Q_1/Q_2 = (T_1/T_2)^4$ where $T_1 \cong 288°K$, the average surface temperature of the earth. If the solar input drops by 5 percent then we have $Q_2 = 0.95Q_1$ so $1.053 = (T_1/T_2)^4$ or $T_2 \cong 288/1.013 \cong 284°K$.

2. Does the particulate buildup cause a sufficient decrease in the average global surface temperature to concern us, assuming that the effect of particulates is, in fact, to decrease the global temperature?
3. Can certain positive feedback mechanisms be activated which would cause drastic climate changes?
   (*a*) Ice has a high albedo (reflectivity) for solar radiation. If a temperature rise occurs such that some of the ice caps melt, will a sufficient decrease in the reflectivity occur so that the increased solar absorption may cause a continued heating trend, and thus a continued reduction of the ice caps?
   (*b*) The opposite positive feedback effect can also be postulated with an increased ice cap layer causing increased reflection, or thus decreased absorption, and a subsequent cooling with increased ice formation.
   (*c*) Another feedback mechanism of concern is the CO_2 balance between the oceans and the atmosphere. The CO_2 solubility is temperature dependent, with CO_2 less soluble at higher temperatures. Consequently, an initial CO_2 atmospheric buildup which might cause a rise in the earth-surface temperature could, in turn, cause a release of CO_2 from the oceans as their temperatures increase, with a further corresponding rise in surface temperature, etc.
   (*d*) Plant life requires carbon dioxide in the photosynthesis process. It is possible that the phytoplankton of the sea would grow faster with an increase in CO_2. Might this lead to a decrease in the CO_2 atmospheric content (for a relatively short time), and therefore decrease the average surface temperature? If so, an increase in CO_2 solubility in the ocean (temperature up, solubility down, and vice versa) would result, and reduce the CO_2 level and temperature even more.

Of course, it may well be that the carbon dioxide and particulate temperature changes will just cancel out. Or it may be that the carbon dioxide available to stimulate phytoplankton growth, using organic wastes deposited in the seas by man, will cause a sufficient release of O_2 (via photosynthesis) to account for that used up in the combustion of fossil fuels. We do not know!

None of the above considerations should suggest that tomorrow we will find ice sheets in the southern United States or all the coastal cities of the world inundated with a rising ocean level. They are meant to suggest that there is a great deal we do not yet understand about the world around us. The time to find out if something is likely to occur (we can never tell exactly what will happen) is before we cause changes in the environment which will create changes on a global scale in the atmosphere.

CARBON DIOXIDE

Carbon dioxide at standard temperature and pressure is a colorless, odorless gas. It can be solidified as the familiar "dry ice." Our interest here in CO_2 exists because it is one of the gases that absorbs thermal radiation. In order to understand the effects CO_2 has on the earth-energy balance, we must briefly discuss thermal radiation and introduce some commonly used terms.

It is convenient to assume that the sun radiates as a blackbody at a single temperature, so that we may make simple calculations concerning the total solar energy output and the spectral distribution from the sun. A solar temperature based on Wien's law ($\lambda_{max}T = 5{,}215\ \mu^\circ R$) is 6,080°K (10,900°R) while one based on the total radiation from the sun is 5,770°K (10,400°R). For the purpose of discussion we shall assume that the sun radiates as blackbody at a temperature of 6,000°K. Thus, the wavelength of maximum output from the sun is given by Wien's law as

$$\lambda_{max} T = 5{,}215 \mu^\circ R$$

so $$\lambda_{max} \cong 0.5\mu$$

The visible radiation to which the eye responds is from the violet at 0.38μ to the red at 0.76μ (3,800 Å to 7,600 Å) so that the maximum solar output is in the visible range. Solar radiation is therefore concentrated as shown in Fig. 2.3 in the ultraviolet, visible, and near infrared wavelength ranges. In contrast, a lower temperature body at, say 255°K (i.e., the effective radiative blackbody temperature of the earth-atmosphere system), would have its maximum radiation at about 11μ. Consequently, the radiation from the surface of the earth is concentrated at the longer wavelengths in the region of infrared radiation. This difference in solar and terrestrial radiation wavelengths is of crucial importance to the energy balance of the earth-atmosphere system.

Carbon dioxide and water vapor are absorbers of radiation but primarily in the infrared, not in the visible. Consequently, they pass solar radiation without much interference but they absorb and reradiate much of the terrestrial radiation emitted from the earth. In particular,

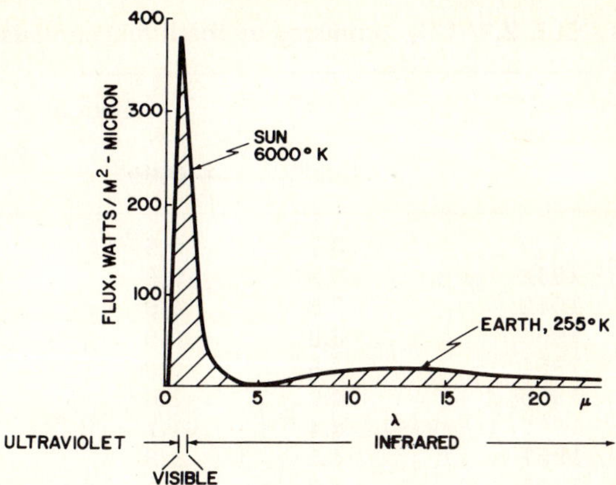

Fig. 2.3 Thermal radiation to and from the earth.

CO_2 has radiation absorption bands about 2.7 microns, 4.3 microns, and in the 12- to 18-micron wavelength range.

Returning now to the effect of CO_2, we know that carbon dioxide is the inevitable result of the combustion process. Consider methane, CH_4, which is the largest component in natural gas. When we burn methane in air we have the chemical reaction for *complete combustion* as

$$\underbrace{CH_4 + 2O_2 + 2(3.76)N_2}_{\text{Reactants}} \rightarrow \underbrace{2(3.76)N_2 + CO_2 + 2H_2O}_{\text{Products}}$$

The burning of 1 pound of CH_4 in air produces about 21,000 Btu of thermal energy in the form of heat and this, of course, is why we burn any fuel—to produce heat. We see also that each mole of methane burned produces a mole of carbon dioxide. Likewise, burning coal in a power plant produces CO_2 as one of the products of combustion. In short, the amount of CO_2 produced by combustion per year is truly prodigious. It is estimated at 9×10^{15} g/yr (Goldberg, in Singer 1970*a*), or 9×10^9 metric tons/yr. The SCEP[2] report estimates a modestly higher figure and its results are shown in Table 2.1. Note, however, that even these amounts are less than *natural* yearly fluctuations caused by the plant cycle. Here plants take up CO_2 in the spring and summer and decay to release CO_2 in the fall and winter. This cycle is evident in Fig. 2.4. If we have incomplete combustion, part of the combustion products are CO (carbon monoxide) and unburned, or partly burned, hydrocarbon compounds from the fuel. Also some of

[2] Study of Critical Environmental Problems.

TABLE 2.1 CO_2 produced by fossil fuel combustion, 1950-1967

Year	Billions of metric tons				
	Coal[a]	Lignite[b]	Refined oil fuels[c]	Natural gas[d]	Total
1950	3.7	0.9	1.4	0.4	6.4
1951	3.8	0.9	1.7	0.5	6.9
1952	3.8	0.9	1.8	0.5	7.0
1953	3.8	0.9	1.9	0.5	7.1
1954	3.8	0.9	2.0	0.6	7.3
1955	4.1	1.0	2.2	0.6	7.9
1956	4.4	1.1	2.4	0.7	8.6
1957	4.5	1.3	2.5	0.7	9.0
1958	4.6	1.4	2.6	0.8	9.4
1959	4.8	1.4	2.8	0.9	9.9
1960	5.0	1.4	3.1	1.0	10.5
1961	4.5	1.5	3.3	1.0	10.3
1962	4.6	1.5	3.5	1.1	10.7
1963	4.8	1.6	3.8	1.2	11.4
1964	5.0	1.7	4.2	1.3	12.2
1965	5.0	1.7	4.5	1.5	12.7
1966	5.1	1.7	4.8	1.6	13.2
1967	4.8	1.7	5.2	1.7	13.4
1980 (est.)[e]	11.1[f]		10.8	4.0	26.0

[a] Assumed carbon content, coal = 75 percent.
[b] Assumed carbon content, lignite = 45 percent.
[c] Assumed carbon content, refined oil fuels = 86 percent.
[d] Assumed carbon content, natural gas = 70 percent.
[e] The 1980 estimate was constructed by multiplying the 1965 emissions for coal (including lignite), refined oil fuels, and natural gas X growth factors (i.e., ratios of 1980 consumption to 1965 consumption for solid and liquid fuels and natural gas)
[f] Coal and lignite combined.
Source: Reprinted from SCEP/Man's Impact on the Global Environment by permission of The M.I.T. Press, Cambridge, Massachusetts. Copyright 1970.

the nitrogen may enter the reaction if the temperatures are sufficiently high, resulting in the production of nitric oxide, NO, and possibly nitrogen dioxide, NO_2. All of these chemical species are undesirable in the atmosphere, as will be discussed later.

What happens to the CO_2? About half of it remains in the atmosphere causing a gradual buildup. About half is absorbed in the oceans or taken up in the biosphere. The oceans now contain about 60 times as much carbon dioxide as the atmosphere. The gradual buildup of CO_2 is shown in Figs. 2.4 and 2.5. Since carbon dioxide is an absorber of radiation in the infrared, particularly in the 12–18 micron range, it acts to put a lid on radiation from the earth to space. Incoming

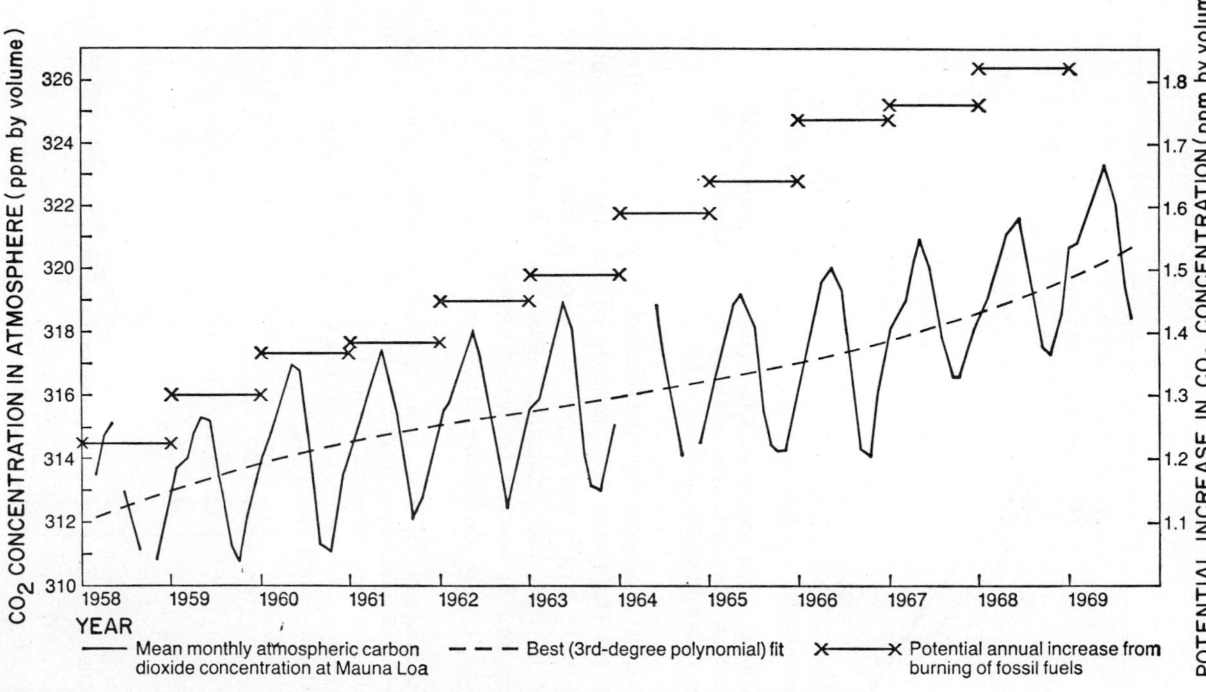

Fig. 2.4 CO_2 concentration from burning fossil fuels. (*Reprinted from SCEP/Man's Impact on the Global Environment by permission of The M.I.T. Press, Cambridge, Massachusetts. Copyright 1970.*)

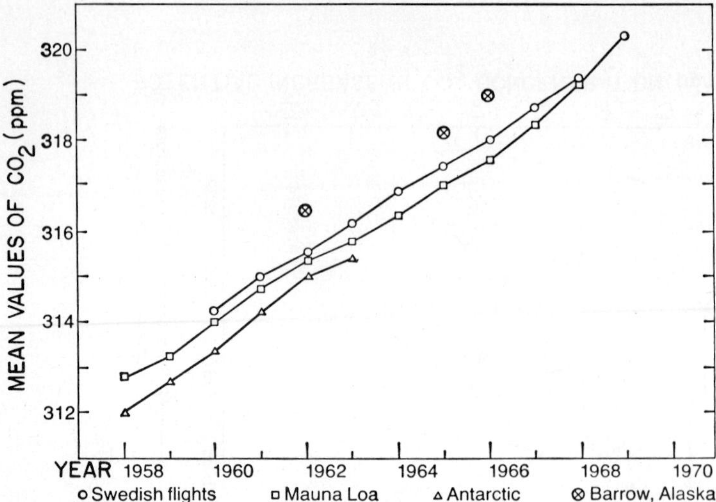

Fig. 2.5 Annual mean values of CO_2. (*Reprinted from SCEP/ Man's Impact on the Global Environment by permission of The M.I.T. Press, Cambridge, Massachusetts. Copyright 1970.*)

solar radiation is essentially unaffected.[3] Thus we have absorption and reemission (at the lower temperature of the upper atmosphere) by the carbon dioxide and water vapor as noted in Fig. 2.6. In point of fact the CO_2 is not the dominant gas for absorption in the lower atmosphere but is second to water vapor. In the stratosphere carbon dioxide and ozone dominate. However, in both regions, as the CO_2

[3] This is the origin of the term, *greenhouse effect*, since the glass in a greenhouse is also transparent to short wavelength solar radiation. But like CO_2, the glass absorbs the long wavelength radiation emitted from inside the greenhouse. However, the glass also reduces convective cooling of the plants by the outside air and this is the dominant effect.

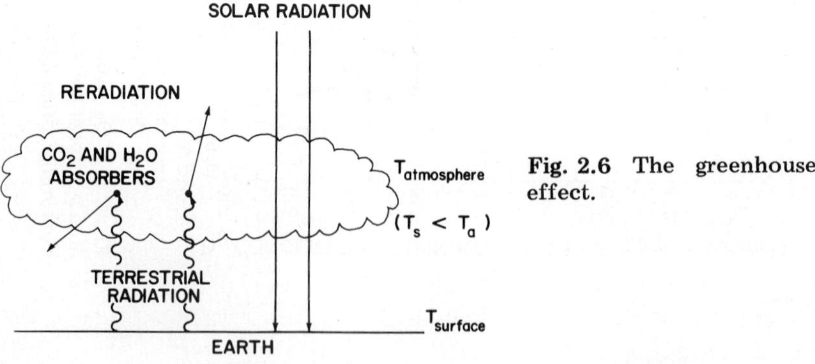

Fig. 2.6 The greenhouse effect.

TABLE 2.2 Change of equilibrium temperature of the earth's surface in °C, corresponding to various changes of CO_2 content of the atmosphere

Change of CO_2 content (ppm)	Fixed absolute humidity		Fixed relative humidity	
	Average cloudiness	Clear	Average cloudiness	Clear
300→150	−1.25	−1.30	−2.28	−2.80
300→600	+1.33	+1.36	+2.36	2.92

Source: Manabe in Singer, 1970*a*.

builds up, absorption increases and it is more difficult for terrestrial radiation to get out into space.

What is the magnitude of this effect? Manabe (Singer, 1970*a*) shows calculations which included the absorption owing to H_2O, O_3, and CO_2, as well as the effects of clouds. His results are shown in Table 2.2. He concludes that with the predicted CO_2 increase to the year 2000 (i.e., 375–385 ppm), the resulting increase in surface temperature would be about 0.5°C. At the present time these results cannot include effects of latitudinal distribution or exchange effects between the ocean and the atmosphere. Consequently, the above numbers are only estimates based on our present ability to model such difficult interactions.

More recently, Rasool and Schneider (1971) have done a set of CO_2 and particulate calculations which indicate both a lesser absolute temperature rise, and more importantly, a *decrease in the rate of change* with increasing CO_2 concentration. Their results are shown in Fig. 2.7.

The important conclusion from their work is not so much the absolute change in temperature, but rather the fact that the rate of temperature increase diminishes with increasing carbon dioxide in the atmosphere. Consequently, the buildup of CO_2 does not cause a runaway temperature rise. They attribute this to the fact that the CO_2 15μ-absorption band in effect saturates, and the addition of more CO_2 does not have a large effect on the infrared opacity of the atmosphere.

What can we now conclude about CO_2 causing an increase in the ambient atmospheric temperature? Mitchell (Singer, 1970*a*) points out that we can only attribute about one-third of the noted worldwide warming trend from 1880 to 1940 to the increase in CO_2. The remaining two-thirds must have been caused by other effects. Thus, even

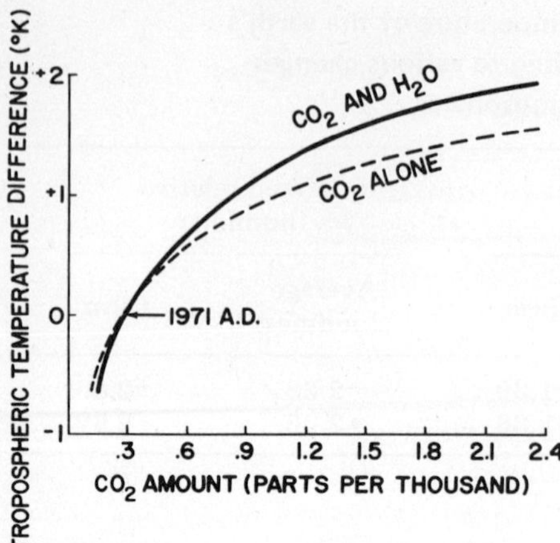

Fig. 2.7 Change in tropospheric temperature as a function of the amount of CO_2 in the atmosphere. The dashed curve is computed for constant surface *absolute* humidity, and the solid curve is for the case in which surface *relative* humidity is maintained constant. Note that the rate of temperature increase diminishes with increasing CO_2 in the atmosphere. (*Reprinted from S. I. Rasool and S. H. Schneider, Science, Vol. 173, pp. 138–141, 9 July 1971. Copyright 1971 by the American Association for the Advancement of Science.*)

though CO_2 may have had only a small effect to date, it appears that the CO_2 buildup coming in the next century may be of sufficient magnitude to cause major concern (even if Rasool and Schneider are correct). This is true even considering the possible effects of particulates.

AEROSOLS

Aerosols are finely divided liquid droplets or solid particles. It is important to recognize that liquid droplets can be considered particles and may act as such in the atmosphere. Phase transformations may occur to produce these aerosols as, for example, when sulfur dioxide gas interacts with moist air to produce droplets of sulfuric acid. Or they may be directly ejected into the atmosphere as solid particulate.

What are the sources of the particles? First, there are large sources of *natural* aerosols, both land and marine sources; secondly, there are man-made aerosols from industry. According to the SCEP (1970) report these sources are:

1. Natural continental sources:
   (*a*) From dust storms and desert areas with particle sizes above 0.3μ radius
   (*b*) From photochemical gas reactions between naturally produced ozone and hydrocarbons resulting in small particles of less than 0.2μ radius
   (*c*) From volcanic eruptions which emit both particulate and gases, in particular SO_2, which can subsequently become an aerosol in the stratosphere

2. Natural oceanic aerosols:
   From the evaporation of oceanic salt spray leaving particulate matter above 0.3μ radius
3. Man-made aerosols:
   (*a*) From the solid particles formed in combustion, i.e., "smoke"
   (*b*) From photochemical reaction between the oxides of nitrogen and the hydrocarbons (smog reactions) which produce small particles of less than 0.2μ radius

In terms of man-made particulates how much is released? Environmental Protection Agency (EPA) figures for the United States (AP-73, 1970) show the particulates to be about 28×10^6 tons per year. Table 2.3 gives these results, including figures for the years 1966, 1967, and 1968. A more detailed breakdown of the sources of particulates is given in the chapter on pollution sources. Note that these figures are estimates based on the best data available, but nonetheless they are approximate. World particulate figures are estimated by

TABLE 2.3 Nationwide emissions of particulates by year

Source	1966	1967	1968	Change from 1966 to 1968
		10^6 tons		
Transportation	1.2	1.1	1.2	N*
Motor vehicles	0.7	0.7	0.8	+0.1
Other	0.5	0.4	0.4	−0.1
Fuel combustion	9.2	8.9	8.9	−0.3
Coal	8.5	8.2	8.2	−0.3
Fuel oil	0.3	0.3	0.3	N
Natural gas	0.1	0.2	0.2	+0.1
Wood	0.3	0.2	0.2	−0.1
Industrial processes	7.6	7.3	7.5	−0.1†
Solid-waste disposal	1.0	1.1	1.1	+0.1
Miscellaneous	9.6	9.6	9.6	N
Man made	2.9	2.9	2.9	N
Forest fires	6.7	6.7	6.7	N
Total	28.6	28.0	28.3	−0.3

*N = Negligible.
†Apparent change.

Goldberg (Singer, 1970*a*) as 2×10^7 metric tons (1 metric ton = 1000 Kg ≅ 2,200 lb) per year for smoke particles; the United States rate is given as 1×10^7 metric tons per year. Since his figures are for smoke particles only (does this include auto exhaust particulate for example?) it is difficult to compare this estimate with the EPA figures.

In any event it is clear that man is putting a lot of particulate into the air. The effect of airborne dust caused by clearing of desert land areas, drought over farmlands, and recent plowing of fields is not even included in these figures. Yet these are also man-made sources, and in many parts of the Midwest and southwestern United States these may be the dominant sources.

Despite all these man-made sources, the production of particulate by natural means is greater. Unfortunately, our knowledge of the magnitudes of both natural and unnatural sources is limited. The SMIC[4] (1971) report estimates that between 5 and 45 percent of all the particulate matter in the atmosphere is produced by man. This wide range reflects our uncertainty about many sources, such as particles formed from natural gaseous emissions.

How long do particulates stay in the air? This time is called the *residence time* and typical values are given in Table 2.4. Note that the product of particulate injection and residence time gives the loading in the atmosphere; that is, an injection rate of 10^7 tons/yr of some material, times a residence time of 0.1 yr, gives a loading of 10^6 tons in the atmosphere.

Just what do all these particles (or aerosols) do? Aerosols will scatter and absorb both solar and terrestrial radiation. Thus they affect the heat balance of the earth. Furthermore they act as nuclei for the condensation of water vapor and thus play a role in cloud formation. They can also act as freezing nuclei. The effect of pollutant particles on rain and snow has been discussed by Schaefer (Singer, 1970*a*) who points out that the massive injection of particulates, from concentrations of autos and industry in large cities, may be initiating and controlling precipitation. This is in contrast to our usual view that these are passively cleaned out of the air by naturally created rain and snow.

Most of the scattering of solar radiation is in the forward direction but a significant proportion, about 10 percent, is backward (SCEP). The effect of backscatter is to increase the reflectivity of the atmosphere. However, the aerosols also reduce the far infrared terrestrial radiation to space and absorb some solar radiation. Which

[4] Study of Man's Impact on Climate.

TABLE 2.4 Aerosol residence times

	SCEP (1970)	Newell (1971)	Removal mechanism
Lower troposphere	6–14 days	6–10 days	rain, gravity fall out for particles $> 10\mu$
Upper troposphere	2–4 weeks	30 days	rain
Lower stratosphere	6 months to 1 year at the equator	4 months	– – –
Upper stratosphere	3–5 years	1 year to 3 years*	– – –
Mesosphere	5–10 years	– – –	– – –

*Residence time is a function of latitude and is greater near the equator.

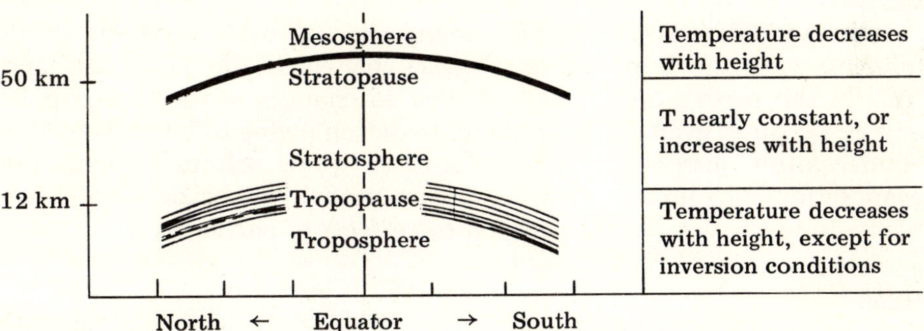

effect is dominant depends on the size, shape, and composition of the particles. Calculations mentioned in the SCEP study indicate that the net effect *could* be one of *warming*.

Rasool and Schneider's calculations, however, indicate a cooling effect. They conclude that the effect of aerosols on visible, i.e., solar, radiation is greater than on the infrared radiation from the earth to space. Thus the net effect is one of cooling of the earth's surface. They find that the rate of temperature decrease is *increased* with increasing aerosol content, such that an increase by a factor of 4 in the global aerosol background *may be* sufficient to decrease the surface temperature by 3.5°K. They conclude, on the basis of Sellers' work (1969), that if this temperature decrease is sustained for some years there will be a change sufficient to trigger an ice age. The SMIC (1971) report also indicates that the net effect of an increase in particulate is to cool the earth. These results, though, depend on many assumptions required to formulate a workable model and point up the need for more information on the properties of particles. Mitchell (Singer, 1970*a*) discusses the effect of particulates, and while agreeing that they act to reduce the earth's surface temperature, he attributes most of the present atmospheric loading to natural sources, in particular of volcanic origin.

Unfortunately, we can conclude little that is definite. In fact, whether particulates have a net cooling effect is still not clear. They do affect the earth's reflectivity, cloud formation, and precipitation, the solar radiation reaching the earth, and that radiation reaching space from the terrestrial infrared radiation. But after accounting for all these individual effects and their interactions, the net result is not clear. Rasool and Schneider's paper (1971) indicates a very significant cooling effect; their conclusion, however, will have to be independently verified.

In addition, a temperature change at the surface of the earth interacts with the troposphere to cause changes in the cloudiness, and thus in the earth's reflectivity, as well as changes in the transport of energy by water evaporation and condensation owing to latent leat. The condensation may occur at a location quite different from the evaporation. So, until a more complete model can be developed, we really do not have a clear answer on the effects of particulates.

HEAT

Is energy released by man in the form of heat properly called a type of air pollution? Yes, for this heat-energy release causes a significant climate change in our cities and may, in the future result in global climate effects. Cities are warmer than their rural surroundings because of energy dissipation and the thermal capacitance of streets and buildings for solar input. Lees, writing in the SCEP report, notes that for 1970 the Los Angeles basin generated heat equivalent to about 6 percent of the solar energy absorbed at the ground. He estimates that the figure will rise to 18 percent by the year 2000!

Energy release occurs both through the evaporation of water, particularly in the generation of electric power, and in the direct heating of the air. As a result, cities are usually warmer, rainier, and foggier than their nonurban surroundings. (Could the combined effects of waste heat and high particulate concentrations have a synergistic[5]

[5] Synergistic $1 + 1 > 2$; anergistic $1 + 1 < 2$.

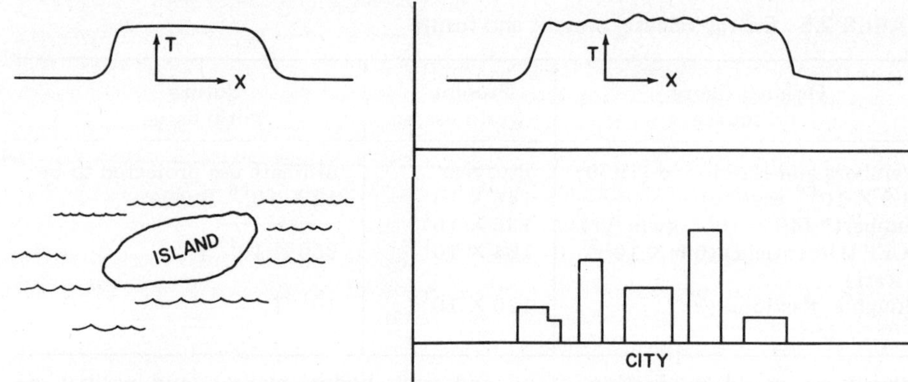

Fig. 2.8 The "heat island" effect.

effect?) Hanna and Swisher (1971) note that in industrial areas the artificial production of energy in watt/m^2 can approach 1,000 (their classification is "super energy center or city") while the solar absorption by the *atmosphere* averages out to only 25 watt/m^2. Thus the city becomes a "heat island" (Fig. 2.8).

What are the effects of this large-scale energy release? We do not know exactly on either a *mesoscale* (1 to 50 kilometers) or on a global scale. As an example, however, Hanna and Swisher's review article notes that city surface temperatures may be as much as 20°F higher than rural areas when the wind is calm. Recently, Erhlich and Holdren (1971) commented that:

> If the present global rate of increase in energy consumption—approximately 5 percent per annum—should persist for another century and a half, man's dissipation then would be equal to 10 percent of the solar energy absorbed over the entire surface of the globe, or one-third of the solar energy absorbed over land. Simple calculations suggest a corresponding mean global temperature increase of about 13°F. Not all climatological authorities agree that thermal input can be translated into temperature change with any such degree of precision. However, all accept the inevitability of climatological and ecological disruption by the growing thermal burden.

Are these authors correct? We can estimate man's present release of energy as heat according to several different references. These are shown in Table 2.5. The rapidly increasing trend in energy production from world oil, coal, and gas production is noted in figures from Table 2.6, taken from the SCEP study. These figures can be used to verify the data of Table 2.5. Coal produces about 13,000 Btu/lb,

TABLE 2.5 Energy release, present and future

Original figure and source	Present world usage	Future world usage
Weinberg and Hammond (1970) (4.9×10^{19} erg/sec)	Btu/year 146×10^{15}	ultimate use projected to be 12×10^{18} Btu/yr
Hubbert* (40×10^{12} kwhr/yr)	136×10^{15}	—————
SCEP (Greenfield) (5.5×10^{12} watt)	164×10^{15}	950×10^{15}‡
Gough & Eastlund (1971) (0.17 Q)†	170×10^{15}	10^{18}‡

*Based on world production of oil and coal; hydro, nuclear, and natural gas sources are not included.
†Q = 10^{18} Btu/yr.
‡In the year 2000.

TABLE 2.6 World production of fossil fuels, 1950–1967

Year	Millions of metric tons			
	Coal	Lignite*	Refined oil fuels†	Natural gas‡
1950	1,340	530	445	155
1951	1,375	550	530	180
1952	1,375	550	585	200
1953	1,380	555	605	210
1954	1,375	550	635	220
1955	1,500	630	705	240
1956	1,595	665	770	260
1957	1,625	765	795	285
1958	1,665	825	830	305
1959	1,730	845	900	345
1960	1,810	875	970	375
1961	1,625	905	1,030	405
1962	1,675	905	1,115	440
1963	1,740	965	1,210	480
1964	1,800	1,005	1,315	525
1965	1,815	1,030	1,410	565
1966	1,845	1,050	1,525	610
1967	1,750	1,040	1,630	655

*After 1962, lignite production figures were given in metric tons coal equivalent. For 1960, 1961, and 1962 (the only years for which there is overlapping data), the apparent conversion factor is 1 metric ton lignite = 0.44 metric ton coal equivalent. This same factor was used in the later data.
†Includes natural gasoline.
‡Assumed density of 8×10^{-4} g cm^{-3} (1,000 m^3 = 0.8 metric ton).
Source: United Nations, *World Energy Supplies.* (Reprinted from SCEP/Man's Impact on the Global Environment by permission of The M.I.T. Press, Cambridge, Massachusetts. Copyright 1970.)

natural gas about 20,000 Btu/lb, and oil about 19,000 Btu/lb. Thus for 1967 we find:

$$\begin{aligned}
&\text{coal } (1{,}750 + 1{,}040) \times 13{,}000 \times 2{,}200 \times 10^6 &&= 80 \times 10^{15} \text{ Btu/yr}\\
&\text{gas } (655) \times 20{,}000 \times 2{,}200 \times 10^6 &&= 29 \times 10^{15} \text{ Btu/yr}\\
&\text{oil } (1{,}630) \times 19{,}000 \times 2{,}200 \times 10^6 &&= 68 \times 10^{15} \text{ Btu/yr}\\
&\qquad\qquad\qquad\qquad\qquad\text{Total} && \ 177 \times 10^{15} \text{ Btu/yr}
\end{aligned}$$

The total figure is in substantial agreement with those given in Table 2.5. The solar input is 17.3×10^{16} watts (Hubbert, 1971) of which about 35 percent is reflected back to space. The input to the earth-atmosphere system is thus about 34.8×10^{20} Btu/yr.[6] If we assume with Ehrlich and Holdren a 5 percent increase compounded yearly, then we find a doubling time of about 14 years, or an increase by a factor of 11.4 in 50 years. Thus in 150 years the increase in energy dissipation is about a factor of 1,500!

If the present usage is taken from Table 2.5 as 170×10^{15} Btu/yr then we find that the fraction of present dissipation to solar input is very small:

$$\frac{170 \times 10^{15}}{340 \times 10^{19}} \cong 0.00005\ (0.005\%)$$

However, if the 5 percent yearly increase were to continue, then in 150 years we would have:

$$\frac{1500 \times 170 \times 10^{15}}{340 \times 10^{19}} = 0.075\ (7.5\%)$$

which is indeed close to the previously quoted figure. Of course, we can continue this exercise and ask what will happen in 200 years at a 5 percent increase per annum. We find the thermal energy dissipation is about equal to the solar input! The conclusion is clear: man cannot continue to increase his energy use at the present rate. Weinberg and Hammond (1970), in fact, predict a leveling off at about 4×10^{21} erg/sec corresponding to a world usage of about 0.5 percent of the solar

[6] The solar constant is about 440 Btu/hr-ft^2 (2 cal/cm^2-min) at the outside of the atmosphere. The projected area of the earth is 1.39×10^{15} ft^2 so the input is 53.6×10^{20} Btu/yr with 35 percent reflected. This leaves 34.8×10^{20} Btu/yr input in agreement with Hubbert's figure.

input. The percentage based on the solar input to the continental land areas is about 1.5 percent. Since the energy dissipation is likely to occur on, or close to, the continents this is a more meaningful figure.

What does an energy dissipation equal to 10 percent of the solar input mean in terms of temperature rise? Sellers (1969) has performed calculations in which a thermal rejection of 50 kly/yr[7] over the continental areas is spread among latitude belts corresponding to the distribution of large cities. His figure corresponds to about 5 percent of the solar input. Even including a correction for the decrease in solar input because of a possible increase in air pollution, he finds a large temperature rise of about 13°C on the average over the whole globe, ranging from 10°C near the equator to 25°C at the North Pole. As Sellers points out, the thermal inertia of the oceans would cause such a rise to take place slowly (perhaps over several hundred years) after the increased heat rejection. As a crude first check on these calculations and on the Ehrlich and Holdren prediction, consider that thermal radiation is proportional to the fourth power of the absolute temperature. For our present situation the average effective black-body radiation temperature of the earth-atmosphere system is 255°K.[8] Then for a heat rejection equal to 10 percent of the solar input we have

$$\frac{Q_1}{Q_2} = \left[\frac{T_1}{T_2}\right]^4$$

$$(1.10)^{1/4} = \frac{T_1}{255}$$

so $$T_1 = (1.024)\,(255) = 261°K$$

or a rise of 6°K (11°R) in agreement with Ehrlich and Holdren. Sellers' more sophisticated calculations, allowing for energy transport by oceanic motion and by evaporation and condensation, predict an even higher figure.

It must follow that, for the long term, it will not be possible to continue the increase in energy usage at our present rate (independent of the availability of energy). This conclusion has serious implications for industrially developed countries such as the United States as the developing countries desire to increase their use of energy.

For the short term, i.e., to the year 2000, we can see that some of the heavily industrialized and urbanized sections of the United States will reach energy releases which are a significant percentage of the solar input, particularly in the winter months.

[7] 1 langley (ly) = 1 cal/cm^2 = 3.6855 Btu/ft^2.

[8] Average surface temperature is about 288°K.

The SCEP report projects that the whole of the northeastern United States (350,000 miles^2) will approach a thermal "waste" of 5 percent of the absorbed solar energy by the year 2000. Smaller areas, but still of significant size, such as the Boston-Washington corridor, will have much higher thermal releases. Such magnitudes will cause some climatic effects, but for better or worse? And will these effects mitigate or enhance those caused by particulate concentrations over the urban areas? Our understanding of the effects produced by both direct heating and increased evaporation is very incomplete. The present regions of elevated temperature will increase in size and number, the heights to which these temperatures are felt will become greater, and increased snow or rain may occur.

SULFUR DIOXIDE

There is one further effect of pollution that, while not yet global, is still noted over large areas. This is the effect of SO_2 in causing acid rain and snowfalls. Gases such as SO_2 and H_2S undergo a transformation in the atmosphere to particulates as sulfates or sulfuric acid droplets. Because of wind motion the sulfur compounds emitted in one location may be removed from the atmosphere in a location many hundreds of kilometers downwind. The removal may be by precipitation, by gravitational fallout, or by direct impact on the ground or on vegetation. Measurements over much of North Central Europe have indicated that the land is gradually becoming more acid (lower pH). Measurements in Sweden show that lakes and rivers are becoming more acid (Sweden, 1971).

The effect of this acidification on plant growth depends on the type of soil. In the Scandinavian countries, where much of the soil is already acid, the additional sulfur compounds reduce the growth rate of forests, for example, and thus the productivity. In lakes and rivers with sufficient acid levels, fish will cease to breed and eventually will leave the area. Figure 2.9 indicates the acidifying trend that has occurred in some Swedish rivers (Sweden, 1971).

As with the previously discussed pollutants, the problem here is one of the long-term potential. It remains to be seen if sufficient international cooperation can be achieved to resolve the difficulties.

CONCLUSION

None of the effects caused by the four pollutants just discussed is yet fully understood. The possible synergistic effects are not even

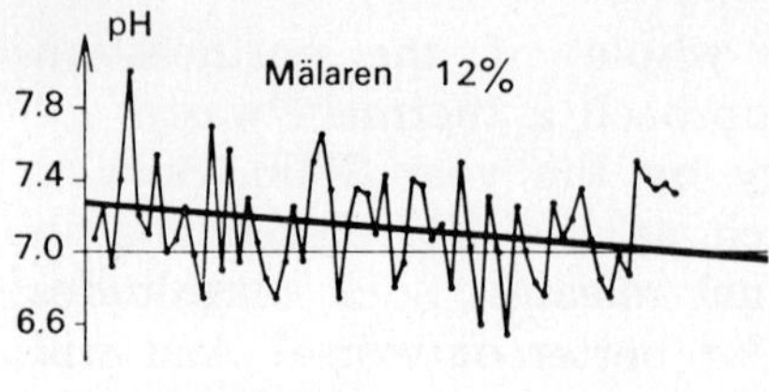

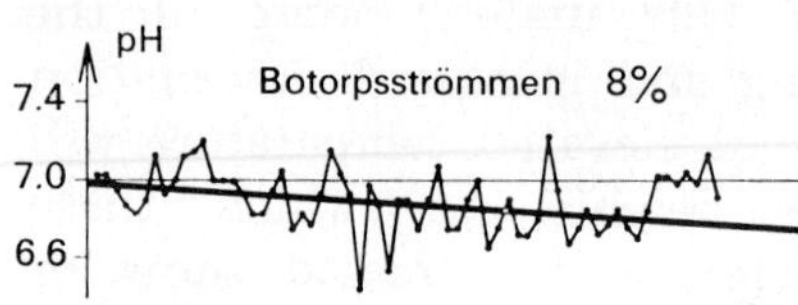

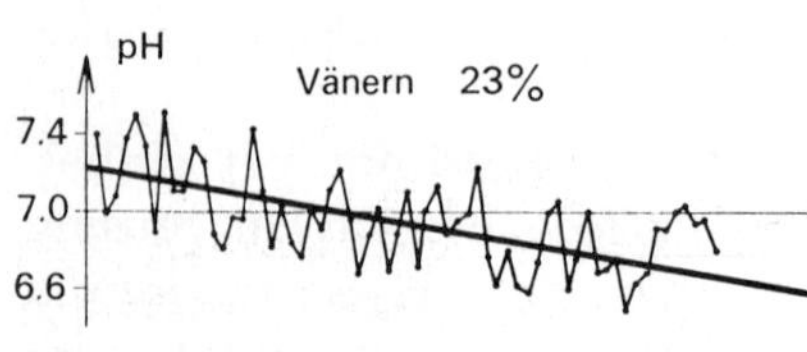

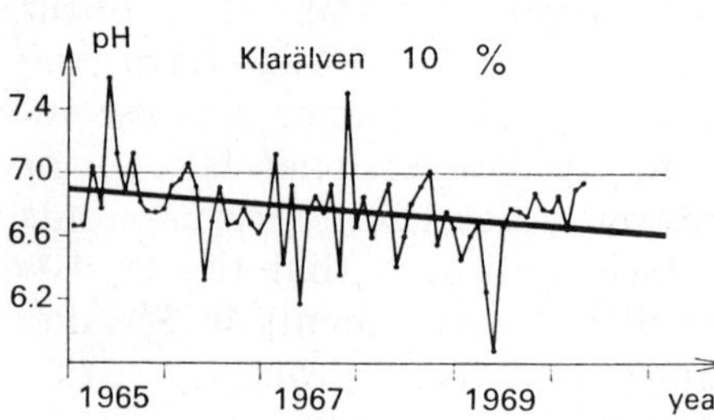

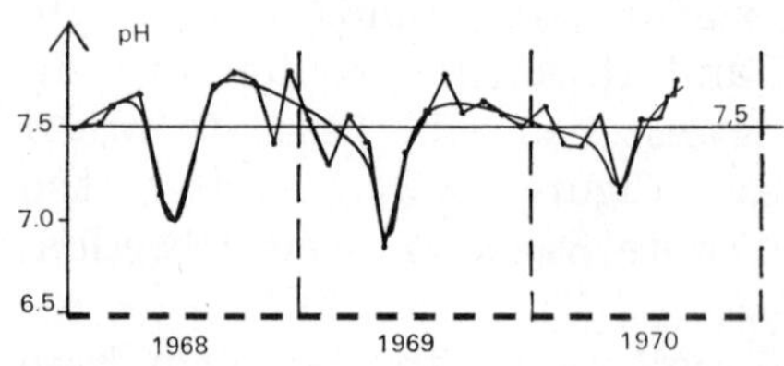

Fig. 2.9 The four upper curves illustrate the pH changes from month to month during 5 years in rivers and lake outlets in southern Sweden. All of them show a negative trend varying from 8 to 23 percent per year, with respect to hydrogen ion concentrations. The lowest curve (data for the river Ljusnan in central Sweden), illustrates the effect of the washout of acid melting water every spring. The rapid pH drop within a short time may be hazardous to the water life.

amenable to crude first analysis. It is clear that the buildup of atmospheric CO_2 acts to increase the ambient temperature. The buildup of particulates probably acts to decrease the ambient temperature, but even this is still under active investigation. In short, there is a great deal about the pollutants' effects that is unknown, and as we sail the good ship earth into the future it would be well to consider these questions now.

REFERENCES

Cloud, P., and A. Gibor: The Oxygen Cycle, *Sci. Am.* September, 1970.

Ehrlich, P. R., and J. P. Holdren: The Heat Barrier, *Sat. Rev.* p. 61, April 3, 1971.

Gough, W. C., and B. J. Eastlund: The Prospects of Fusion Power, *Sci. Am.* February, 1971.

Hanna, S. R., and S. D. Swisher: Meteorological Effects of the Heat and Moisture Produced by Man, *Nucl. Safety*, vol. 12, pp. 114–122, 1971.

Hubbert, M. K.: The Energy Resources of the Earth, *Sci. Am.* September, 1971.

"Inadvertent Climate Modification," Report of the Study of Man's Impact on Climate (SMIC), M.I.T. Press, Cambridge, Mass., 1971.

"Man's Impact on the Global Environment," Report of the Study of Critical Environmental Problems (SCEP), M.I.T. Press, Cambridge, Mass., 1970.

Newell, R. E.: The Global Circulation of Atmospheric Pollutants, *Sci. Am.* January, 1971.

"Nationwide Inventory of Air Pollutant Emissions 1968," National Air Pollution Control Administration Publication AP-73, 1970.

Oort, A. H.: The Energy Cycle of the Earth, *Sci. Am.* September, 1970.

Rasool, S. I., and S. H. Schneider: Atmospheric Carbon Dioxide and Aerosols: Effect of Large Increases on Global Climate, *Science*, vol. 173, pp. 138–141, 1971.

Sellers, W. D.: A Global Climatic Model Based on the Energy Balance of the Earth-Atmosphere System, *J. Appl. Meteorology*, vol. 8, pp. 392–400, 1969.

Singer, S. F. (ed.): "Global Effects of Environmental Pollution," Springer-Verlag, New York, 1970. (a)

Singer, S. F.: Human Energy Production as a Process in the Biosphere, *Sci. Am.* September, 1970. (b)

Sweden: Air Pollution across National Boundaries, the Impact on the Environment of Sulfur in Air and Precipitation. Sweden's National Report to the United Nations Conference on the Human Environment, Utrikesdepartementet, Stockholm, 1971.

Weinberg, A. M., and R. P. Hammond: Limits to the Use of Energy, *Am. Sci.*, vol. 58, pp. 412–418, 1970.

PROBLEMS

2.1 Estimate the percent of land area covered by roads in your city, state, and the United States. What potential effects does this have on the climate? Estimate the percent of land devoted to the automobile (roads, parking lots, etc.) in your city.

2.2 How much might the oceans of the world rise if the Arctic Sea ice were to melt completely? (The ice is several meters thick, SMIC, 1971.)

2.3 How much might the oceans of the world rise if the ice caps on Greenland and the Antarctic were to melt? (Antarctic ice is about 4,000 meters thick; Greenland has a sufficient ice cap to raise the oceans by several meters.)

2.4 Estimate the consumption of oxygen from the burning of fuel in a recent year. How much oxygen will be consumed in fuel combustion from 1970 to 2000? Is this a significant amount?

2.5 Discuss the possible effects of cloud cover on the earth's energy balance. How do man's activities affect the cloud cover?

2.6 Calculate the present ambient earth-atmosphere temperature (i.e., the effective radiative temperature). Estimate what effect a change of albedo of 10 percent would have.

2.7 What is the earth's albedo a function of? How do man's activities affect it?

2.8 Calculate the fraction of the solar input that the United States reaches in energy use per year. For your own city, what is this fraction in the summer, in the winter?

2.9 Estimate the particulate production in your city. What are the major sources? Can they be controlled?

2.10 Consider a 1,000-megawatt coal-fired power plant. Estimate the CO_2 production per year for this plant if the coal is 75 percent carbon and produces 12,000 Btu/lb.

2.11 Assume that gasoline is isooctane, C_8H_{18}. Calculate the CO_2 produced per day from 4×10^6 cars, each burning 2 gal/day.

2.12 Noting that the albedo (reflectance) of the combined earth-atmosphere system is about 0.35 and the solar constant is 2 cal/cm^2-min, determine the effective temperature of the earth-atmosphere system.

2.13 The simplest model for the greenhouse effect is to assume that an absorbing gas forms a completely absorbing layer in the atmosphere for long wavelength radiation. Assuming that this layer absorbs all terrestrial radiation and passes all solar radiation, calculate the effective temperature of the earth and of the layer.

2.14 A rainstorm produces 0.2 inches of water over an area 10 miles square. Calculate the energy release in condensing this quantity of water. How does this energy requirement compare to that of a 1,000-megawatt power plant?

2.15 Water vapor acts as an absorber of terrestrial radiation and, in fact, is the main absorber in the atmosphere. One model of atmospheric behavior is that the *relative humidity* remains constant with perturbations in global temperature. Assuming this fact is correct, discuss how it would act as a positive feedback to further change in the global temperature.

2.16 The total recoverable carbon containing fuel is estimated as 2×10^{12} tons. If all this fuel is burned, estimate the maximum CO_2 concentration that would be reached in the atmosphere. (The present concentration is 320 ppm CO_2.)

3

POLLUTION SOURCES AND EMISSION INVENTORIES

> **The pollution of the air we breathe, ironically enough, is an indirect result of our pursuit of an even higher standard of living. Air pollution derives from the burning of fuel for heat and power, from the processing of materials, and from the disposal of wastes. Air pollution, in short, comes from those everyday activities which are so integral a part of this modern advanced nation.**
>
> *Department of Health, Education, and Welfare, 1966*

Before we can determine what enters the air as pollution we need to know what air normally contains. The composition of clean, dry air is given in Table 3.1. Note that many of the compounds which we consider to be pollutants are minor constituents of clean air, that is, NO_2, O_3, SO_2, CO, and NH_3. Though their concentrations are very low in unpolluted air, under polluted conditions they are greatly increased. For example, concentrations of 0.50 ppm for ozone, O_3, occur in the Los Angeles basin, and CO levels of 20 to 50 ppm for 30 minutes are common along busy urban streets. Thus it is not the mere presence of these gases that causes problems, but it is the greatly increased concentrations that occur from man's activities.

What are sources of air pollution in the United States? In this chapter we shall look at the estimated air pollution emissions in the United States for 1968, the year of the last available complete statistics. The figures are taken from "Nationwide Inventory of Air Pollutant Emissions, 1968 (AP-73)." Simply listing these figures

TABLE 3.1 Composition of clean, dry air near sea level

Component	Content % by volume	Content ppm	Component	Content % by volume	Content ppm
Nitrogen	78.09	780,900	Hydrogen	.00005	0.5
Oxygen	20.94	209,400	Methane	.00015	1.5
Argon	.93	9,300	Nitrogen dioxide	.0000001	0.001
Carbon dioxide	.0318	318	Ozone	.000002	0.02
Neon	.0018	18	Sulfur dioxide	.00000002	.0002
Helium	.00052	5.2	Carbon mon-oxide	.00001	0.1
Krypton	.0001	1	Ammonia	.000001	.01
Xenon	.000008	0.08			
Nitrous oxide	.000025	0.25			

Note: The concentrations of some of these gases may differ with time and place, and the data for some are open to question. Single values for concentrations, instead of ranges of concentrations, are given above to indicate order of magnitude, not specific and universally accepted concentrations.
Source: Reprinted by permission from "Cleaning Our Environment—The Chemical Basis for Action," American Chemical Society, 1969.

may seem unexciting but it is necessary in order to get an idea both of the magnitude of the problem and the source of the emissions.

PARTICULATES

We have already seen in Chap. 2 the compilation of particulate emissions for 1968 by general categories. A more detailed listing in Table 3.2 indicates the nationwide emissions of particulates by source. Note that these figures are in the millions of tons per year. The figure for industrial processes of 7.5×10^6 tons/yr is broken into specific industrial sources in Table 3.3.

It may be surprising to see that the Environmental Protection Agency (EPA) estimates that forest fires are accountable for almost one-quarter of particulate emissions. Of this about one-third results from controlled burning and two-thirds from uncontrolled fires. Coal burning accounts for 29 percent of the particulate emission and 5.6 million tons of the 8.2 million tons is emitted from power plants. Power plants represent concentrated sources and control technology exists to clean up a large percentage of the particulate from those plants burning coal or oil. In spite of available technology they account for 20 percent of the particulate emissions in the United States (5.6 tons/28.3 tons). Note that at the present time solid waste disposal is a small contributor to the emissions of particulates. One must always recognize the limitations of gross figures such as these which are based on best estimates. Some figures may indicate only a minimal emission source when viewed in relation to a national or global scale. But to those living close to the source, the emission is a big problem.

CARBON MONOXIDE

Table 3.4 gives figures for the 1968 emissions of carbon monoxide in millions of tons per year. The largest fraction by far is caused by gasoline-fueled motor vehicles. Most emissions occur in urban areas and this often means high concentrations along downtown streets where concentrations of 10–15 ppm for 8-hour periods are common during the driving day; even higher concentrations have been measured in major cities. Exposure to such levels for an eight-hour day results in a

TABLE 3.2 Nationwide emissions of particulates, 1968

Source	Emissions, 10^6 tons/yr	Percent of total
Transportation	1.2	4.3
Motor vehicles	0.8	2.8
Gasoline	0.5	1.8
Diesel	0.3	1.0
Aircraft	N*	N
Railroads	0.2	0.7
Vessels	0.1	0.4
Non-highway use of motor fuels	0.1	0.4
Fuel combustion in stationary sources	8.9	31.4
Coal	8.2	29.0
Fuel oil	0.3	1.0
Natural gas	0.2	0.7
Wood	0.2	0.7
Industrial processes	7.5	26.5
Solid waste disposal	1.1	3.9
Miscellaneous	9.6	33.9
Forest fires	6.7	23.7
Structural fires	0.1	0.4
Coal refuse burning	0.4	1.4
Agricultural burning	2.4	8.4
Total	28.3	100.0

*N = negligible.

Table 3.3 Particulate emissions from industrial processes, 1968

Industry	Emissions, tons/yr
Iron and steel	1,910,000
Other primary metals	40,000
Grey-iron foundries	170,000
Other secondary metals	50,000
Cement	870,000
Stone, sand, rock, etc.	870,000
Coal cleaning	185,000
Phosphate rock	205,000
Lime	450,000
Asphalt batching	540,000
Other mineral products	180,000
Oil refineries	100,000
Other chemical industries	90,000
Grain handling and storage	800,000
Pulp and paper	720,000
Flour and feed milling	320,000
Other	30,000

carboxyhemoglobin level of about 2.5 percent, that is, 2.5 percent of the red blood cells have CO attached instead of O_2. CO has about 200 times the affinity of O_2 for hemoglobin and thus readily displaces O_2 from the hemoglobin molecule.

Clearly then, great reduction in urban CO levels can only be made by removing the auto or by controlling the combustion process in its engine. Agricultural burning and forest fires are nonurban sources of carbon monoxide, and are therefore less dangerous to man. Air pollution from solid waste disposal can be controlled through better combustion design.

SULFUR OXIDES

Table 3.5 shows emissions of sulfur oxides, mainly SO_2, for 1968. Again we find one category predominating—fuel combustion in stationary sources. Power plants were responsible for 16.8×10^6 tons of SO_2, or 50 percent of the nation's total. The burning of coal and fuel oil is not spread evenly throughout the country, however, and the Northeastern states had almost 50 percent of the emissions. Large sources can also exist locally in all parts of the United States. In southern Arizona copper smelters are a major SO_2 source and emit an estimated 5,000 tons of SO_2 per operating day or about 1,700,000 tons per year. Thus, to someone living close to a copper smelter the fact that power plants are the major national source of SO_2 is not as important as the existence of a local SO_2 source of a different nature, but of a very high emission capability.

Typically coal (or residual fuel oil) contains from 1 to 3 percent sulfur and the combustion of coal produces SO_2 at the rate of 38 times the percent sulfur in units of pounds of SO_2 per ton of coal burned. (One percent sulfur coal has 20 pounds of sulfur per ton. Since SO_2 has a molecular weight of 64 and sulfur's molecular weight is 32, we can at most produce twice as much SO_2 by weight as the sulfur content of the fuel.) Thus 2 percent sulfur coal will produce about 76 lb of SO_2 per ton of coal burned. A modern coal-fired plant uses about 10^4 tons of coal per 1,000-megawatt day and with 2 percent sulfur coal will therefore produce about 76×10^4 lb of

TABLE 3.4 Nationwide carbon monoxide emissions, 1968

Source	Emissions, 10^6 tons/yr	Percent of total
Transportation	63.8	63.8
Motor vehicles	59.2	59.2
Gasoline	59.0	59.0
Diesel	0.2	0.2
Aircraft	2.4	2.4
Railroads	0.1	0.1
Vessels	0.3	0.3
Non-highway use of motor fuels	1.8	1.8
Fuel combustion in stationary sources	1.9	1.9
Coal	0.8	0.8
Fuel oil	0.1	0.1
Natural gas	N*	N
Wood	1.0	1.0
Industrial processes	9.7	9.6
Solid waste disposal	7.8	7.8
Miscellaneous	16.9	16.9
Forest fires	7.2	7.2
Structural fires	0.2	0.2
Coal refuse burning	1.2	1.2
Agricultural burning	8.3	8.3
Total	100.1	100.0

*N = negligible.

TABLE 3.5 Nationwide sulfur oxides emissions, 1968

Source	Emissions, 10^6 tons/yr	Percent of total
Transportation	0.8	2.4
Motor vehicles	0.3	0.9
Gasoline	0.2	0.6
Diesel	0.1	0.3
Aircraft	N*	N
Railroads	0.1	0.3
Vessels	0.3	0.9
Non-highway use of motor fuels	0.1	0.3
Fuel combustion in stationary sources	24.4	73.5
Coal	20.1	60.5
Distillate fuel oil	0.4	1.2
Residual fuel oil	3.9	11.8
Natural gas	N	N
Wood	N	N
Industrial processes	7.3	22.0
Solid waste disposal	0.1	0.3
Miscellaneous	0.6	1.8
Forest fires	N	N
Coal refuse	0.6	1.8
Total	33.2	100.0

*N = negligible.

SO_2 per day at a 1,000-megawatt rating. It is easy to see why emissions of SO_2 have been rising rapidly as the nation's energy requirements have risen. From 1966 to 1968 there was an estimated 2.4×10^6 ton increase in SO_2 emissions from steam-electric power plants.

NITROGEN OXIDES

Emissions of oxides of nitrogen are given in Table 3.6. The two primary sources of nitrogen oxides, NO_x, are motor vehicles and power generation stations. Thus control techniques must include both those for the individual car and for the individual power station, the respective combustion chambers of which are orders of magnitude apart in size. Note that the burning of natural gas which produces negligible SO_x, CO, and HC does produce a very significant quantity of the oxides of nitrogen. Different burner design to decrease peak temperatures and time that nitrogen spends at high temperature will help to reduce emissions of NO_x from power plants. The reduction in auto-produced emissions is difficult since some techniques to reduce NO_x increase CO and HC emissions.

HYDROCARBONS

Table 3.7 lists hydrocarbon emissions for which, again, the motor vehicle is the primary source. Organic solvent evaporation comes from sources such as the paint and varnish industry and the local dry-cleaning plant. We shall see in Chap. 5 that hydrocarbons are a very important component of photochemical smog.

TABLE 3.6 Nationwide nitrogen oxides emissions, 1968

Source	Emissions, 10^6 tons/yr	Percent of total
Transportation	8.1	39.3
Motor vehicles	7.2	34.9
Gasoline	6.6	32.0
Diesel	0.6	2.9
Aircraft	N*	N
Railroads	0.4	1.9
Vessels	0.2	1.0
Non-highway use of motor fuels	0.3	1.5
Fuel combustion in stationary sources	10.0	48.5
Coal	4.0	19.4
Fuel oil	1.0	4.8
Natural gas†	4.8	23.3
Wood	0.2	1.0
Industrial processes	0.2	1.0
Solid waste disposal	0.6	2.9
Miscellaneous	1.7	8.3
Forest fires	1.2	5.8
Structural fires	N	N
Coal refuse burning	0.2	1.0
Agricultural burning	0.3	1.5
Total	20.6	100.0

*N = negligible.
†Includes LPG and kerosene.

TABLE 3.7 **Nationwide hydrocarbon emissions, 1968**

Source	Emissions, 10^6 tons/yr	Percent of total
Transportation	16.6	51.9
Motor vehicles	15.6	48.8
Gasoline	15.2	47.5
Diesel	0.4	1.3
Aircraft	0.3	0.9
Railroads	0.3	0.9
Vessels	0.1	0.3
Non-highway use of motor fuels	0.3	1.0
Fuel combustion in stationary sources	0.7	2.2
Coal	0.2	0.6
Fuel oil	0.1	0.3
Natural gas	N*	N
Wood	0.4	1.3
Industrial processes	4.6	14.4
Solid waste disposal	1.6	5.0
Miscellaneous	8.5	26.5
Forest fires	2.2	6.9
Structural fires	0.1	0.3
Coal refuse burning	0.2	0.6
Agricultural burning	1.7	5.3
Organic solvent evaporation	3.1	9.7
Gasoline marketing	1.2	3.7
Total	32.0	100.0

*N = negligible.

NATURAL SOURCES

Before taking up the question of how these emissions were estimated we shall look at a comparison between the size of natural sources of some of these pollutants and the size of man-caused sources. Much of this material is taken from Robinson and Moser (1970), Robinson and Robbins (in Singer, 1970), and Kellogg et. al. (1972). For SO_2 the worldwide man-caused emissions are estimated to be between 100×10^6 tons/year and 146×10^6 tons/year of which 70 percent is due to combustion of sulfur containing coal. Over 90 percent of these emissions are in the Northern Hemisphere, most in the belt between 30° and 60° north latitude. In terms of total *sulfur* emissions, natural sources are about twice as great as the man-caused sources. SO_4 sea salts (44×10^6 tons), marine sources of H_2S (30×10^6 tons), and land sources of H_2S (70×10^6 tons), combined with the 73×10^6 tons of sulfur in the SO_2 and 3×10^6 tons of additional sources add to a total estimated sulfur emission of 220×10^6 tons/year. However, by the year 2000 it is calculated that man-caused sources will be approximately equal to the natural sources taken over the entire globe. In the Northern Hemisphere man will have overwhelmed the natural sources.

Man-caused nitrogen oxide emissions as equivalent NO_2 are estimated as 53×10^6 tons/year over the world, 95 percent of which are emitted in the Northern Hemisphere. Half of these are from the combustion of coal. Natural sources of NO_x are estimated as between 15 times (Robinson and Moser, 1970) and 7 times (Robinson and Robbins, 1970) the pollutant emissions.

Carbon monoxide world sources are estimated as 304×10^6 tons/year from man-caused activity, again 95 percent of which is emitted into the Northern Hemisphere. Until recently natural sources were taken to be smaller and were estimated between 30×10^6 and 80×10^6 tons/year depending upon the amount resulting from photochemical reactions involving the naturally emitted hydrocarbon compounds. Since there is no buildup of CO in the atmosphere there must be a finite half-life for the gas and thus a sink to act as a removal mechanism. The residence time of CO in the atmosphere had been estimated to be as short as 1 month to as long as 1 to 2 years. The discrepancy can be eliminated by finding a much larger natural production of CO, in which case man's activities are not producing the majority of atmospheric CO. Recent evidence (Weinstock and Niki, 1972) indicates that, in fact, natural sources of CO are some 10 times larger than the man-caused sources so the shorter estimates of residence time are thought to be correct.

Natural emissions of hydrocarbons far outweigh man-caused emissions. However, much of the natural emissions is methane, which does not participate in the photochemical smog cycle. Emissions of heavier hydrocarbons have been estimated between 170×10^6 tons/year (Went, 1960) to 10^9 tons/year (Went, 1966). Man-caused hydrocarbon emissions are about 90×10^6 tons/year, 70 percent of which are from petroleum-related activities and 95 percent of which are again released into the Northern Hemisphere. An estimated one-third of these compounds are *reactive*, a term which will be discussed in Chap. 5.

The mechanisms by which some of these gaseous emissions are transformed to particulates, that is, sulfates and nitrates, are discussed in the SMIC (1971) report. In general, we see that natural sources of emission are greater than man-caused sources taken on a worldwide basis. This is no reason for complacency, however, since emissions from natural sources are overwhelmed in urban areas by the local man-caused sources.

EMISSION FACTORS

How were the previous tables developed? Clearly every factory or power plant was not inspected and its effluent measured. Rather individual processes have been studied and *emission factors* developed for these. An estimate is then made of the number of such processes. The total emission then equals the product of the emission factor times the number of such sources. Emission factors are only estimates, however, since they represent at best the values for the particular processes studied. Other similar processes may not give the same emissions. In some cases very large samples have been studied as, for example, cars in southern California, and here the emission factors may be more accurate for average-type computations.

Power Plants

Consider Table 3.8, taken from "Compilation of Air Pollutant Emission Factors" (AP-42, 1968), which gives particulate emission factors for coal-combustion equipment without any controls. Using such a table one can make estimates of uncontrolled particulate emission for a given plant once the type of unit is specified. (An example will follow shortly.) The particulates from coal consist largely of carbon, silica (SiO_2), alumina (Al_2O_3) and iron oxide (Fe_2O_3 or Fe_3O_4), emitted as *fly ash*. Gaseous material is also emitted from coal-fired power plants. These emission factors are given in Table 3.9 as average values since the type of combustion equipment and size is not specified. The type of unit refers to size with units greater than 100×10^6 Btu/hr taken to be a power plant, units from 10×10^6 to 100×10^6 taken as industrial plants, and units less than 10×10^6 as domestic and commercial plants. Since high-quality coal produces about 13,000 Btu/lb we can convert these to a Btu basis using 26×10^6 Btu/ton of coal burned.

Other fuels used in power plants are natural gas and fuel oil. Emission factors for natural gas are given in Table 3.10. As previously noted, the combustion of natural gas produces a significant quantity of NO_x given here as an equivalent amount of NO_2. The heating value of gas is 1,000 Btu per scf (standard cubic foot) at a density of 0.052 lb/scf. Thus the heating value is 19,250 Btu/lb. The particulate produced here is less than 5μ in size. Fuel-oil emission factors are given in Table 3.11. Power plants burn residual fuel oil which is the lowest

TABLE 3.8 Particulate emission factors for coal combustion without control equipment

Type of unit	Particulate per ton of coal burned*, lb	Percent 44 microns or greater	Percent 20 to 44 microns	Percent 10 to 20 microns	Percent 5 to 10 microns	Percent less than 5 microns
Pulverized						
General	16A	25	23	20	17	15
Dry bottom	17A	25	23	20	17	15
Wet bottom without fly ash reinjection	13A	25	23	20	17	15
Wet bottom with fly ash reinjection†	24A	25	23	20	17	15
Cyclone	2A	10	7	8	10	65
Spreader stoker:						
without fly ash reinjection	13A	61	18	11	6	4
with fly ash reinjection†	20A	61	18	11	6	4
All other stokers	5A	70	16	8	4	2
Hand-fired equipment	20	–	–	–	–	100

*The letter A on all units other than hand-fired equipment indicates that the percent ash in the coal should be multiplied by the value given. Example: If the factor is 17 and the ash content is 10 percent, the particulate emission before the control equipment would be 10 × 17, or 170 lb of particulate per ton of coal.

†Values should not be used as emission factors. Values represent the loading reaching the control equipment always used on this type of furnace.

TABLE 3.9 Gaseous emission factors for coal combustion
(lb/ton of coal burned)

Pollutant	Type of unit		
	Power plant	Industrial	Domestic and commercial
Aldehydes (HCHO)	0.005	0.005	0.005
Carbon monoxide	0.5	3	50
Hydrocarbons (CH_4)	0.2	1	10
Oxides of nitrogen (NO_2)	20	20	8
Oxides of sulfur (SO_2)	38S*	38S*	38S*

*S = % sulfur in coal, e.g., if sulfur content is 2%, the oxides of sulfur emission would be 2 X 38 or 76 lb of sulfur oxides per ton of coal burned.

quality oil and is the leftover after the lighter hydrocarbons have been distilled off, that is, gasoline, kerosene, distillate fuel oil, and so on. Since sulfur is one of the residues to the distillate process, it is concentrated in the resid. Sulfur contents of resid oil vary from less than 1 percent for some domestic United States resids and below 0.5 percent for resid from crude oils of Indonesia, Libya, and Nigeria, to 2-3 percent for Venezuelan resid, and 3-4 percent for resids from some Arabian oils. Fuel oil has a heating value of about 18,000 Btu/lb.

TABLE 3.10 Emission factors for natural gas combustion
(pounds per million cubic feet of natural gas burned)

Pollutant	Type of unit		
	Power plant	Industrial process boilers	Domestic and commercial heating units
Aldehydes (HCHO)	1	2	N
Carbon monoxide	N*	0.4	0.4
Hydrocarbons	N	N	N
Oxides of nitrogen (NO_2)	390	214	116
Oxides of sulfur (SO_2)	0.4	0.4	0.4
Other organics	3	5	N
Particulate	15	18	19

*N = negligible.

TABLE 3.11 Emission factors for fuel oil combustion *(pounds per 1,000 gallons of oil burned)*

Pollutant	Type of unit			
	Power plant	Industrial and commercial		Domestic
		Residual	Distillate	
Aldehydes (HCHO)	0.6	2	2	2
Carbon monoxide	0.04	2	2	2
Hydrocarbons	3.2	2	2	3
Oxides of nitrogen (NO_2)	104	72	72	72
Sulfur dioxide	157S*	157S*	157S*	157S*
Sulfur trioxide	2.4S*	2S*	2S*	2S*
Particulate	10	23	15	8

*S = % sulfur in oil, e.g., if the sulfur content is 2%, the sulfur dioxide emission would be 2 × 157 or 314 lb of sulfur dioxide per 1,000 gallons of oil burned.

Example. Consider the design of a new power plant consisting of three 750-megawatt units. The following fuels are commercially available: a low sulfur coal, residual fuel oil, and natural gas. The analyses are

coal	8% ash	0.5% sulfur	11,000 Btu/lb
resid oil	—	1 % sulfur	18,000 Btu/lb
natural gas	—	—	19,000 Btu/lb (1,000 Btu/scf)

What will be the emissions of particulate, NO_x, SO_x for each fuel? The coal-fired boiler units are to be of the "general-pulverized" type and the thermal efficiency of the plant is estimated to be 38 percent. Which fuel would you recommend?

Thermal Analysis: There will be 2,250 megawatts in all. The energy input required is

$$2{,}250/0.38 = 5{,}930 \times 10^6 \text{ watts} = 20{,}200 \times 10^6 \text{ Btu/hr}$$

so the coal required is

$$\frac{20{,}200 \times 10^6}{11{,}000} = 1{,}834 \times 10^3 \text{ lb/hr} = 917 \text{ tons/hr} = 22{,}000 \text{ tons/day}$$

Coal Emissions: Particulate emissions are taken from Table 3.8 and gaseous emissions from Table 3.9 as:

Particulate: 16(8% ash) × 917 = 117,300 lb/hr
NO_2: 20 × 917 = 18,340 lb/hr
SO_2: 38 (0.5) × 917 = 17,400 lb/hr

Gas Required:

$$\frac{20{,}200 \times 10^6}{1{,}000 \text{ Btu/scf}} = 20.2 \times 10^6 \text{ scf/hr}$$

Emissions:

particulate	$15 \times 20.2 = 303$ lb/hr
NO_2	$390 \times 20.2 = 7{,}890$ lb/hr
SO_2	$0.4 \times 20.2 = 8$ lb/hr

Fuel Oil Required: The emissions factors are given on the basis of 1,000 gallons of oil. We shall assume a density of 0.95× (62.4). We have

$$\frac{20{,}200 \times 10^6}{18{,}000 \text{ Btu/lb}} = 1{,}120 \times 10^3 \text{ lb/hr}$$

or

$$\frac{1{,}120 \times 10^3}{7.9 \text{ lb/gal}} = 142 \times 10^3 \text{ gal/hr}$$

Emissions:

particulate	$10 \times 142 = 1{,}420$ lb/hr
NO_2	$104 \times 142 = 14{,}800$ lb/hr
SO_2	$157 \times 1 \times 142 = 22{,}300$ lb/hr

A summary of these results is given in Table 3.12 in pounds per hour.

The results of this table indicate why natural gas is the best fuel to use from the point of view of emissions. Unfortunately, it is also in the shortest supply and may be the most expensive. Let us continue this example with a brief look at the economics of different fuels. Consider that natural gas costs 35 cents per million Btu, residual oil 25 cents per million Btu, and coal 18 cents per million Btu. The latter figure includes the cost of ash disposal but includes no transportation costs. Using these figures, calculate the cost of

TABLE 3.12 Emissions from a power plant (*lb/hr*)

	Particulate	SO_2	NO_2
Coal	117,300	17,400	18,340
Gas	303	8	1,890
Oil	1,420	22,300	14,800

each fuel to operate the plant for 1 year. If the plant has a 30-year life, how much money would be "available" for air-pollution control for each fuel, assuming that the emissions are to be limited to those that result from the burning of gas with no controls. The costs are

gas
$0.35 \times 20{,}200 \times 8{,}760$ hr/yr = \$62,000,000

oil
$0.25 \times 20{,}200 \times 8{,}760$ hr/yr = \$44,200,000

coal
$0.18 \times 20{,}200 \times 8{,}760$ hr/yr = \$31,800,000

For a 30-year plant life, using gas as the base line for economic calculations, we have a differential cost for oil of $18{,}000{,}000 \times 30$ year = 540×10^6 dollars and for coal $30{,}000{,}000 \times 30$ year = 900×10^6 dollars.[1]

Naturally this example is not meant to indicate that your local power company agrees it has almost a billion dollars available over the next thirty years for pollution control! But the numbers do offer some interesting speculation. Note that to bring the coal-fired unit down to the particulate level of a gas-fired plant requires a collector efficiency greater than 99.7 percent. We can conclude from this example that natural gas is going to continue to be in demand if pollution controls for electric utilities are required. At present many utilities use gas when high pollution levels are predicted, that is, when "episodes" are possible. Most coal is significantly higher than 0.5 percent sulfur so the SO_2 emissions would be considerably higher than calculated here. The coal used in this example was based on that available in the southwestern United States.

Automobiles

Emission factors for uncontrolled automobiles are listed in Table 3.13. Note that these are based on an average route speed of 25 miles per hour (40 km/hr) in urban areas. In the United States an urban vehicle is estimated to drive 3.25 trips per day of 8 miles each, getting 14.4 miles to the gallon of gas. These figures indicate a usage of 1.8 gallons per day. Los Angeles and Tucson, Arizona figures indicate that 2.0 gallons per day are used.

[1] Actual fuel costs may be substantially different from these figures, but the general rule is that gas is most expensive and coal is least expensive. The imported Russian natural gas will cost about \$1.00 per million Btu, but presumably will not be used for power production. In certain parts of the country, for example, the Southwest, gas may be cheaper than oil.

TABLE 3.13 Emission factors for uncontrolled automobile exhaust

Type of emission	Emissions		
	Pounds per 1,000 vehicle-miles	Pounds per 1,000 gallons of gas	Pounds per vehicle-day
Aldehydes (HCHO)	0.3	4	0.007
Carbon monoxide	165.0	2,300	4.160
Hydrocarbons (C)	12.5	200	0.363
Oxides of nitrogen (NO_2)	8.5	113	0.202
Oxides of sulfur (SO_2)	0.6	9	0.016
Organic acids (acetic)	0.3	4	0.007
Particulates	0.8	12	0.022

As noted by Duprey in AP-42, automobile emissions factors are highly variable depending on the altitude of the city and the local driving patterns, that is, freeway driving versus stop-and-go city driving. Emission factors accounting for traffic patterns are indicated in Table 3.14. Cars built starting with the 1968 model have some form of HC and CO control (in addition to the PCV valve). Nitrogen oxide controls were required in California beginning in 1971 models. Thus the emission factors given here need to be adjusted to account for new cars when applied to the general auto population. Cars are considered to have a life of 10 years but this too is a function of the section of the country.

Diesel engine emission factors are given in Table 3.15. In some categories diesels are better than gasoline engines. But they are generally worse on particulates and NO_2.

Other emission factors may be found in Duprey, AP-42, 1968. Emission factors for certain industrial processes are given in the problems at the end of this chapter.

TABLE 3.14 Emission factors for uncontrolled automobile exhaust (*pound per vehicle-mile*)

Route type	Average route speed, mph	Hydrocarbons*	Carbon monoxide
Business	10	0.023	0.35
Residential	18	0.015	0.21
Arterial	24	0.013	0.17
Rapid transit	45	0.0085	0.10

*Expressed as carbon as measured by flame ionization detector.

TABLE 3.15 Emission factors for diesel engines (*pounds per 1,000 gallons of diesel fuel*)

Type of emission	Emission factor
Aldehydes (HCHO)	10
Carbon monoxide	60
Hydrocarbons (C)	136
Oxides of nitrogen (NO_2)	222
Oxides of sulfur	40
Organic acids (acetic)	31
Particulate	110

SUMMARY

We see that motor vehicles and electric power production are the two main sources of air pollution in the United States. On a global basis natural sources of what we call pollutants are still far greater than man-caused sources. Man, however, is catching up, particularly for sulfur, and man overwhelms nature when industry and population are concentrated. Emission factors, based on typical processes, can be used to *estimate* the amount of pollutant resulting from particular processes.

REFERENCES

Duprey, R. L.: "Compilation of Air Pollutant Emission Factors," National Air Pollution Control Administration, PHS Publication 999-AP-42, 1968.

"Inadvertent Climate Modification," Report of the Study of Man's Impact on Climate (SMIC), M.I.T. Press, Cambridge, Mass., 1971.

Kellogg, W. W., R. D. Cadle, E. R. Allen, A. L. Lazrus, and E. A. Martell: The Sulfur Cycle, *Science*, vol. 175, pp. 587–596, 1972.

"Nationwide Inventory of Air Pollutant Emissions," National Air Pollution Control Administration, Publication AP-73, 1970.

Robinson, E. and C. E. Moser: Global Gaseous Pollutant Emissions and Removal Mechanisms, *Proc. Second International Clean Air Congress, Washington, D. C.*, December, 1970.

Singer, S. F., (ed.): "Global Effects of Environmental Pollution," Springer-Verlag, New York, 1970.

Weinstock, B. and H. Niki: Carbon Monoxide Balance in Nature, *Science*, vol. 176, pp. 290–292, 1972.

Went, F. W.: Organic Matter in the Atmosphere, *Proc. Nat. Acad. Sci.*, vol. 46, p. 212, 1960.

——: On the Nature of Aitken Condensation Nuclei, *Tellus*, vol. 18 (203), pp. 549–555, 1966.

PROBLEMS

3.1 A 1,000-megawatt pulverized coal-fired unit (general type) of 40% thermal efficiency using 1.7% sulfur coal is to be built. The ash content of the coal is 10% and the heating value is 12,000 Btu/lb. Calculate the emissions from this plant for particulate, NO_x, and SO_x if no controls are used. If uncontrolled, how much emission would occur of particulate of less than 5μ in diameter? Now assume that an electrostatic precipitator with an efficiency as noted below is purchased for use with the plant. How much fly ash must be disposed of each year? How much particulate of less than 5μ diameter is released per day and per year?

Efficiency of precipitator

particles size	0–5μ	5–10	10–20	20–44	> 44
efficiency, %	75	94.5	97	99.5	100

Emitted particulate size range

particles size	0–5	5–10	10–20	20–44	> 44
% by weight	15	17	20	23	25

3.2 Develop a table comparing energy alternatives on the basis of fuel costs, particulate emission, and SO_2 emitted. Consider hydro, nuclear, gas, oil, coal (90% fly-ash removal) and coal (99.5% fly-ash removal). The coal will be 10% ash and 2% sulfur; the oil 2% sulfur; the gas as given in Table 3.10.

3.3 Make up an emissions inventory for your city, or if you live in a large city, do so for just particulates and SO_x. You may have to consult such sources as the Chamber of Commerce, yellow pages, or the industrial development office to find out what industry exists. Compare your results with the existing inventory if one is available.

3.4 An open burning dump emits about 17 lb of particulate/ton of refuse burned. Calculate the yearly particulate emissions from this source for a city of 100,000 if the average individual generates 5 lb of refuse/day for a 6-day week.

3.5 Duprey (AP-42) gives emission factors for a 1968 vintage fanjet aircraft with 4 engines as 7.4 lb of particulate/flight; 9.2 lb of NO_2/flight; 29.0 lb general HC/flight; and 2.2 lb of aldehydes/flight. If a large airport has 200 four-engine flights and 200 two-engine flights per day (a flight is a takeoff and a landing but no idle or taxi time), calculate the total emissions at the airport from this source. How many automobile-days does this source represent? (See Table 3.12.)

3.6 A safety analysis for an ammonia storage tank requires knowing the source strength of ammonia in g/sec released if the tank breaks. The tank is to hold 20,000 metric tons which will spread over an enclosure of 10,000 m^2. If the boiloff rate is an estimated 1cm/hr, calculate the emission and the time for all the ammonia to boil off. Would these results cause you to be concerned in the event of an accident?

3.7 A copper smelter is producing about 500 tons of copper per day from the copper ore chalcopyrite, $CuFeS_2$. Estimate the SO_2 emissions from this smelter if no controls are applied and all the sulfur goes up the stack. Your result is probably low since the ore will contain some iron pyrites, FeS_2.

3.8 A western city of 300,000 people has 0.67 cars/person. Each car uses about 2 gal gas/day. Calculate the emissions of particulates, NO_x as equivalent NO_2, and CO per day, in this city, from cars. Assume that no pollution controls at present exist on the autos in order to get a worst-case estimate.

3.9 A 10,000 lb/hr continuous coffee bean roaster with a recirculation system exhausts about 4,000 scf/min from the roaster and 120 scf per lb coffee from the cooler. The emission factors for particulate are 4.2 lb/ton of beans from the roaster and 1.4 lb m/ton of beans from the "stoner"-cooler (a stoner removes any heavy objects from the coffee before grinding). Calculate the particulate loading in grains/scf in the two exhausts if no controls are applied (7,000 grains = 1 lb). If a simple cyclone collector is purchased with an overall efficiency of 70% for the particulate from this plant, what will the emissions be in gr/scf?

3.10 Certain emissions of particulates are considered "minor." Assuming that automobile tires are consumed as particulate which is lost into the air, estimate the total particulate from this source per year in the United States. If 1,000 cigarettes produce 1 lb of smoke particles, estimate the total particulate production from this source per year in the United States. Can you think of any other such "minor" sources?

3.11 Estimate the amount of hydrocarbon emitted into the air from the filling of an automobile gas tank. Is this a significant amount?

3.12 Write a summary of the recent work in determining sources and sinks for CO and estimate the residence time for this gas in the atmosphere. References of interest are: *Science*, vol. 176, pp. 290-292, 1972; *J. Air Poll. Control Assoc.*, p. 824, October, 1972; *Science*, vol. 172, pp. 943-945 and pp. 1229-1231, 1971; *Science*, vol. 167, pp. 984-986, 1970.

3.13 Develop a diagram showing the natural and man-caused sources of sulfur and their circulation through the environment.

4

THERMODYNAMICS, KINETICS AND AIR POLLUTION

Good gardeners know that garden refuse need never be burned. Twigs and other prunings which are not easily composted can always be left to dry before burning, and they make a good basis for generating a hot fire in which potatoes may be roasted and fun had by all.

R. S. Scorer, 1968

Most air pollution is created by emission of the products of combustion. You are already familiar with certain of its aspects since everyone has heard of smog and probably of the sulfur oxides. We know that photochemical smog is the product of a complex series of reactions starting with the combustion process in the internal combustion engine of the automobile. Sulfur oxides result from the burning of coal and oil and belch out in huge quantities in the generation of electricity at the large power plants currently coming into use. Thus a basic understanding of combustion is of importance in understanding combustion-generated air pollution. The information in this chapter will be useful in later chapters when learning about photochemical smog and the internal combustion engine as well as in the sections on NO_x and SO_x.

We begin with a few general comments on combustion, followed by information on reacting mixtures, chemical equilibrium, and chemical kinetics. Chapter 5 deals with photochemical smog and draws upon the information presented here.

COMBUSTION

Gasoline is a mixture of hydrocarbon compounds which have been distilled from petroleum. The composition of raw petroleum varies among oil fields. However, the saturated hydrocarbons dominate. Sulfur is also a constituent in oil (as well as in coal); low-sulfur oil is now in great demand as a fuel because the SO_2 emission is reduced. Low-sulfur oil is called "sweet."

Natural gas contains the lighter aliphatic compounds and is largely methane, CH_4. A typical gas used in the Southwest would be 80 to 90 percent methane, 5 to 10 percent ethane, and the rest other compounds. Other natural gases contain significant amounts of propane.

While gasoline is a mixture it is convenient to consider it a single hydrocarbon usually taken as octane or isooctane, C_8H_{18}. If we burn gasoline in the cylinder of a car we expect to have the following reaction:[1]

$$C_8H_{18} + XO_2 + X(3.76)N_2 \rightarrow a\ CO_2 + b\ H_2O + (3.76X)N_2$$

$$\text{Reactants} \rightarrow \text{Products} \tag{4.1}$$

In order to balance the equation we find that the X, a, and b must be such to give the reaction:

$$C_8H_{18} + 12.5O_2 + 12.5(3.76)N_2 \rightarrow 8\ CO_2 + 9\ H_2O + 47.0\ N_2 \tag{4.2}$$

If the reaction were to occur in this manner it would be a *stoichiometric* combustion process, that is, all the oxygen atoms in the oxidizer react chemically to appear in the products. The nitrogen appears to go along for the ride.

What is the correct air/fuel ratio, AFR, to provide stoichiometric combustion? We have (N is the number of moles)

$$AFR = N\text{ air}/N\text{ fuel} = (12.5 + 3.76(12.5))/1$$

$$AFR = 59.5\text{ moles air/mole fuel}$$

On a weight basis this becomes

$$AFR = 59.5\ (28.95)/114 = 15\text{ units air/unit fuel}$$

[1] Air can be treated as a mixture of 1 mole of O_2 to 3.76 moles of N_2.

This is a very typical AFR for hydrocarbons. Thus we need to use about 15 lb of air for each pound of fuel burned, or 15 grams for each gram of fuel burned. An American car driving 30 miles per day, a typical round-trip commute, will use 2 gallons of gas and therefore burn about 13 lb of gasoline. With an AFR of 15 this means that the car takes in about 195 lb of air. When one considers that Los Angeles County has over 4 million cars, we find the astounding figure of $(4 \times 10^6)(195)$ or 780×10^6 lb of air per day going through automobile engines. If only 1 percent of this is a pollutant then the Los Angeles basin must take up 7 *million* lb a day of pollution. In fact, including CO as a pollutant and assuming no emission controls, more like 3 percent of this exhaust is a form of air pollution. This provides all kinds of chemicals in the atmosphere to form reactants for complex chemical reactions, the products of which may not come from automobiles at all!

Why do we burn hydrocarbons either as gas in cars, natural gas in home furnaces, or JP-4 in jet planes? We do this, of course, to extract energy from the combustion process. Burning 1 lb of octane provides 19,100 Btu of energy, assuming that the water produced is in the vapor form. The heating values of hydrocarbons can be found in most thermodynamic books.

Equation (4.1) assumes that the reaction is complete, that is, all of the reactants are used up and no CO or NO is produced in the combustion process. The real fact is that the combustion process is usually incomplete and therefore noxious products can result. The unburned hydrocarbons, or partially burned ones, along with the oxides of nitrogen and carbon monoxide are the major emission problem from automobiles.

To understand why reactions do not go to completion requires knowledge of both the equilibrium conditions for reactions and the kinetics of reactions. A reaction at a certain temperature and pressure and given a long time (long compared to the residence time, say, in the piston cylinder of the automobile engine) will reach an *equilibrium state* which may include products of combustion other than just CO_2 and H_2O. High temperature combustion processes can thus produce CO even if given time to reach equilibrium since the equilibrium state may include CO as one of the products. Usually, however, the combustion process does not have sufficient time to reach equilibrium and thus the process becomes *kinetically limited*. That is, it is limited not by equilibrium considerations but by the fact that reactions take time to proceed to completion, whether the completion state is one of stoichiometric combustion or not.

EQUILIBRIUM CALCULATIONS

This section assumes a knowledge of thermodynamics at the level of Wark, "Thermodynamics," Reynolds and Perkins, "Engineering Thermodynamics," or Van Wylen and Sonntag, "Fundamentals of Classical Thermodynamics." The student should understand the material of Chap. 11 and especially Chap. 12 in the text by Reynolds and Perkins. Chapter 11 deals with thermodynamics of reacting mixtures and is obviously important in terms of combustion-generated pollution. Chapter 12 is concerned with equilibrium calculations.

As an example of a reacting mixture let us look again at the combustion of isooctane. We have

$$C_8H_{18} + 12.5O_2 + 12.5(3.76)N_2 \rightarrow 8CO_2 + 9H_2O + 47.0\ N_2$$

The air/fuel ratio for this reaction is

$$\text{AFR} = 59.5 \text{ moles air/mole fuel}$$

or

$$\text{AFR} = 15.0 \text{ lb air/lb fuel}$$

For stoichiometric combustion the carbon-to-oxygen ratio is 8 carbon atoms to every 25 oxygen atoms, or 1 carbon to 3.125 oxygen atoms. According to this reaction only CO_2 is produced, not CO. However, equilibrium calculations show that CO is, in fact, produced. We shall now see why.

At 1 atmosphere let us find the equilibrium state of CO, CO_2, and O_2 for this ratio of carbon atoms to oxygen atoms.[2] We then have for this single reaction:

$$CO_2 \rightleftharpoons CO + \tfrac{1}{2}O_2 \tag{4.3}$$

Now using P_o as 1 atmosphere and χ as the mole fraction, we have for the equilibrium condition

[2] Note that this equilibrium composition is not the same as would occur in the combustion of C_8H_{18} in air where many competing reactions involving oxygen and carbon would be taking place. Products of combustion would include, but not be limited to, H, CH, OH, H_2, H_2O, C, CO, CO_2, O, O_2, N, NH, N_2, NH_3, NO, NO_2. For illustrative purposes, however, we shall use the isooctane stoichiometric C/O ratio.

$$K(T) = \frac{\chi_{CO}{}^{\nu_{CO}} \chi_{O_2}{}^{\nu_{O_2}}}{\chi_{CO_2}{}^{\nu_{CO_2}}} \left(\frac{P}{P_o}\right)^{\nu_{CO} + \nu_{O_2} - \nu_{CO_2}}$$

where $K(T)$ is the equilibrium constant and the ν's are the stoichiometric coefficients. Then for this reaction we have

$$K(T) = \frac{\chi_{CO} \chi_{O_2}{}^{1/2}}{\chi_{CO_2}} \left(\frac{P}{P_o}\right)^{1 + \frac{1}{2} - 1} \tag{4.4}$$

or since $P = P_O$,

$$K(T) = \frac{\chi_{CO} \chi_{O_2}{}^{1/2}}{\chi_{CO_2}} \tag{4.5}$$

As a side condition we know that the mole fractions must total unity:

$$\chi_{O_2} + \chi_{CO_2} + \chi_{CO} = 1 \tag{4.6}$$

A third relation follows from the given initial atomic composition.

$$\frac{\chi_{CO_2} + \chi_{CO}}{2\chi_{CO_2} + \chi_{CO} + 2\chi_{O_2}} = \frac{1}{3.125} \tag{4.7}$$

To solve these three simultaneous equations let us set $\chi_{CO_2} = a$, which shall then be our unknown.

From the equations we find

$$\chi_{CO} + \chi_{O_2} = 1 - a$$

or

$$\chi_{O_2} = 1 - a - \chi_{CO}$$

Substituting

$$\frac{\chi_{CO} + a}{2a + \chi_{CO} + 2\chi_{O_2}} = \frac{1}{3.125}$$

$$\frac{\chi_{CO} + a}{\chi_{CO} + 2 - 2\chi_{CO}} = \frac{1}{3.125}$$

or

$$\chi_{CO} = 0.484 - 0.758a$$
$$\chi_{O_2} = 0.516 - 0.242a$$

Using the remaining relation

$$K(T) = \frac{(0.484 - 0.758a)(0.516 - 0.242a)^{1/2}}{a} \tag{4.8}$$

Let us now solve this equation for "a" for several different temperatures but at a pressure of 1 atmosphere. The equilibrium constant is given usually as $\log_{10} K$ so we have the following information

$\log_{10} K$	T, °K	K
−45.043	298	9.15×10^{-46}
−10.199	1000	6.34×10^{-11}
− 2.863	2000	1.37×10^{-3}
− 0.469	3000	0.340
− 0.699	4000	4.99

Solving by trial and error we find the concentration *at equilibrium* to be as noted in Table 4.1. Some additional results are also to be found in the table. What can we conclude from this exercise? First at low

TABLE 4.1 Equilibrium concentration for the single reaction: $CO_2 \rightleftharpoons CO + \frac{1}{2}O_2$

P = 1 atm	T, °K	χ_{CO}	χ_{CO_2}	χ_{O_2}
	298	negligible	0.637	0.362
C/O = 1/3.12	1000	<0.001	0.637	0.362
	2000	0.003	0.635	0.363
	3000	0.198	0.378	0.425
	4000	0.437	0.063	0.501
C/O = 1/2	3000	0.363	0.454	0.182
C/O = 1/5	3000	0.113	0.264	0.623
P = 10psia C/O = 1/5	3000	0.128	0.247	0.625
P = 30psia C/O = 1/5	3000	0.089	0.296	0.615

temperatures only CO_2 and O_2 are present. The dissociation of CO_2 does not begin until about 2000°K. The reaction $CO_2 \rightleftharpoons CO + \frac{1}{2}O_2$ is endothermic (requires energy to proceed) when going from left to right. It is a general conclusion that endothermic reactions are more complete at higher temperatures and this is shown by the results in the table.

The equilibrium state is also affected by the pressure. An increased pressure tends to reduce the number of moles (and hence the number of molecules) in the equilibrium mixture. Consequently the effect here of increased pressure is to increase the CO_2 concentration. The carbon/oxygen ratio must, of course, also affect the equilibrium concentration. We see that an initial state with an excess of oxygen atoms, that is, fuel lean, results in lower concentrations of carbon monoxide.

Conversely, excess fuel or, as here, an increased ratio of carbon to oxygen atoms, promotes the formation of carbon monoxide and the concentration increases. At 3000°K we find for C/O ratios of 1/2, 1/3.12, 1/5, carbon monoxide mole fractions, respectively, of 0.363, 0.198, 0.113. A quick conclusion might be simply to run all engines air rich to reduce carbon monoxide. Unfortunately, this causes the formation of other problem pollutants.

The above example is a particularly simple one involving only three species. The more general combustion equation for isooctane would be

$$C_8H_{18} + XO_2 + YN_2 \rightarrow aCO_2 + bH_2O + cO_2 + dH_2 + eCO +$$

$$fH + gOH + hO + iN_2 + jN + kNO + \text{- - - - - - - - -}$$

To work out the coefficients would require knowledge of all of the possible reactions such as

$$H_2 + CO_2 \rightleftharpoons CO + H_2O$$

$$N_2 \rightleftharpoons 2N$$

$$O_2 \rightleftharpoons 2O$$

$$NO \rightleftharpoons \frac{1}{2}O_2 + \frac{1}{2}N_2$$

$$H_2O \rightleftharpoons H + OH$$

$$H_2 \rightleftharpoons 2H$$

$$2H_2O \rightleftharpoons 2H_2 + O_2$$

To find the equilibrium composition with this large number of possible reactions would clearly require the use of a computer. However, the composition of the products of combustion can be predicted from equilibrium considerations just as we have done for our single reaction. Knowing which reactions to consider is, however, not an easy task and requires experience. Since we can write one equation of conservation of atoms for each element, the number of independent equilibrium equations which must be written is equal to the number of species considered less the number of elements contained in these species. In this example we need seven equilibrium reactions and four equations for conservation of species, that is, C, O, N and H.

The results of some studies done by Sawyer (1971) will follow. We must first define the *equivalence ratio*. In our CO_2 example we have specified a C/O ratio in terms of the number of atoms of carbon divided by the number of atoms of oxygen initially present. Usually the initial reaction mixture is specified as

$$\varphi = \frac{\text{fuel/oxidizer}}{(\text{fuel/oxidizer})_{\text{Stoichiometric}}}$$

and φ is the equivalence ratio, a nondimensional quantity. $\varphi = 1$ corresponds to a stoichiometric mixture, $\varphi < 1$ means that we have specified a mixture which has less fuel/oxidizer than a stoichiometric mixture and is thus fuel lean (air rich). Likewise, $\varphi > 1$ means a mixture which is fuel rich (air lean).

The final equilibrium composition of a combustion mixture will depend greatly on whether the reactants were fuel rich or lean, that is, whether $\varphi > 1$ or $\varphi < 1$. Table 4.2 presents the results for the combustion of heptane, C_7H_{16}, and air at $T = 1500°K$ and $P = 1$ atm as a function of the equivalence ratio. Note that *equilibrium considerations predict no unburned hydrocarbons.* Carbon monoxide builds up only for $\varphi > 1$, that is, fuel rich-air lean, and nitric oxide occurs only for $\varphi < 1$, fuel lean.

The effect of temperature is shown in Fig. 4.1 where the equilibrium composition from the combustion of heptane is presented for $\varphi = 1$, stoichiometric combustion. Carbon monoxide builds up at high temperature as the CO_2 dissociates and as our simple previous example indicated. Nitric oxide also builds up at high temperature.[3]

[3] The adiabatic flame temperature for a stoichiometric heptane-air mixture at 298°K is about 2250°K so results beyond this point would require either an external input of energy or preheated reactants.

TABLE 4.2 Product composition for the combustion of heptane and air. $P = 1$ [atm], $T = 1500°K$, varying equivalence ratios. Composition in mole fractions

ϕ	Fuel lean				Stoichio-metric	Fuel rich
	.25	.33	.50	.67	1.00	2.00
Ar	.00971	.00911	.00900	.00889	.00868	.00686
CO_2	.03271	.04334	.06422	.08459	.12381	.03370
H_2O	.03736	.04951	.07337	.09665	.14155	.07819
N_2	.76596	.76124	.75200	.74299	.72576	.57354
NO	.00111	.00104	.00089	.00072	.00002	.00000
O_2	.15364	.13569	.10046	.06609	.00005	.00000
OH	.00005	.00005	.00006	.00006	.00001	.00000
CO	.00000	.00000	.00000	.00000	.00009	.16211
H_2	.00000	.00000	.00000	.00000	.00004	.14559
H	.00000	.00000	.00000	.00000	.00000	.00001

Source: R. F. Sawyer, 1971.

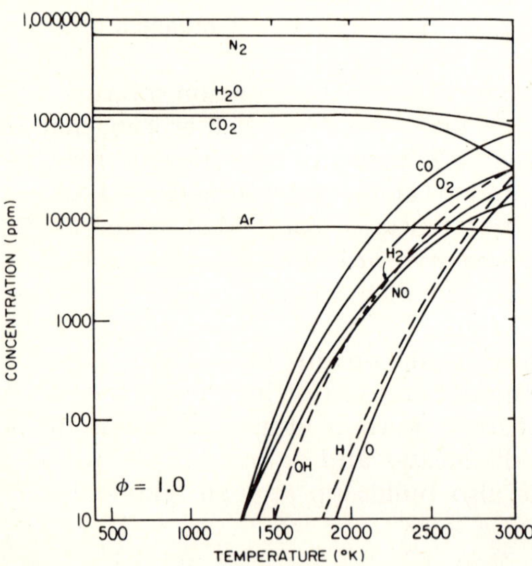

Fig. 4.1 Effect of temperature on equilibrium composition. Stoichiometric heptane/air at a pressure of 1 atm.

Reading some approximate results from this graph we can develop Table 4.3, which indicates that at equilibrium conditions the carbon monoxide level at 2500°K (4500°R) is about 3.5 percent of the products of combustion. The NO level is about 0.6 percent. However, below about 1500°K (2700°R) the products are almost exclusively CO_2, H_2O, and N_2. Since automobile exhaust, which is obviously well below 2700°R at the tailpipe, has CO and NO in it, there must be something lacking in our analysis.

We may conclude from the above results that:

1. The effect of φ, or alternatively, the air/fuel ratio, is very important. Air-rich mixtures, $\varphi < 1$, will tend to produce minimal CO but will produce NO.
2. Keeping combustion temperatures down tends to reduce CO concentrations and NO formation. One way to do this in the internal combustion engine is to reduce the compression ratio.
3. No unburned hydrocarbons are predicted from equilibrium calculations but we know that there is a large variety of such compounds in the products of combustion. Thus something beyond equilibrium considerations must be occurring. What else is happening that we have not accounted for in our analysis?
4. Below about 1500°K equilibrium calculations indicate little pollutant formation. Since we know this is not the case, our analysis is incomplete.

TABLE 4.3 Equilibrium products from combustion of heptane $\phi = 1$

Species	1500 °K		2000 °K		2500 °K	
	2700 °R		3600 °R		4500 °R	
	ppm	%	ppm	%	ppm	%
CO	100	0.01	4,000	0.40	35,000	3.5
CO_2	120,000	12	115,000	11.5	95,000	9.5
H_2	50	0.005	1,000	0.10	8,000	0.8
H_2O	140,000	14	135,000	13.5	12,000	12
N_2	730,000	73	73,000	73	700,000	70
O_2	60	0.006	2,000	0.20	14,000	1.4
O	—	—	30	0.03	2,000	0.2
OH	10	0.001	800	0.08	9,500	0.95
H	—	—	50	0.005	2,000	0.2
NO	22	0.0022	800	0.08	6,000	0.60
Ar	—	≅1	—	≅1	—	≅1

CHEMICAL KINETICS

We can use thermodynamics to learn the equilibrium concentrations for the products of combustion. However, thermodynamics cannot tell us the *rate* at which the reaction proceeds to this equilibrium concentration, or in a real process, whether we ever reach an equilibrium at all. *Reaction kinetics* is used to study the rate of formation of compounds. Many of the products of combustion which are air pollutants result from incomplete reactions in which there is insufficient time for the reaction to reach the equilibrium state. Thus the kinetics of reactions is an important topic for the engineer interested in combustion-generated air pollution.

If we consider a black-box chemical reactor system as a combustion chamber, then thermodynamics can tell us about states 1 and 2, if state 2 is an equilibrium state. It is reaction kinetics, however, which will show how the system got from 1 to 2.

Fig. 4.2 The combustion chamber.

The stoichiometric equation (or chemically balanced equation if not stoichiometric combustion) relates state 1 (the reactants) to state 2 (the products). Intermediate steps are not involved in writing the chemical balance. However, the real processes are not as simple as a single equation. Consider, for example, the combustion of hydrogen-oxygen. We have the stoichiometric equation

$$2\,H_2 + O_2 \rightarrow 2\,H_2O$$

And the equilibrium composition can be calculated, for example, at 2000°K, $\log_{10} K = -3.531$ for the right to left reaction. What is the real process? The postulated mechanism is the following set of reactions:

$$(a) \quad M + H_2 \rightarrow 2\,H + M \tag{4.9}$$

This is the *initiating* step and produces radicals which, in combustion processes, are an important part of the sequence of reactions. The radicals, also called free radicals, are a species that have an unpaired electron and thus are very reactive. M is a molecule which can act as an energy absorber in the process but is not affected chemically. The initiating step is followed by the propagating chain of reactions postulated as:

(*b*) $H + O_2 \rightleftharpoons OH + O$

$O + H_2 \rightleftharpoons OH + H$ Propagating chain (4.10)

$OH + H_2 \rightleftharpoons H_2O + H$

Note that here H is both a reactant and a product so it is generated over and over again, leading to a chain reaction. The propagating chain is in turn followed by the terminating step.

(*c*) $OH + H + M \rightleftharpoons H_2O + M$ (4.11)

This is the *terminating* step in which the radicals, here OH and H, are removed. Thus the simple combustion process represented by the stoichiometric equation turns out to be not so simple after all.

It is easy to see now why the whole *process is time dependent* and the reaction can be cut off if it leaves the reaction chamber too soon. A nonequilibrium concentration may easily result if this occurs. As a further complication, catalysts and impurities may change the rate of reaction and can, in fact, completely stop it. Thus the reaction-rate constant must be experimentally determined for each specific reaction with specific side conditions.

Reactions are classified according to the order of reaction, which is determined by experiment. In a first order reaction the rate of reaction is proportional to the concentration of only one of the reacting substances. Thus if the reaction

$$\nu_1 A + \nu_2 B \rightarrow \text{products} \tag{4.12}$$

can be characterized as having the change of species A concentration given by

$$-\frac{d(A)}{dt} = k_A (A) \tag{4.13}$$

then we say it is a first-order reaction. The minus sign indicates that species A is being consumed. The rate of reaction is just the change of concentration of A with time, where concentration is denoted as (A) and k_A is the rate constant (not the same as the equilibrium constant).

The solution to this particularly simple differential equation is

$$\frac{d(A)}{(A)} = -k_A \, dt$$

$$\ln (A) = -k_A t + C$$

or

$$(A) = (A_0)e^{-k_A t} \tag{4.14}$$

where A_0 is the initial concentration. Thus for this first-order reaction the change in concentration of species A is exponential in time and the rate constant k_A determines how rapidly the concentration of A changes. For a first-order reaction the units of k are clearly sec^{-1}.

A second-order reaction would have the change in concentration proportional to a concentration to the second power. Consider

$$\nu_1 A + \nu_2 B \rightarrow \text{products} \tag{4.15}$$

where

$$-\frac{d(A)}{dt} = k_A(A)^2 \tag{4.16a}$$

or

$$-\frac{d(A)}{dt} = k_A(A)(B) \tag{4.16b}$$

Either of these are second-order reactions. Both of these types of equations can be solved simply for the time dependence of species A. The first yields

$$\frac{1}{(A)} - \frac{1}{(A_0)} = k_A t \tag{4.17}$$

The second requires noting that the change in A is related to the change in B via the stoichiometric equation. For a second-order reaction the rate constant must have the units of 1/(concentration)-time, which in the engineering literature is usually written as cm^3/(g-mole)(sec) or liters/(g-mole)(sec). However, the rate constant is also given as cm^3/(molecule)(sec) and the word molecule is often not written explicitly. To convert, one needs the number of molecules per g-mole, that is, Avogadro's number.

In a third-order reaction the rate of reaction is proportional to the third power of concentration, either of an individual species or a combination of concentrations of species adding up to the power three.

Reactions are not necessarily of integer order nor are they always as simple as noted above. For example, consider the case where the products of a reaction can react to reproduce the original reactants. Then it follows that as the reaction approaches equilibrium the rate, as we see it, will decrease, since some of the reactants will be produced from the products. An equilibrium state will be reached with equal forward and backward rates to the reaction. This is the equilibrium state which we have already studied. Consider the case where we have

$$A + B \underset{k_2}{\overset{k_1}{\rightleftharpoons}} C + D \tag{4.18}$$

where the rate constant k_1 refers to the forward direction and k_2 to the backward direction. Then *near equilibrium* the reaction is second-order overall and first-order with respect to each concentration.[4]

We have

$$-\frac{d(A)}{dt} = k_1(A)(B) - k_2(C)(D) \tag{4.19}$$

That is, the decrease in the concentration of A depends on both the forward and backward reactions. At equilibrium the rates must be equal, $d(A)/dt = 0$, and so

$$k_1(A)(B) = k_2(C)(D) \tag{4.20}$$

However, we also know that at equilibrium

$$K(T) = \frac{\chi_3^{\nu_3}\chi_4^{\nu_4}}{\chi_1^{\nu_1}\chi_2^{\nu_2}}\left(\frac{P}{P_o}\right)^{\nu_3+\nu_4-\nu_1-\nu_2} \tag{4.21}$$

At a pressure of P_o and for the reaction of interest where all the stoichiometric coefficients equal 1 we thus have

$$K(T) = \frac{(C)(D)}{(A)(B)} = \frac{k_1}{k_2} \tag{4.22}$$

Note it is only true at equilibrium that the equilibrium constant $K(T)$ is related in this manner to the reaction constants k_1 and k_2. This

[4] Consider that any collision between an A and a B molecule removes one A molecule and any collision between a C and a D molecule produces one A molecule.

relationship can be used to determine one of the reaction constants if the other has been measured.

As another example we will consider the case of consecutive reactions, in particular two first-order consecutive reactions. The example will indicate what happens in the initiation-chain-termination sequence of reactions. Consider the stoichiometric sequence of reactions

$$A \xrightarrow{k_A} B$$

$$B \xrightarrow{k_B} C$$

where B is produced as a product of reaction 1 and removed as a reactant in reaction 2. The overall reaction is

$$A \rightarrow C$$

We have

$$\frac{d(A)}{dt} = -k_A(A)$$

$$\frac{d(B)}{dt} = k_A(A) - k_B(B) \qquad (4.23)$$

$$\frac{d(C)}{dt} = k_B(B)$$

Note that adding these equations we have

$$\frac{d}{dt}(A + B + C) = 0, \quad \text{or } (A + B + C) = \text{constant}$$

The first equation is solved as

$$A = A_0 e^{-k_A t} \qquad (4.24)$$

Then the second equation becomes

$$\frac{d(B)}{dt} + k_B(B) = k_A A_0 e^{-k_A t} \qquad (4.25)$$

The solution to this differential equation requires both a general and particular solution which are given as

$$\text{general solution} \quad (\text{B}) = a_1 e^{-k_\text{B} t}$$

$$\text{particular solution} \quad (\text{B}) = a_2 e^{-k_\text{A} t}$$

Thus

$$(\text{B}) = a_1 e^{-k_\text{B} t} + a_2 e^{-k_\text{A} t} \tag{4.26}$$

where a_1 and a_2 must be determined from the initial conditions. Let us assume that both (B) and (C) are initially 0. Substituting our solution for (B), Eq. (4.26), back into the differential equation, Eq. (4.25), we find that

$$-a_1 k_\text{B} e^{-k_\text{B} t} - a_2 k_\text{A} e^{-k_\text{A} t} + a_1 k_\text{B} e^{-k_\text{B} t} + a_2 k_\text{B} e^{-k_\text{B} t}$$
$$= k_\text{A} \text{A}_0 e^{-k_\text{A} t}$$

At $t = 0$ we then have

$$-a_2 k_\text{A} + a_2 k_\text{B} = k_\text{A} \text{A}_0$$

or

$$a_2 = \frac{k_\text{A} \text{A}_0}{k_\text{B} - k_\text{A}}$$

Thus

$$(\text{B})_{t=0} = 0 = a_1 + \frac{k_\text{A} \text{A}_0}{k_\text{B} - k_\text{A}}$$

and so

$$(\text{B}) = \frac{k_\text{A} \text{A}_0}{k_\text{B} - k_\text{A}} (e^{-k_\text{A} t} - e^{-k_\text{B} t}) \tag{4.27}$$

We could now turn to the solution of the third equation; however, note that the chemical equation provides us with some further information, namely

$$\text{A} + \text{B} + \text{C} = \text{constant} = \text{A}_0 \tag{4.28}$$

so

$$C = A_0 - A_0 e^{-k_A t} - \frac{k_A A_0}{k_B - k_A}\left(e^{-k_A t} - e^{-k_B t}\right)$$

$$C = A_0 (1 - e^{-k_A t}) - \frac{k_A A_0}{k_B - k_A}\left(e^{-k_A t} - e^{-k_B t}\right) \tag{4.29}$$

At this point we can introduce the idea of a *rate-determining step*. Our solution for the change of species C with time can be simplified if one of the reactions is much faster than the second. If $k_B \gg k_A$ (the second reaction is much faster than the first) then the above expression reduces to

$$C \cong A_0(1 - e^{-k_A t}) \tag{4.30}$$

and species B is rapidly converted to species C. Alternatively, if $k_A \gg k_B$ then

$$C \cong A_0 + A_0(-e^{-k_B t})$$

or

$$C = A_0(1 - e^{-k_B t}) \tag{4.31}$$

This solution represents the first-order reactions as governed by the rate constant k_B, the slower step in the two-step reaction. In short, if one or more of the rate constants is larger than the others, then we can simplify the mechanism accordingly. This idea is often used in dealing with multiple reactions such as occur in chemical combustion.

Our particular two-step reaction can be qualitatively plotted as noted in Fig. 4.3. The concentration of A falls off exponentially; B rises to a maximum and falls off to zero; and C rises continuously with a maximum slope where B is a maximum. This example shows how even a simple two-step, first-order analysis can exhibit the possibility of a *kinetically limited* reaction. If we should stop the reaction before sufficient time has elapsed for it to go to completion, we would have species A and B as well as the expected species C. In combustion processes, A and B may prove to be contaminates.

A last class of reactions important in understanding the combustion process is the group of consecutive reactions involving propagation chains. The previous example in this section of hydrogen/oxygen is such a case. We can classify such systems as follows:

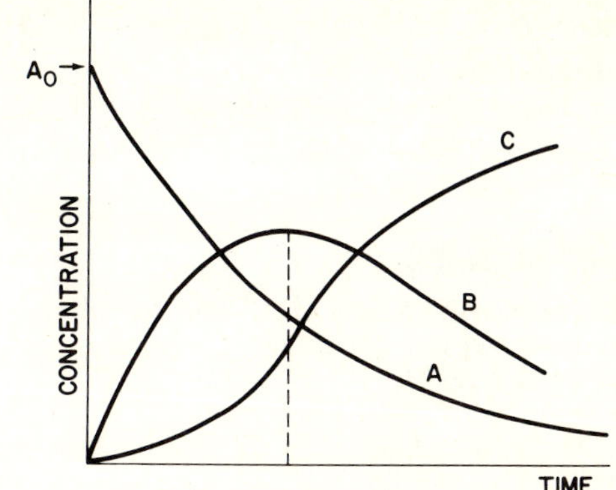

Fig. 4.3 Concentration as a function of time.

1. *Initiation* reaction in which the production of species as radicals gives rise to further reactions.
2. *Propagation* chain reactions which produce intermediate products and often involve a product that may become a reactant in a further reaction.
3. *Termination* reactions which remove the radicals and result in final products.

In analysing such systems the idea of a rate-limiting reaction often leads to simplification. Such chain reactions are important in the formation of carbon monoxide and nitric oxide in the products of combustion from both internal combustion engines and external combustion engines.

CONCLUSION

We can now see why the process of chemical combustion can produce some partially burned or unburned hydrocarbons and significant CO and NO concentrations. The combustion process is usually kinetically limited and does not reach an equilibrium state. Rather, some concentration is reached which represents that of previous temperature and pressure conditions and not the exhaust conditions. This "frozen" concentration is the one which contains the air pollutants. As a particular example of this we shall see in Chap. 12 for NO_x how important the ideas of equilibrium and reaction kinetics are in understanding the formation and control of that pollutant.

The fundamental idea to be retained from this chapter is that understanding combustion-generated air pollution has three requirements, an understanding of stoichiometric combustion processes, an understanding of the approach to an equilibrium concentration and how it is calculated, and an understanding of the effect of kinetic limitations on the final concentration. Even a qualitative appreciation of these three subjects, as outlined in this chapter, is a great aid in following why various control measures do or do not work.

REFERENCES

Reynolds, W. C. and H. C. Perkins: "Engineering Thermodynamics," McGraw-Hill Book Company, New York, 1970.

Sawyer, R. F.: "Combustion Generated Air Pollution," E. S. Starkman (ed.), Plenum Press, New York, 1971.

Scorer, R. S.: "Air Pollution," Pergamon Press, London, 1968.

Van Wylen, G. J., and R. E. Sonntag: "Fundamentals of Classical Thermodynamics," John Wiley & Sons, Inc., New York, 1965.

Wark, K.: "Thermodynamics," 2d ed., McGraw-Hill Book Company, New York, 1971.

PROBLEMS

4.1 What is a stoichiometric mixture? What do the terms enthalpy of formation and heat of formation mean? What do the terms excess air and theoretical air mean?

4.2 Write out the chemical equation for the reaction of octane (C_8H_{18}) with a stoichiometric reaction with air. Determine the theoretical air/fuel ratio for this reaction on both a molal and mass basis.

4.3 Write out the chemical equation for the reaction of octane (C_8H_{18}) with (*a*) 100 percent excess air and (*b*) 250 percent theoretical air.

4.4 Write the chemical equation for the reaction of *n*-butane (C_4H_{10}) with a stoichiometric reaction with air. Determine the theoretical air/fuel ratio for this reaction on both a molal and mass basis.

4.5 An oil fuel contains 84 percent C and 16 percent H_2 by mass. Find the stoichiometric air for complete combustion of 1 lb of fuel. Determine the air/fuel ratio.

4.6 Compute the air/fuel ratio by mass used in an engine if the exhaust gas dry analysis in percent by volume is CO_2, 0.124; O_2, 0.032; CO, 0.001; H_2, 0.002.

4.7 Liquid benzene (C_6H_6) is burned in a stoichiometric reaction with dry air. Determine the air/fuel ratio and the dew point of the products if the total pressure is 13.5 psia. Redo the problem if 200 percent excess air is used.

4.8 How much energy is released when 1 mole of propane (C_3H_8) reacts with 80 percent theoretical air when both the inlet and exit states of the combustion are at the standard reference state? How much energy is released in a stoichiometric reaction?

4.9 Determine the enthalpy of combustion at 77°F for ethane (C_2H_6) when gaseous ethane reacts, assuming liquid water in the products. Do the same for acetylene (C_2H_2). Determine your answer in Btu/lb fuel.

4.10 Estimate the adiabatic flame temperature for a stoichiometric reaction of *n*-butane (C_4H_{10}) with air.

4.11 Redo the previous problem for the adiabatic flame temperature if the reaction occurs with 400 percent theoretical air.

4.12 Determine the adiabatic flame temperature that would be obtained from complete combustion of C_8H_{18} (octane) with theoretical air if (*a*) the inlet conditions are 1 atm at 77°F, and (*b*) if the inlet conditions are 1000°R. (Note an average $c_p = 7.90$ Btu/lb mole-°R for C_8H_{18} in this temperature range, $\hat{h}^{\circ}_{C_8H_{18}} = -89{,}680$ Btu/lb mole).

4.13 Estimate the adiabatic flame temperature when methane (CH_4) at 77°F reacts with 80 percent theoretical air, also at 77°F. (Note that CO will be present in the products since the reaction is one of incomplete combustion.)

4.14 A laboratory analysis shows that CO_2 becomes 10 percent dissociated into CO and O_2 at 4300°R when the total pressure is 1 atm. Find the equilibrium constant from this information and compare to that given in your thermodynamics text.

4.15 Determine the degree of dissociation of CO_2 at 5000°R, 1 atm if 100 percent excess oxygen is supplied to the reaction.

4.16 Find the mole fractions of O present in equilibrium oxygen at 1000°K and at 5000°K ($P = 1$ atm).

4.17 Find the mole fractions of N present in equilibrium nitrogen at 1000°K and 5000°K ($P = 1$ atm).

4.18 1 mole of CO reacts with 1 mole of O_2 in a steady-flow process. Both inlet conditions are 77°F at 1 atm. The final products are a mixture of CO, CO_2, O_2 at 1 atm. Determine the equilibrium compositions at 3000°K and 2800°K.

4.19 The simplest model for the burning of methane is the four-step reaction noted below. If this process is cut off before completion is reached, what undesirable products will result?

$$CH_4 + O_2 \rightarrow CH_2O + H_2O$$
$$CH_2O + O_2 \rightarrow CO + H_2O_2$$
$$CO + \frac{1}{2}O_2 \rightarrow CO_2$$
$$H_2O_2 \rightarrow H_2O + \frac{1}{2}O_2$$

5

LOS ANGELES

What to do When the Smog is Bad

1. Plan to stay indoors as much as possible.
2. Avoid driving and areas of heavy traffic.
3. Avoid smoking, dust, fumes, sprays.
4. Use properly filtered air conditioning.
5. Avoid strenuous physical activity.
6. If breathing is difficult, call your doctor.
7. Keep informed on daily air pollution conditions.

Daily announcements are also broadcast by all radio stations in this area. During an alert situation, more frequent bulletins are broadcast.

Riverside, California, City Council and Community Relations Commission

PHOTOCHEMISTRY

In the previous chapter we discussed combustion, equilibrium, and the rates of reaction. Let us now turn to photochemistry so that we can, in turn, study the evolution of photochemical smog, that wondrous mixture of chemicals which southern Californians know so well.

In photochemical reactions the initiating step is absorption of a photon by an atom, molecule, free radical or ion. The result of this absorption depends on the energy and therefore the wavelength of the photon. To be initiated photochemical *reactions* require light in the visible and ultraviolet portion of the spectra. The absorption can produce an excited species, dissociation, internal rearrangement, or fluorescence. We are interested here in species which absorb a photon and then dissociate since these are fundamental in the formation of photochemical smog. The initial effect of photon absorption and dissociation is called the primary photochemical reaction. Subsequent reactions caused by these primary products are called secondary photochemical reactions and these too are critically important in the formation of smog. We can summarize the process with a species denoted by A as

$$A + h\nu \rightarrow A'$$

followed by dissociation of A′ into two products,

$$A' \rightarrow D_1 + D_2$$

This is the reaction of interest to us but there is an incredible number of possible processes that can occur after the absorption of a photon. The reader interested in other processes is referred to the text by Calvert and Pitts (1966).

It should be clear that the number of photochemical reactions occurring in the atmosphere will depend on the intensity of solar radiation and that the particular reaction will depend on the energy (wavelength) of the photon absorbed. We know that

$$E = h\nu$$

and

$$c = \lambda\nu$$

so

$$E = \frac{hc}{\lambda}$$

Photochemists refer to a number of photons equal to Avogadro's number as an Einstein. Thus

$$1 \text{ Einstein} = 6.0225 \times 10^{23} \text{ photons}$$

Substituting for Plank's constant and the velocity of light we find

$$E = 2.859 \times 10^5/\lambda \quad \text{kilocal/Einstein} \quad (\lambda \text{ in Å})$$

$$E = 28.59/\lambda \quad \text{kilocal/Einstein} \quad (\lambda \text{ in } \mu)$$

The kilocal unit is usually used since chemical reactions, for example dissociation, are presented in terms of the energy in kilocalories required to cause 1 g-mole of species to react. Table 5.1

TABLE 5.1 Photon energy as a function of wavelength

Region of spectrum	Microns λ	E, kilocalories per g-mole
Ultraviolet	0.3	95.3
Violet, visible limit	0.38	75.1
	0.4	71.4
Visible	0.7	40.8
Red-visible limit	0.76	37.6
Infrared	1.0	28.6

gives the energy content of an Einstein of photons as a function of wavelength.

The intensity of solar radiation is a function of latitude, time of year, time of day, cloud cover, amount of pollution in the atmosphere, and so on. The solar zenith angle is defined as noted in Fig. 5.1. The variation of solar zenith angle and time of day is shown on Fig. 5.2 for the latitude of Los Angeles. Other latitudes have a similar behavior. Qualitatively this figure indicates how the diurnal variation in solar input will go. The insolation (solar input) depends on the amount of air through which the radiation must travel given as the air mass, m. The air mass is the ratio of the length of path of the direct solar radiation through the atmosphere relative to a vertical path. Thus $m = 1$ for a zenith angle of 0° and $m = 2$ for a zenith angle of 60°. Details for obtaining solar inputs are given in Leighton (1961) and Robinson (1966). It is clear that the summer months in the United States will tend to bring high insolation and therefore produce high smog levels, although the local meteorology plays a dominant role in determining when smog levels reach high concentrations. The period of high solar insolation is long at latitudes typical of the United States (20 to 50°N) and lasts approximately from April to September.

It is the combination of light intensity, light duration, availability of reactants, and stability of the atmosphere which controls the

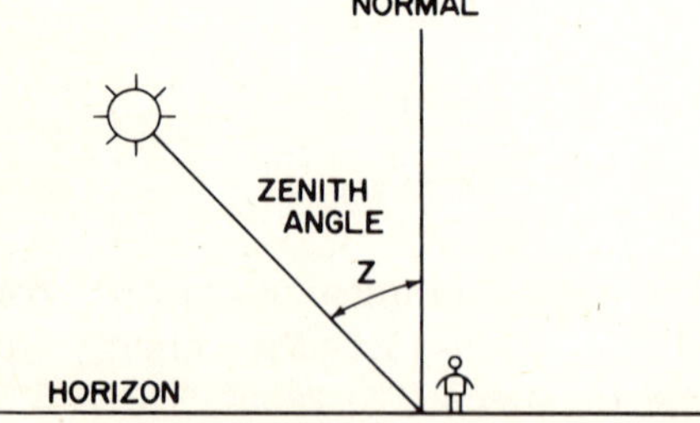

Fig. 5.1 Solar zenith angle.

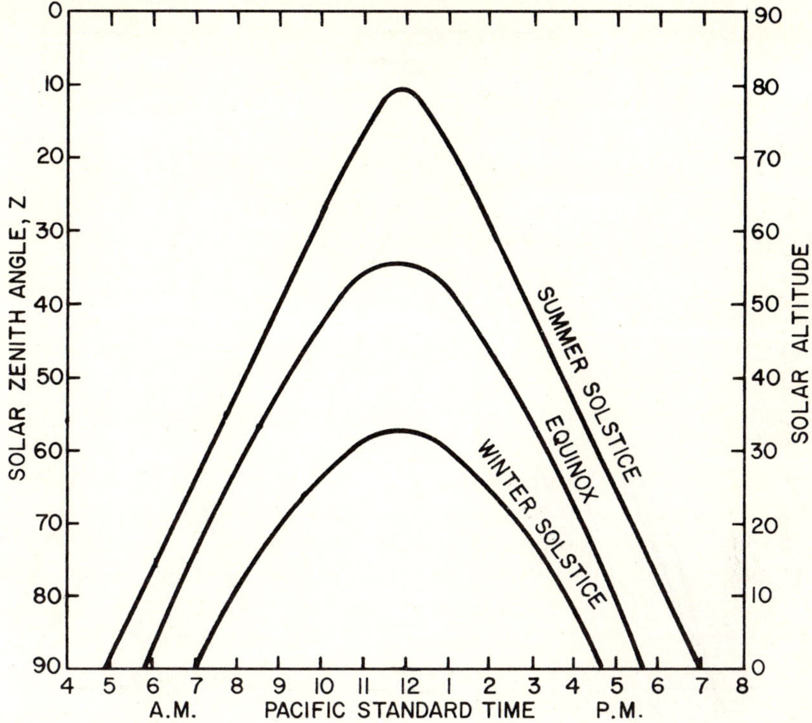

Fig. 5.2 Relation between solar zenith angle and time of day at Los Angeles. (*After Leighton.*)

concentration of photooxidized compounds. In the United States the maximum light intensity is approximately constant with latitude for the summer months. In the wavelength range from 3000 to 4000 Å the maximum total intensity is 2×10^{16} photons/cm^2-sec and this level is a broad maximum lasting for 4 to 6 hours per day. Winter values range from 0.7×10^{16} to 1.5×10^{16} photons/cm^2-sec depending on the latitude. Values near this maximum last for about 2 to 4 hours.

Corrections for atmospheric scattering and absorption by molecular species are required in order to calculate the actual solar energy input at a given location. However, the calculation is not even this "simple" because of the effects of the air pollutants themselves. Figure 5.3 from Leighton (1961) shows the attenuation of *direct* solar input by photochemical smog over a 10-Å bandwidth centered at 3235 Å. Note again that this figure shows attenuation of direct radiation. A significant portion (under some conditions more than half) of radiation from the sun under unpolluted conditions reaches the surface of the earth indirectly, only after being scattered, and is called

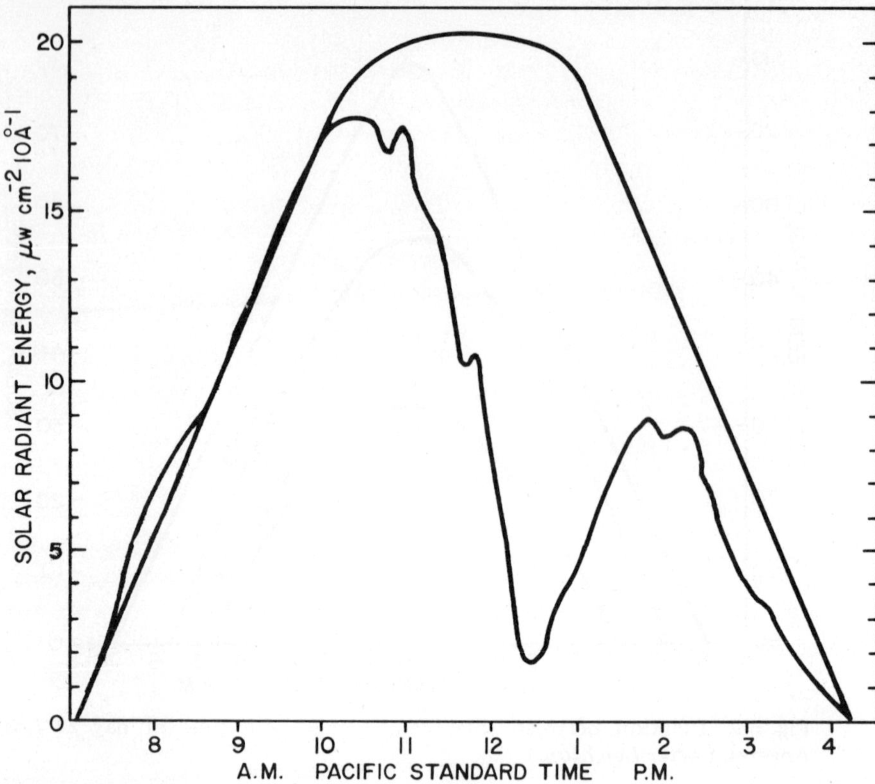

Fig. 5.3 Attenuation of direct solar energy by photochemical smog. Both lines represent the direct solar energy over a 10-Å bandwidth centered at 3235 Å, incident on a normal surface at 800-ft elevation. The smooth line is calculated. The irregular line is the observed energy, from Stair's data for October 18, 1954 at Pasadena. (*Leighton, 1961.*)

the indirect or sky radiation. It is clear from the figure, however, that smog does reduce the solar input significantly in this bandwidth which corresponds to wavelengths absorbed by NO_2.

NO_2

A specific example of a photochemical reaction is the photodissociation of NO_2, nitrogen dioxide. NO_2 absorbs over virtually the whole of the visible and ultraviolet range of the solar spectrum with a decrease in absorption in the longer wavelength visible portion. Thus the gas appears to have a reddish-brown color. This dissociation reaction shown as

$$NO_2 + h\nu \rightarrow NO + O$$

requires sufficient energy to break the bond between the NO and NO_2, which is about 72 kilocalories/g-mole at 25°C. From Table 5.1 we see this requires light of wavelength less than about 0.4μ (4000 Å). The dissociation is a function of wavelength increasing with wavelengths below 4000 Å. Above 4200 Å there is insufficient energy to cause dissociation and other photochemical effects such as fluorescence occur. Below about 3700 Å the *primary quantum yield* (percent of molecules undergoing dissociation per photon absorbed) is over 90 percent. How fast is this reaction? In "typical" sunlight the half-life of NO_2 is about 2 minutes (Stephens, 1969*b*). This particular fast reaction is one of very great importance in the formation of photochemical smog, as we shall soon see.

PHOTOCHEMICAL SMOG

What do we put into air to cause photochemical smog? Here we will discuss primarily what happens in the Los Angeles basin and what goes into that giant chemical reactor. It is most important, however, to remember that photochemical smog is not exclusive to Los Angeles. It has been observed in most major cities of the world, and particularly in those with high solar input. Thus in the United States, the San Francisco area, Phoenix, Denver, and Salt Lake City show evidence of photochemical smog during the months of high insolation. Keep in mind that Philadelphia, St. Louis, and virtually all major United States cities have had occurrences of this form of air pollution. When we use Los Angeles as our example let us not forget all the other vulnerable cities.

Contaminants in tons per day from major sources in Los Angeles County are noted in Tables 5.2 and 5.2*a*. In all, these sources total up to over 13,000 tons per day of gases and particulates emitted into the natural atmospheric basin. The great majority of these are from motor vehicles (68 percent of the total contaminants). Specific emissions from motor vehicles as a function of year are noted in Figs. 5.4, and 5.5.[1] The increase in NO_x in the late 1960s is due to use of pollution controls designed to reduce hydrocarbons! Note that, from the vantage point of 1969, the pollution controls expected to be used in the future would only drop the nitrogen oxide to about the levels existing in 1950 before concentrations rise again as the number of vehicles in the county

[1] Profile of Air Pollution Control in Los Angeles County, Los Angeles County Air Pollution Control District, 1969.

TABLE 5.2 Profile of air pollution in Los Angeles County, 1969

Item	Quantity
Population	7,300,000
Los Angeles County land area	4,083 Sq. Miles
Los Angeles Basin land area	1,250 Sq. Miles
Solvents used (emitted)	1,000,000 Lbs./Day
Refinery crude throughput	731,000 Bbls./Day
Fuels burned during Rule 62 period[(a)]	
Oil containing more than 0.5% sulfur	193,200 Gals./Day[(c)]
Oil containing 0.5% or less sulfur	59,430 Gals./Day
Natural gas	1,894,400 MCF/Day
Refinery make gas	272,400 MCF/Day
Fuels burned during Rule 62.1 period[(b)]	
Oil containing more than 0.5% sulfur	487,400 Gals./Day[(c)]
Oil containing 0.5% or less sulfur	2,828,100 Gals./Day
Natural gas	2,128,200 MCF/Day
Refinery make gas	277,200 MCF/Day
Gasoline-powered vehicle registration	3,950,000
Gasoline consumed	8,000,000 Gals./Day
Diesel-powered vehicle registration	14,500
Diesel fuel consumed by motor vehicles	150,000 Gals./Day
Jet-powered aircraft flights/day	1,505
Jet fuel consumed (within L.A. County)	330,000 Gals./Day
Piston-driven aircraft flights/day	9,650
Helicopter flights/day	590
Aviation gasoline consumed (within L.A. County)	110,000 Gals./Day

(a) Rule 62 period is from April 15 through November 15.
(b) Rule 62.1 period is from November 16 through April 14.
(c) Burned under variance granted by Hearing Board.
Source: Los Angeles County Air Pollution Control District.

increases. These astounding statistics led to the legislation of 1970 designed to force auto emissions to 10 percent of the 1970 level.

In short, every day a great quantity of hydrocarbons and a lesser but very important amount of nitrogen oxides (largely as nitric oxide, NO) are funneled into the Los Angeles basin. What happens to these gases? They participate in a long chain of chemical reactions which produce an incredible variety of chemical garbage. We have thus created in the Los Angeles atmosphere a chemistry laboratory with reactions uncontrolled by man and occurring only as nature dictates.

Some hydrocarbons emitted are "worse" than others. Figure 5.6 notes which organics are of high or low reactivity. This simply means that certain classes of organic hydrocarbons are more likely to take part in these chemical reactions than other classes. One must, of course, be careful to define the basis for reactivity. If we use the nitric oxide photooxidation rate as our basis, then the reactivity of various exhaust products is as noted on the figure. The photooxidation rate means the rate at which the hydrocarbons cause NO to be oxidized to NO_2, given in parts per billion per minute. To determine this, nitric oxide is irradiated in the presence of specific hydrocarbon compounds and the buildup of NO_2 is noted (Glasson and Tuesday, 1970).

To control photochemical smog we need to control the emissions of the more reactive hydrocarbon species, if this is in fact possible. In general these are the internally double-bonded olefins, the diolefins, and the cycloalkenes. The important point here is that not all hydrocarbon compounds are equally bad in the air, at least in terms of participating in the photochemical smog mechanism. We can carry this one step further by noting (in Table 5.3) the hydrocarbon concentrations measured in air samples in Riverside, California, which is east of Los Angeles but connected to the basin by a broad valley. Table 5.3 (AP-64) emphasizes two important points. First, the number of hydrocarbon compounds in the air is truly incredible. Second, there is a change in the individual species concentration with time of day. Why? The change is attributable not just to further emissions during the day, but also to the fact that concentrations of

TABLE 5.2*a* Contaminants, in tons per day, from major sources within Los Angeles County

Major source	Organic gases			Particulates	NO_x	SO_2	CO	Total
	Reactivity							
	High	Low	Total					
Motor vehicles	1,255	475	1,730	45	645	30	9,470	11,920
Org. solvent usage	100	400	500	17	–	1	–	520
Petroleum	55	165	220	4	45	55	30	355
Aircraft	45	45	90	12	15	3	190	310
Combustion of fuels	–	9	9	15	235	40	1	300
Chemical	–	–	–	–	–	90	–	90
Other	–	3	3	16	11	3	4	35
Total (rounded)	1,455	1,095	2,550	110	950	225	9,695	13,530

Source: See Table 5.2.

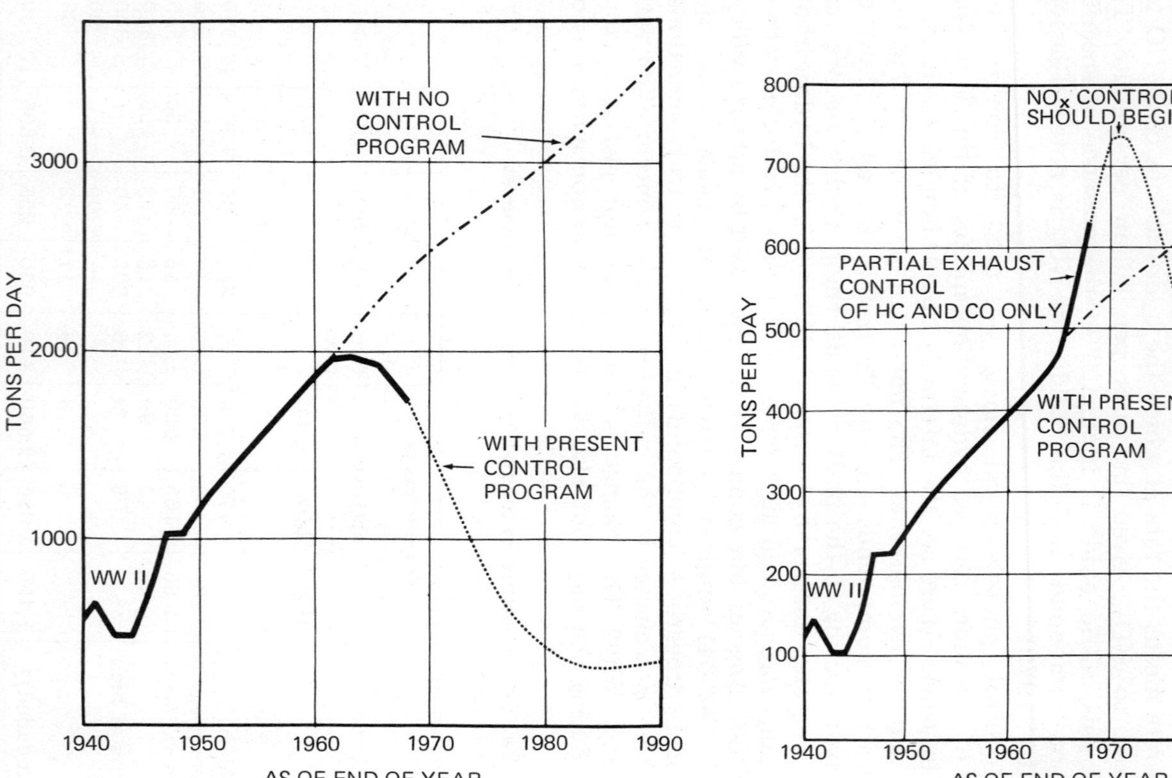

Fig. 5.4 Emissions of hydrocarbons and other organic gases in tons per day from motor vehicles in Los Angeles County.

Fig. 5.5 Emissions of oxides of nitrogen in tons per day from motor vehicles in Los Angeles County.

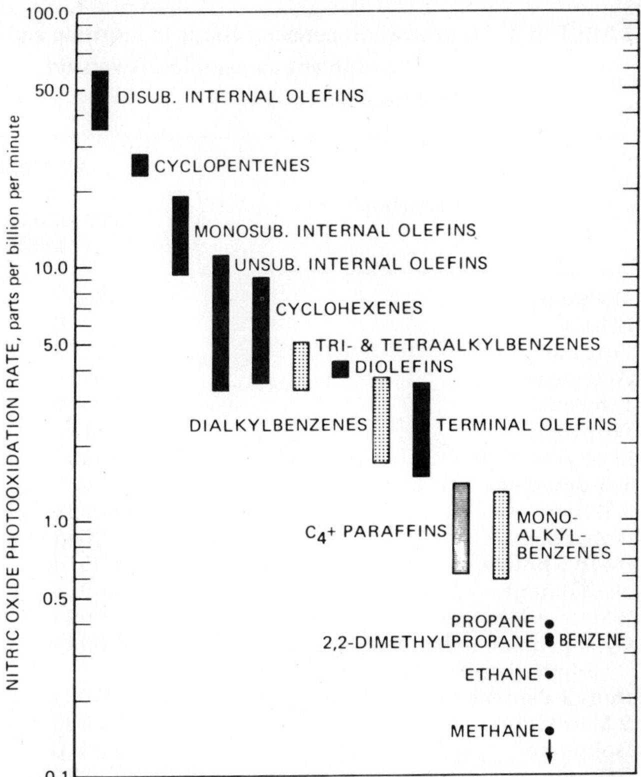

Fig. 5.6 Reactivity of various exhaust products. (*Reprinted by permission from Environmental Science and Technology, Vol. 4, p. 922, 1970. Copyright © 1970 by the American Chemical Society.*)

the more reactive compounds will be reduced during the day. The 7:30 AM figures thus represent "new smog," the 4:10 PM concentrations "well aged smog." Note that the olefins and diolefins (alkenes and alkadienes) do indeed decrease in concentrations, for example, ethylene, propylene, butadiene, butene, etc. Also the cycloalkenes concentration such as cyclopentene has dropped markedly. However, the alkanes (saturated chain hydrocarbons) are not very reactive. Thus the concentrations of methane and ethane have risen as they represent the stable compounds which do not react and which can therefore build up in time as the end products of other reactions.

In meteorological conditions different from Los Angeles the emission of hydrocarbons and nitric oxide into the atmosphere, while certainly undesirable, would not result in the concentrations of photochemical smog which occur in this basin. We shall soon devote

TABLE 5.3 Hydrocarbon concentrations in morning and afternoon ambient air samples, Riverside, California, Fall 1968

Hydrocarbon	Concentration, ppm	
	7:30 AM, P.S.T. Sept. 24, 1968	4:10 PM, P.S.T. Oct. 24, 1968
Methane	2.355	2.530
Ethane	0.0636	0.0722
Propane	0.0188	0.0499
Acetylene	0.0770	0.0420
Ethylene	0.0656	0.0179
Propylene	0.0192	0.0013
Methyl acetylene	0.0026	0.0012
1,3-Butadiene	0.0036	0.0003
1-Butene	0.0026	0.0003
Isobutene	0.0052	0.0020
trans-2-Butene	0.0014	<0.0002
cis-2-Butene	0.0014	<0.0002
2-Methyl-1-butene	0.0024	0.0004
Cyclopentene (with 2-methyl-1, 3-butadiene)	0.0044	0.0008
trans-2-Pentene	0.0024	<0.0002
2-Methyl-2-butene	0.0026	<0.0002
Isobutane	0.0080	0.0197
Butane	0.0276	0.0620
Isopentane	0.0392	0.0412
Pentane (with 3-methyl-1-butene)	0.0224	0.0214
Cyclopentane	0.0032	0.0024
2,2-Dimethyl butane (with 1-pentene)	0.0016	0.0009
2,3-Dimethyl butane	0.0038	0.0019
2-Methyl pentane (with *cis*-2-pentene)	0.0124	0.0096
3-Methyl pentane	0.0084	0.0061
Hexane	0.0090	0.0083

Source: AP-64.

several sections to meteorology but a few brief comments are necessary here. Los Angeles, in addition to having a great concentration of over 4 million cars, has a limited volume of air in which to dilute the products of combustion. Any location has a limited volume of the atmosphere into which to dump its waste, but the problem is particularly acute in Los Angeles because of a very stable temperature inversion which limits the vertical mixing. Further, typical low wind speeds limit the horizontal transport of pollutants away from the city. The inversion at about 1,500 to 3,000 feet (500 to 1,000 meters) is caused by the presence of a high pressure area, the Pacific High, off the coast of southern California. The air in this region is sinking or subsiding and thus is compressed and warmed. The region of highest temperature is not at the ground but is at a modest altitude dependent upon several parameters such as the time of day, the strength of the ocean breeze which is usually cool, and the strength of the high pressure region. The effect of the inversion is to place a lid over the basin which is walled in by mountains to the east and north. Thus the vertical height available for mixing is greatly limited and the emitted pollutants have no place to go. The weak breezes are partly an effect of the geography since the coastline takes an east-west bend at that point in California. So it is a combination of concentrated emissions, limited air volume, and strong insolation which produces smog conditions.

The photochemical smog mechanism is treated in the next section. It is a complex subject and not yet completely understood, although the general details are now firmly in hand. One of the significant facts from the study is how secondary pollutants are produced from interactions among the primary emissions. In fact, some of the secondary pollutants represent new compounds unique to the photochemical smog process. We have learned in short, that the whole may actually be greater than the sum of the parts.

PHOTOLYTIC CYCLE

We have found that the Los Angeles basin is full of reactive hydrocarbons, nitric oxide, and sunlight. What happens to this chemical mess? The reactions of interest start with one we have already discussed, namely, the photolysis of NO_2 as

$$NO_2 + h\nu \rightarrow NO + O$$

Ozone, O_3, is then formed by reacting very quickly with the oxygen *atom* in the reaction

$$O + O_2 + M \rightarrow O_3 + M$$

where M is a third body needed to absorb the energy of the reaction. Without the M body there would just be an exchange of oxygen atoms within an oxygen molecule. Even with this requirement of a triple collision, the reaction is kinetically fast. In air the M body will probably be N_2 or O_2 since these account for most of the available molecules. The third reaction is one which completes the cycle and is

$$O_3 + NO \rightarrow NO_2 + O_2$$

This reaction is also rapid and thus a constant level of each species NO, NO_2, and O_3 could be expected if this were all that happened. We can consider that these three reactions form a cycle as noted in Fig. 5.7.

The steady-state ozone concentration from these reactions can be calculated as a function of the NO_2 initial concentration. Stephens indicates (1969*b*) that at 10 pphm (parts per hundred million) NO_2 we expect about 2.7 pphm (0.027 ppm) ozone. Most of the nitrogen oxide emitted into the air from the combustion process is NO, and the NO_2 levels are usually not above about 10 pphm. However, the ozone levels are well above 2.7 pphm and in fact, often run to 50 pphm (0.5 ppm) for 1 hour peak averages. Thus there must be some other mechanism involved which affects the nitrogen oxide-ozone cycle so that ozone can build up.

We need to find some way in which NO can be oxidized to NO_2 without consuming the ozone. Thus this species could build up and the nitric oxide level would be reduced as, in fact, occurs in the atmosphere. The required mechanism is produced by the hydrocarbon compounds in the atmosphere. The very reactive oxygen *atoms* can attack a hydrocarbon, say an olefin (a hydrocarbon with one or more carbon-carbon double bonds), and split it into two parts at the double bond. Olefins are reactive because the double bond can be readily oxidized. Ozone can also do this but the oxygen atom reaction is faster (Cadle and Allen, 1970). The result is a free radical, that is an incomplete hydrocarbon, which is highly reactive and continues to get involved in other reactions. We can denote the oxygen atom-olefin reaction as

$$O + \text{olefin} \rightarrow R + R'O$$

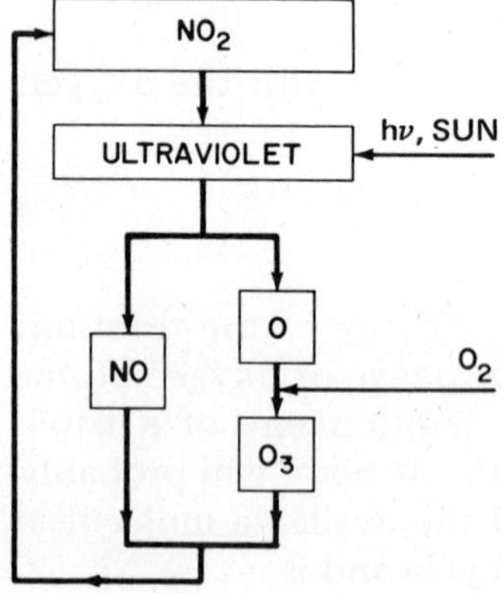

Fig. 5.7 The photolytic cycle for the oxides of nitrogen.

where R and R′ are organic radicals. The organic free radicals react with O_2 to form peroxy radicals as

$$R + O_2 \rightarrow RO_2 \text{ (or R OO)}$$

Species noted as ending in OO are called peroxy radicals and can yield an oxygen atom to nitric oxide to form nitrogen dioxide and thus interfere with the previously discussed NO_2, NO, O_3 cycle. Since each hydrocarbon molecule requires one oxygen atom to initiate its oxidation (which could react to convert an NO to NO_2 via the O_3 process) more than one NO molecule must be oxidized to NO_2 per hydrocarbon molecule broken up to permit ozone to accumulate. This cycle is noted on Fig. 5.8 and shows two points at which NO is oxidized to NO_2 by the action of the organic radicals. Thus these radicals are responsible for much of the oxidation of NO to NO_2 and consequently the buildup of ozone.

The other radical formed in the initial process denoted as R′O might be an aldehyde (an oxygenated hydrocarbon) since aldehydes are among the products known to form from the reaction of O and the olefins. The aldehydes are eye irritants and formaldehyde, CH_2O, is noted on the figure. As a side effect of the peroxyradical we may have

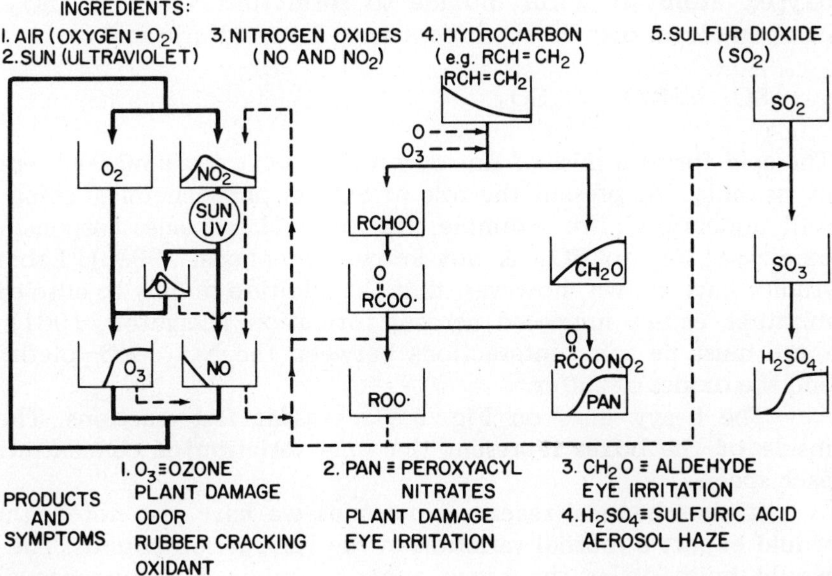

Fig. 5.8 An outline of the photochemistry of smog. (*Stephens, 1969a.*)

the formation of peroxyacetyl nitrate (PAN), which has the chemical form

$$CH_3\overset{\overset{\displaystyle O}{\|}}{C}OO\,NO_2$$

In effect the conversion of the last part of the NO by the radicals is difficult since the concentration is low. Thus in place of the oxidation reaction

$$R\overset{\overset{\displaystyle O}{\|}}{C}OO + NO \rightarrow NO_2 + R\overset{\overset{\displaystyle O}{\|}}{C}O$$

we may have the reaction

$$R\overset{\overset{\displaystyle O}{\|}}{C}OO + NO_2 \rightarrow R\overset{\overset{\displaystyle O}{\|}}{C}OO\,NO_2$$

where R can be the methyl group CH_3. PAN is a very potent eye irritant as well as a compound which causes damage to plants.

This figure also notes that the peroxy radical can also donate an oxygen atom to sulfur dioxide to form sulfur trioxide, SO_3. This compound can then react with water to form sulfuric acid as

$$SO_3 + H_2O \rightarrow H_2SO_4$$

The acid forms a mist of fine aerosols which cause a marked reduction in visibility. At present the role of SO_2 in photochemical smog is not well understood; for example, the particular species responsible for oxidizing SO_2 to SO_3 is not known (Stephens, 1969*a*). Laboratory studies have shown, however, that the addition of SO_2 to auto exhaust mixtures causes increased aerosol formation (Leighton, 1961). Thus there must be some interactions between the NO_2—NO—olefin cycle and the oxides of sulfur.

The heavy lines on Fig. 5.8 represent fast reactions. The lines inside of the boxes represent the time variation of concentration of each species.

If the chemical reactions occur as we have just noted, then we would expect a diurnal variation in the various compounds. The ozone would build up as the nitric oxide is oxidized to nitrogen dioxide through the action of the hydrocarbons. We would also expect the

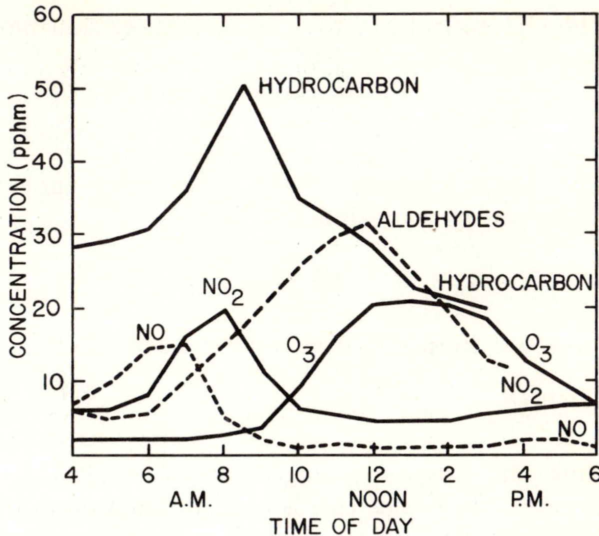

Fig. 5.9 Variation of the components of photochemical smog with time in the Los Angeles basin. (*From Leighton.*)

aldehydes to build up since they are a product of the breakup of the hydrocarbons by oxygen atom attack. Figure 5.9 notes the variation of concentrations with time of day and is taken from Leighton. The NO and NO_2 data are from 1958 and the other data are from 1953-54 so only qualitative comparison should be made. Note that the ordinate is in parts per hundred million. See if you can explain the time variation in terms of the reactions noted in Fig. 5.8.

MECHANISM

We have outlined the overall mechanism for the formation of photochemical smog as it is presently thought to occur. Details of the specific reactions are difficult to nail down since incredible numbers of possible reactions exist. However, several authors have postulated simulation models. The NO_2—NO—O_3—NO_2 cycle is a key element in all the models. After that set of reactions, the various authors differ in details of the models. We shall comment here on only a few of the possible additional complications.

It is known that the aldehydes are attacked by atomic oxygen to produce hydroxyl radicals (here we use acetaldehyde)

$$O + CH_3CHO \rightarrow CH_3CO + OH$$

The OH radical can then react with carbon monoxide to produce CO_2 and H as

$$OH + CO \rightarrow CO_2 + H$$

The hydrogen can then react with molecular oxygen to form hydroperoxy radicals as

$$H + O_2 + M \rightarrow HO_2 + M$$

However, these can then oxidize nitric oxide to nitrogen dioxide as

$$HO_2 + NO \rightarrow NO_2 + OH$$

Thus the original nitrogen-oxygen cycle can also be affected in this manner. Note that this is a chain reaction in which the OH is first used as a reactant and then produced as a product. The oxidation of NO, which causes the buildup of O_3, is thus affected by relatively inert carbon monoxide which is readily available as a pollutant. The OH is regenerated and the CO is available so the reaction can go on and on.

Further, many of the chemical species in the atmosphere, other than NO_2, can absorb sunlight. Other absorbers include nitric acid, nitrates, aldehydes, peroxides, acylnitrates and the particulates. The photodissociation caused by some of these absorptions will contribute reactive radicals which will enter the mechanisms outlined here. Still other reactions involve complications in the nitric oxide cycle. For example, we know that

$$O_3 + NO_2 \rightarrow NO_3 + O_2$$

$$NO_3 + NO_2 + M \rightarrow N_2O_5 + M$$

$$(\text{or } 2NO_2 + O_3 \rightarrow N_2O_5 + O_2)$$

and

$$N_2O_5 + H_2O \rightarrow 2HNO_3 \text{ (nitric acid)}$$

Thus we can produce nitric acid in the atmosphere. We have already noted the formation of sulfuric acid. In short, photochemical smog represents a reacting salad bowl fueled by the sun with a variety of resulting noxious products.

A somewhat more complete reaction chain is listed in Table 5.4 and illustrates some of the numerous chemical species which can be

TABLE 5.4 A photochemical model

	Reaction	
NO_x cycle	1. $NO_2 + h\nu \rightarrow NO + O$	
	2. $O + O_2 + M \rightarrow O_3 + M$	
	3. $O_3 + NO \rightarrow NO_2 + O_2$	
	4. $O_3 + NO_2 \rightarrow NO_3 + O_2$	produces *nitric acid*
	5. $NO_3 + NO_2 \xrightarrow[H_2O]{} 2HNO_3$	
	6. $NO + NO_2 \xrightarrow[H_2O]{} 2HNO_2$	nitrous acid
	7. $HNO_2 + h\nu \rightarrow NO + OH$	
	8. $CO + OH \xrightarrow[O_2]{} CO_2 + HO_2$	CO effect produces an OH chain
	9. $HO_2 + NO \rightarrow NO_2 + OH$	
	10. $HC + O \rightarrow \alpha\, RO_2\cdot$	aldehyde and peroxy radicals
	11. $HC + O_3 \rightarrow \beta\, RO_2\cdot + \gamma\, RCHO$	
	12. $HC + OH \rightarrow \delta\, RO_2\cdot + \epsilon\, RCHO$	
	13. $HC + RO_2 \rightarrow \zeta\, RO_2\cdot + \eta\, RCHO$	
	14. $RO_2 + NO \rightarrow NO_2 + \varphi\, OH$	oxidation of NO to NO_2
	15. $RO_2 + NO_2 \rightarrow PAN$	production of peroxyacetyl nitrate

HC = hydrocarbon
$RO_2\cdot$ = peroxy radical

$\alpha \rightarrow \varphi$ stoichiometric coefficients

Source: Seinfeld, 1971. See also T. A. Hecht, and J. H. Seinfeld, *Environ. Sci. Technol.* vol. 6, pp. 47–57, 1972; and R. G. Lamb, and J. H. Seinfeld, *Environ. Sci. Technol.* vol. 7, pp. 253–261, 1973.

produced. Note particularly the products which are acids, hydrocarbon radicals, and PAN. One can quickly see that modeling the photochemical smog reactions is not an easy task. Specific reaction-rate constants are given in Westburg and Cohen (1970) and Cadle and Allen (1970). To simplify handling the large number of possible reactions only the faster reactions are included in the more complex models. Furthermore it is important to realize that the previous discussion represents what we think is true. Future laboratory research and investigation into the atmosphere will improve our understanding and the proposed models will have to be changed accordingly.

OXIDANTS

"Oxidation" generally denotes the loss of one or more electrons by an atom, ion, or molecule. Thus carbon is oxidized to carbon dioxide when burned in air. Reduction is therefore the gain of one or more electrons. In the $C + O_2 \rightarrow CO_2$ reaction the oxygen is reduced, the carbon oxidized. Photochemical smog includes several pollutants which are called oxidants or photochemical oxidants, in particular ozone, O_3, and PAN, peroxyacetyl nitrate. These are substances which will oxidize certain reagents not readily oxidized by oxygen and thus might be considered "more oxidizing than oxygen". PAN is known to be a potent eye irritant and ozone causes adverse health effects in plants. The properties of ozone and PAN are listed in Tables 5.5 and 5.5*a*.

The concentrations of these oxidants are important since both are known to be hazards to health. It will also be of interest to compare federal ambient air quality standards to present-day concentrations; Table 5.6 gives a summary of the maximum oxidant concentrations (AP-63). The hourly average here is the concentration averaged over each clock hour. The maximum hourly average is the highest value recorded for any one-hour average. The peak concentration is either a five-minute average, that is, the peak five-minute value, or an "instantaneous" average. The test used to measure these does not

TABLE 5.5 Physical properties of ozone

Physical state	Colorless gas
Chemical formula	O_3
Molecular weight	48.0
Melting point	$-192.7 \pm 0.2^\circ C$
Boiling point	$-111.9 \pm 0.3^\circ C$
Specific gravity relative to air	1.658
Vapor density	
At $0^\circ C$, 760 mm Hg	2.14 g/liter
At $25^\circ C$, 760 mm Hg	1.96 g/liter
Solubility at $0^\circ C$ (Indicated volume of ozone at $0^\circ C$, 760 mm Hg)	0.494 ml/100 ml water
Conversion factors	
At $0^\circ C$, 760 mm Hg	1 ppm = 2141 $\mu g/m^3$ 1 $\mu g/m^3$ = 4.670×10^{-4}
At $25^\circ C$, 760 mm Hg	1 ppm = 1962 $\mu g/m^3$ 1 $\mu g/m^3$ = 5.097×10^{-4}

Source: AP-63.

TABLE 5.5a Physical properties of peroxyacetyl nitrate

Physical state	Colorless liquid
Chemical formula	$CH_3C(=O)OONO_2$
Molecular weight	121
Boiling point	No true boiling point, compound decomposes before boiling
Vapor pressure at room temperature	About 15 mm Hg
Conversion factors	
At 0°C, 760 mm Hg	1 ppm = 5398 $\mu g/m^3$ 1 $\mu g/m^3$ = 1.852×10^{-4} ppm
At 25°C, 760 mm Hg	1 ppm = 4945 $\mu g/m^3$ 1 $\mu g/m^3$ = 2.022×10^{-4} ppm

Source: AP-63.

discriminate between oxidant species. Further, reducing agents such as SO_2 will act to lower the oxidant reading. These results are thus a measure of the *net* oxidizing properties of the atmospheric pollutants. An atmosphere which has a high concentration of both oxidizing and reducing agents will not show a meaningful reading using this technique (colorimetric analysis using neutral phosphate-buffered potassium iodide).

As one would expect, the values for oxidant concentration differ greatly among different cities. AP-63 notes that the "high peak concentration reported for the St. Louis air monitoring station is one of a series of extraordinarily high readings which usually occur late at night and are of short duration. It is suspected they result from emissions from a nearby large chemical complex rather than from an atmospheric photochemical reaction." This observation suggests that one must be careful in locating a sampling station if one wants to get true atmospheric averages. However, to the people downwind of the chemical complex the location of the sampling station is not of interest.

A word of caution is appropriate here regarding averaging. High ozone levels are associated with photochemical processes which require sunshine. Thus daily averages include a weighting factor of two-thirds of the time in which one would expect low ozone levels since the insolation is low or zero. Yearly averages are even less meaningful since inclusion of winter months will bring average ozone levels down almost to the rural background level.

A particular monthly variation in oxidant level is indicated in Fig. 5.10 for Phoenix, Denver, and Los Angeles. The average here is the mean by month of the daily maximum 1-hour average concentrations.

TABLE 5.6 Summary of maximum oxidant concentrations recorded in selected cities, 1964–1967

Station	Total days of available data	Number and percent of total days with maximum hourly average equal to or greater than concentration specified						Maximum hourly average, ppm	Peak concentration, ppm
		0.15 ppm		0.10 ppm		0.05 ppm			
		Days	Percent of days	Days	Percent of days	Days	Percent of days		
Pasadena	728	299	41.1	401	55.1	546	75.0	0.46	0.67
Los Angeles	730	220	30.1	354	48.5	540	74.0	0.58	0.65
San Diego	623	35	5.6	130	20.9	440	70.6	0.38	0.46
Denver*	285	14	4.9	51	17.9	226	79.3	0.25	0.31
St. Louis	582	14	2.4	59	10.1	362	62.2	0.35	0.85
Philadelphia	556	13	2.3	60	10.9	233	41.9	0.21	0.25
Sacramento	711	16	2.3	104	14.6	443	62.3	0.26	0.45
Cincinnati	613	10	1.6	55	9.0	319	52.0	0.26	0.32
Santa Barbara	723	11	1.5	76	10.5	510	70.5	0.25	0.28
Washington, D.C.	577	7	1.2	65	11.3	313	54.2	0.21	0.24
San Francisco	647	6	0.9	29	4.5	185	28.6	0.18	0.22
Chicago	530	0	0	24	4.5	269	50.8	0.13	0.19

*11 months of data beginning February, 1965.
Source: AP-63.

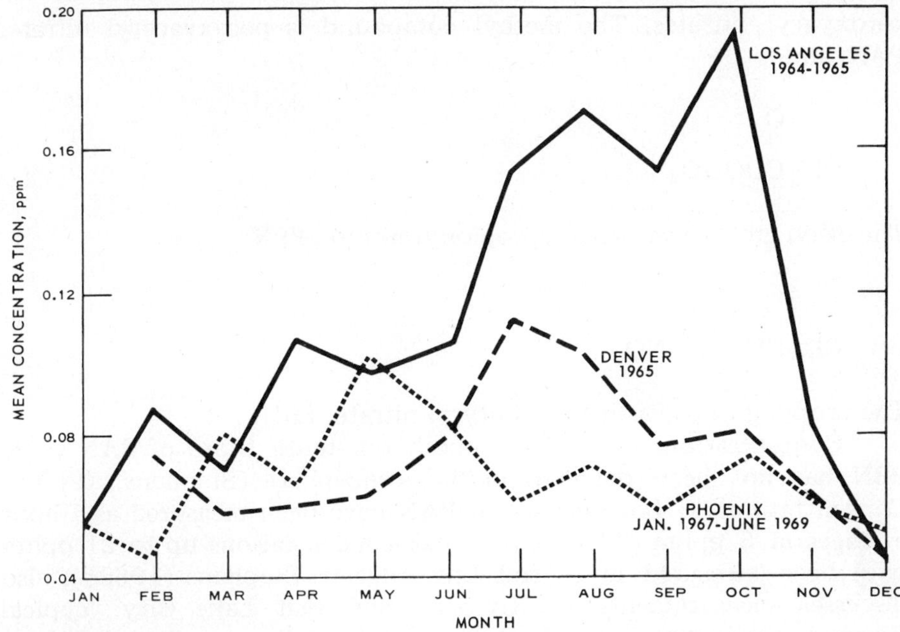

Fig. 5.10 Monthly variation of mean daily maximum 1-hour average oxidant concentrations for three selected cities. (*From AP-63.*)

These represent therefore a measure of the worst conditions averaged out first over 1-hour periods and then over a monthly period. Since Los Angeles tends to have stable atmospheric conditions and fewer clouds in the late summer and autumn, we find peak concentrations high in August, September, and October. In Phoenix, July, August, and early September are rainy months with frequent afternoon thunderstorms and clouds. Here, the clear, sunny months of May and June have the peak oxidant levels. Denver shows a "normal" summer peak.

PEROXYACETYL NITRATE

Leighton's (1961) book discusses in detail the isolation of PAN and the numerous experiments required to discover this particular product of photochemical smog. It is an interesting story of chemical detective work, particularly since the book was written just when the relevant experiments had been completed. Before the compound was identified it was called compound X and many people attributed all of the atmospheric ills of the Los Angeles basin to the mysterious compound X. Several years of work were necessary before it was decided that a new family of oxidants was manufactured in the atmosphere, the peroxyacyl nitrates. The methyl compound is peroxyacetyl nitrate, PAN, and is

$$CH_3\overset{\overset{\displaystyle O}{\|}}{C}OONO_2$$

The ethyl group gives peroxypropionyl nitrate, PPN:

$$CH_3CH_2\overset{\overset{\displaystyle O}{\|}}{C}OONO_2$$

The propyl group gives peroxybutyryl nitrate, PBN.

Concentrations of PPN are about one-tenth those of PAN while PBN has not been detected in the atmosphere (Stephens, 1969*b*). Concentrations in Los Angeles of PAN have been measured as 1-hour averages at 6 pphm (AP-63). But peak concentrations up to 21 pphm have been measured in central Los Angeles. Stephens (1969*b*), also discusses measurements in Riverside and Salt Lake City. Typical Riverside measurements, 1 pphm with peaks to 5 pphm, are substantially lower than those of downtown Los Angeles. The compound is unstable and disappears under irradiation and by chemical reactions.

Let us now compare the new 1971 Environmental Protection Agency ambient air quality standards with the present concentrations. The federal standard calls for photochemical oxidants, corrected for SO_2 and NO_2, not to exceed 160 $\mu g/m^3$ (0.08 ppm) for a 1-hour averaging time. This limit is not to be violated more than once per year. Oxidant levels noted in Table 5.6 show that Pasadena exceeded this level at least 55 percent of the days. San Diego was over the limit more than 20 percent of the days. Every city listed had a maximum hourly average equal to or greater than 0.13 ppm. Thus every city is over the limit and every city except Chicago is at least 100 percent over the new standard. We clearly have a long way to go!

REFERENCES

Air Quality Criteria for Hydrocarbons, National Air Pollution Control Administration, publication AP-64, 1970.

Air Quality Criteria for Photochemical Oxidants, National Air Pollution Control Administration, publication AP-63, 1970.

Cadle, R. D. and E. R. Allen: Atmospheric Chemistry, *Science*, vol. 167, pp. 243–249, 1970.

Calvert, J. G. and J. N. Pitts: "Photochemistry," John Wiley & Sons, Inc., New York, 1966.

Glasson, W. A. and C. S. Tuesday: Hydrocarbon Reactivities in the Atmospheric Photooxidation of Nitric Oxide, *Environ. Sci. Technol.*, vol. 4, pp. 916-924, 1970.

Leighton, P. A.: "Photochemistry of Air Pollution," Academic Press, New York, 1961.

Los Angeles County Air Pollution Control District: Profile of Air Pollution in Los Angeles County, 1969.

Robinson, N.: "Solar Radiation," Elsevier, Amsterdam, 1966.

Seinfeld, J.: lecture, May 6, 1971, University of Arizona, Chemical Engineering Department.

Stephens, E. R.: Photochemistry of Smog, *Calif. Air Environment*, vol. 1, Apr-June, 1969 (*a*).

——: Chemistry of Atmospheric Oxidants, *J. Air Pollution Control Assoc.*, vol. 19, pp. 181-185, 1968 (*b*).

Westberg, K. and N. Cohen: The Chemical Kinetics of Photochemical Smog as Analyzed by a Computer, AIAA paper 70-753, AIAA Third Fluid and Plasma Dynamics Conference, July, 1970.

6
THE AUTOMOBILE

> It still appears that a cleaned up internal combustion engine, possibly with extensive modifications, will be the overall best automotive power plant for at least the next decade . . .
>
> *W. M. Brehob, 1971*

We have seen that motor vehicles are a major source of several of the air pollutants. We have also discussed photochemical smog which results from reactions among the chemical species emitted from motor vehicles, especially gasoline-burning cars. In this chapter we shall learn more about these emissions and the control measures that have been and are being taken. In order to understand the methods of control it is necessary to consider the thermodynamics of internal and external combustion engines so we may understand typical operating conditions.

First of all, however, we should consider why pollution from cars is a problem in 1970 when it was not so in 1940. Though engine design has improved to reduce pollutants, why does pollution increase?

Briefly, the number of cars is increasing at a rate which is even more rapid than the population rise. Table 6.1 lists the population and the number of motor vehicles in the United States. More importantly, the number of vehicles per capita is also listed. We see that recent vehicle pollution problems result not merely from a rising population but more so from the increase in vehicles/person. Considering that some

TABLE 6.1 Automobile ownership

Year	Population	Motor vehicles*	Vehicles/person
1900	75,994,000	8,000	—
1910	91,972,000	468,000	0.005
1920	105,710,000	9,231,000	0.087
1930	122,775,000	26,545,000	0.22
1940	131,669,000	32,025,000	0.24
1950	150,697,000	49,161,000	0.33
1955	—	62,689,000	0.38‡
1960	178,464,000	73,869,000	0.41
1965	—	90,360,000	0.47‡
1970	204,000,000	104,702,000†	0.51

*Cars, trucks, and buses.
†1969 data.
‡Estimated.

67 million of the 1970 population were 15 or younger and another 7.5 million were 75 or older we find that the number of vehicles per eligible driver is fast approaching 1. This may not be the ultimate figure but it does represent the maximum for pollution since one can drive only one vehicle at a time. Figures 5.4 and 5.5, which present HC and NO_x emissions in Los Angeles County, have already indicated what happens to emissions as the population and per capita car ownership increase.

OPERATING CONDITIONS

The operating conditions for both internal and external combustion engines may be modeled as "pattern cycles" with thermodynamic coordinates such as pressure-volume, p-v, or temperature-entropy, T-s.

The *Otto cycle* is the pattern cycle for reciprocating spark-ignition engines. The pressure of the gas within the cylinder of an idealized spark-ignition engine is shown as a function of the piston position in Fig. 6.1. With the piston at top dead center (tdc), the intake valve opens and a fresh charge of fuel-air mixture is sucked in. At bottom dead center (bdc) the intake valve closes, and the return stroke causes the gas to be compressed. In the idealized system ignition occurs instantaneously at tdc, causing a rapid rise in temperature and pressure. The gas is then expanded on the outstroke, until at bdc the exhaust valve opens, and the gas "blows down" through the exhaust port. With a fourth stroke the gases are purged out. In the idealized Otto cycle the compression and expansion processes are considered reversible and

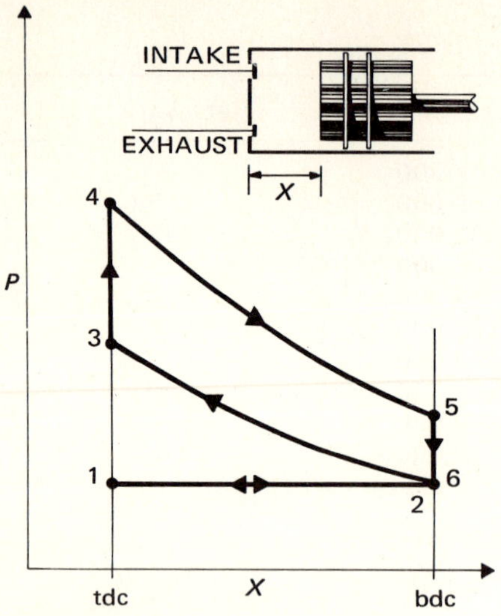

Fig. 6.1 Pressures in an idealized spark-ignition engine.

adiabatic, that is, isentropic, and it is assumed that the pressure within the cylinder during the intake and exhaust strokes is equal to the atmospheric pressure. The work done by the piston on the gas inside the cylinder during the exhaust stroke is exactly equal to the work done on the piston by the gas during the intake stroke, so that useful work output results only from the excess of the work done by the gas during the expansion stroke over that done on the gas during the compression stroke.

The process representation for the fluid during the compression, ignition, and expansion parts of the cycle is shown in Fig. 6.2. The combustion process is idealized in terms of a simple energy addition (as

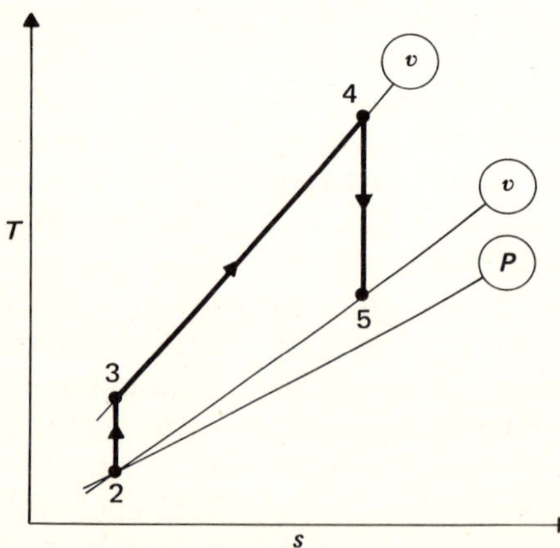

Fig. 6.2 Idealized Otto-cycle process representation.

heat), and the changes in the chemical composition of the mixture are neglected. If it is further idealized that the gas is a perfect gas with constant specific heats, appropriate thermodynamic analysis leads to a simple algebraic expression for the efficiency of the Otto cycle in terms of the compression ratio,

$$\eta = \frac{W_{\text{net}}}{Q_{34}} = 1 - \frac{1}{r^{k-1}} \tag{6.1}$$

where the compression ratio r is

$$r = \frac{v_2}{v_3} = \frac{v_5}{v_4}$$

This relationship for $k = 1.4$ is shown in Fig. 6.3.

It is clear that by increasing the compression ratio we can improve the efficiency of the ideal engine. (In the real engine, however, there is a maximum compression above which efficiency does not increase.) However, this would also increase the peak temperatures in the engine which, for example affects the production of NO_x. There is therefore a trade off between efficiency and pollution.

A real spark-ignition engine will not meet the performance of the highly idealized Otto cycle. Combustion takes time, and for this reason it is initiated before tdc by "advancing the spark." Furthermore, there will be a pressure drop across the valve during intake and exhaust; the piston must do work on the air to get it out, and this is more than the work done on the piston by the cylinder gases during the intake stroke. Heat transfer is involved, so the compression and expansion processes are not isentropic. The pressure-displacement diagram of a realistic spark-ignition engine is shown in Fig. 6.4.

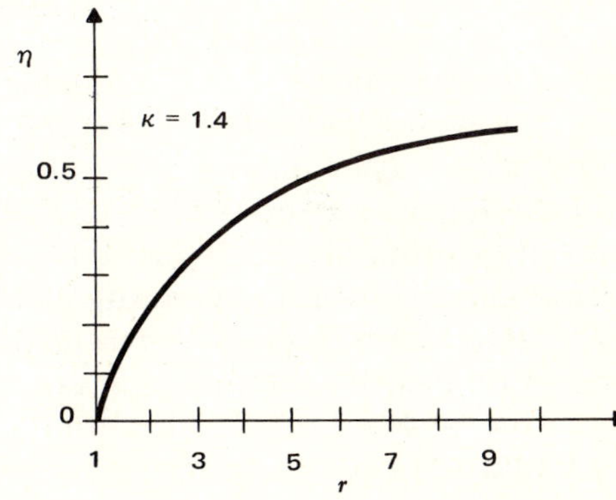

Fig. 6.3 Otto-cycle efficiency.

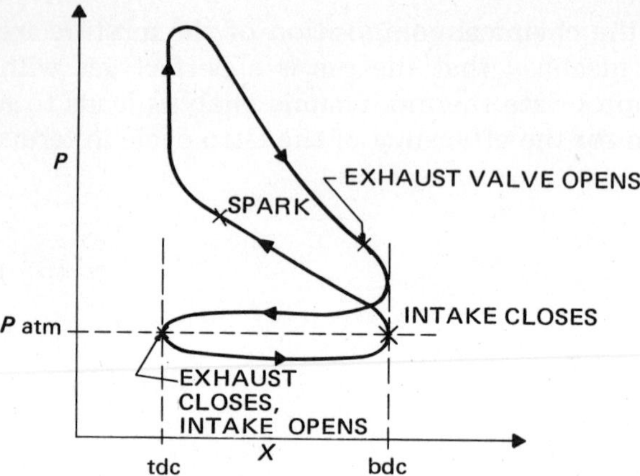

Fig. 6.4 Pressures in a real spark-ignition engine.

The peak pressures in such an engine will be 30–40 atmospheres and the peak temperatures 4500–5000°R (2500–2800°K). In the ideal Otto cycle the peak operating conditions are, of course, higher. But in the real cycle the peak (point 4 in Fig. 6.1) is cut off since the combustion process is not a heat addition at fixed volume, that is at tdc, but is a chemical reaction occurring during a period of time while the compression and the power strokes of the engine take place. The pressure at the engine exhaust is above atmospheric and the temperature is about 1500°R (not at the tail pipe!).

At part load or idle the peak cycle pressures and temperatures are lower and consequently the pollution formation mechanisms and concentrations are quite different.

From the curve of efficiency versus compression ratio it is clear that one would like to operate at high compression ratio, v_2/v_3. However, in the spark-ignition engine this is often difficult to do, because the fuel-air mixture may undergo *detonation* in which the fuel begins to burn beyond the area ignited by the spark. This causes high-pressure waves in the combustion chamber ("knocking") which may cause severe damage if allowed to continue over a period of time. Fuels with higher octane ratings have higher detonation temperatures, and the higher compression ratios of the automobiles of the late 1960s were due primarily to improved fuel rather than to improved engine technology. However, operating at high compression ratio increases HC and NO_x formation so present engines are not operated at the maximum possible compression ratio.

One method of avoiding the detonation during compression with its limitation on the compression ratio is to inject the fuel *after* the

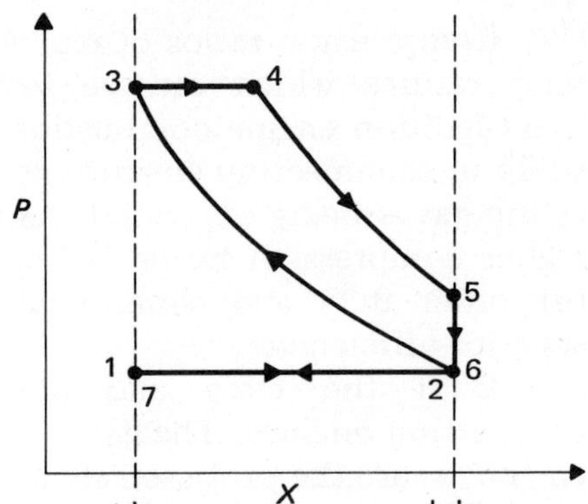

Fig. 6.5 Pressures in an idealized diesel engine.

compression process. Since there is no fuel in the cylinder during the compression process, higher compression ratios (and therefore higher air temperatures) can be reached without detonation. Then, upon injection of the fuel, it ignites spontaneously owing to the high air temperature. This ignition tends to occur fairly uniformly over the cylinder, without harmful blast waves. Such an engine is called a *compression-* (rather than spark-) *ignition* engine.

The *diesel cycle* is the pattern cycle for reciprocating compression-ignition engines. In the idealized system air is compressed to tdc, at which time fuel is injected, and it is idealized that the combustion process takes place at constant pressure for part of the expansion stroke. The remainder of the expansion stroke and the compression stroke are idealized as isentropic. The pressures within the cylinder of an idealized diesel engine are shown in Fig. 6.5. The process representation is shown in Fig. 6.6.

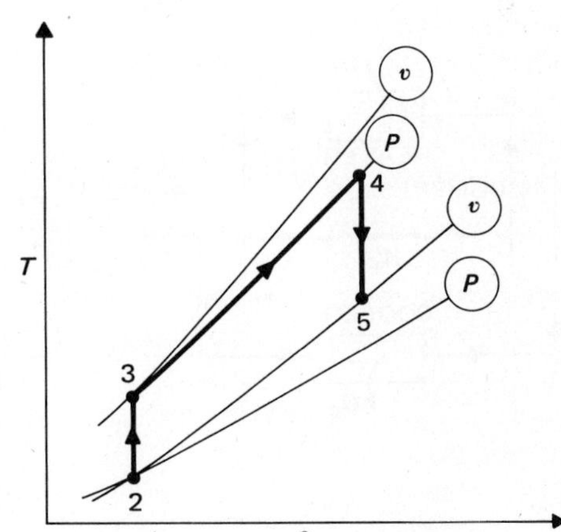

Fig. 6.6 Idealized diesel-cycle process representation.

Compression ratios of real engines are limited by the pressures and temperatures which can be tolerated within the cylinders. In the spark-ignition engine combustion occurs as the gas is being compressed, while in the injection engines (compression-ignition) combustion begins as the gas is being expanded. As a result, diesel engines can operate at higher compression ratios (of order 18:1) than spark-ignition engines (of order 9:1) and consequently, can obtain comparable and even superior efficiencies.

Both the Otto and diesel cycles are models for internal combustion engines. The gas turbine engine is a continuous combustion engine (as are the proposed steam engines for motor vehicles).

The pattern cycle for the gas-turbine power system is known as the *Brayton cycle*. The hardware schematic and process representation for a closed Brayton-cycle system are shown in Fig. 6.7. In the idealized Brayton cycle the compressor and turbine processes are isentropic, and pressure drop across the heat exchangers is neglected. If the working fluid is treated as a perfect gas with constant specific heats, a simple expression for energy-conversion efficiency may be obtained in terms of the pressure ratio

$$\eta = 1 - \left[\frac{1}{P^*}\right]^{(k-1)/k} \tag{6.2}$$

where

$$P^* = \frac{P_2}{P_1} = \frac{P_3}{P_4}$$

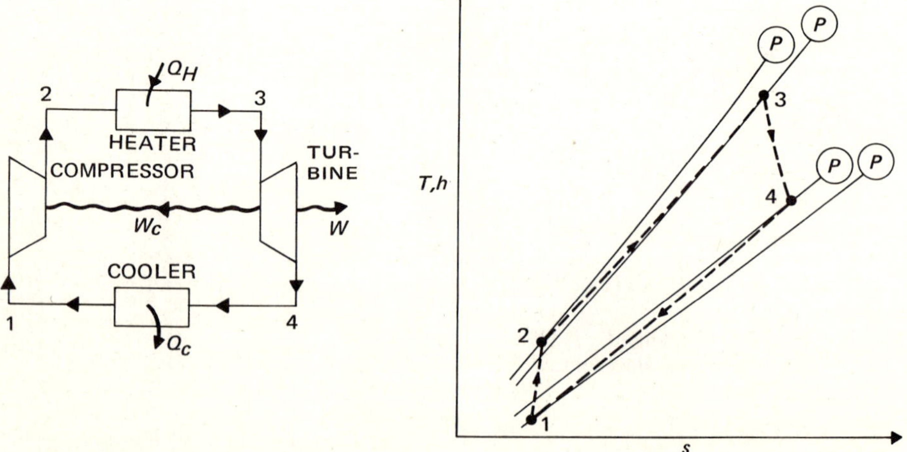

Fig. 6.7 The Brayton cycle.

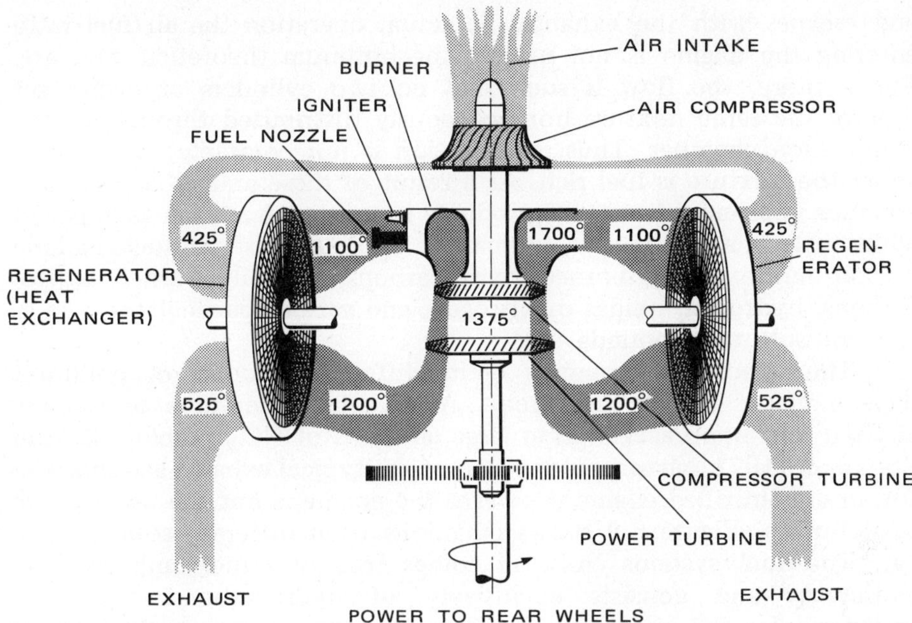

Fig. 6.8 Schematic of an automotive gas-turbine power system.

The development of compact and efficient regenerators may allow the gas-turbine engine to move into the competitive automotive market. Of particular importance is the *rotary regenerator*, an unusual type of heat exchanger. A disk of porous metal is rotated through two semicircular ducts. Hot gases in one duct heat up the metal, which is then moved by the disk rotation to the other duct, where the energy is transferred as heat from the metal to the cooler gas. Such exchangers can be made very compact and highly efficient. Figure 6.8 shows a schematic of one manufacturer's system. Note particularly the dual turbine disks, and the operating temperatures, in °F, revealed in the diagram.

EMISSION SOURCES FROM CARS

In this section we shall be most interested in emissions from automobiles since these are the largest source of motor-vehicle pollutants. Actual engine operation involves a very rapid batch-burning process. After ignition, the flame progresses in the combustion chamber until it cools and stops burning or is quenched as it nears the chamber wall. This process leaves a layer of unburned hydrocarbon next to the wall, a portion of which subsequently mixes with the burned charge and escapes with the exhaust. In actual operation the air/fuel ratio entering the engine is not usually the optimum theoretical mixture. Furthermore, the flow is such that no two cylinders or cycles get exactly the same mixture homogeneously distributed throughout the combustion chamber. Thus, combustion is not complete, particularly when the mixture is fuel rich. As a result of these and other factors a complex mixture of exhaust products is emitted from the tailpipe. In addition to water, nitrogen, and carbon dioxide, these products include carbon monoxide, unburned hydrocarbons, partially burned hydrocarbons, hydrogen, oxides of nitrogen, and various particulates such as lead and sulfur compounds.

The uncontrolled engine emits different amounts of pollutant depending on the driving mode. At idle large amounts of CO are emitted and upon acceleration large amounts of NO_x. Table 6.2 from Starkman (1971) gives an indication of the typical exhaust gas emission for an uncontrolled engine. However the engine is not the only source of pollution. Figure 6.9 shows emissions from different sources on a car. The fuel systems emission comes from the fuel tank and the carburetor and consists exclusively of hydrocarbons. Crankcase emission, also HC, comes mainly from the gas-air mixture which blows by the piston rings. Exhaust gas contains CO, NO_x, and additional HC as well as the lead compounds emitted. An indication of the variety of HC compounds from the products of combustion of an automobile engine is given in Fig. 6.10 from Hafstad (1969). We see that there are an incredible variety of HC compounds which can be emitted. We can also see that effective control techniques must include ways to reduce emissions from the crankcase, the fuel tank and carburetor, and the engine itself.

TABLE 6.2 Typical exhaust gas compositions

Mode of operation	Unburned hydrocarbons* ppm	Carbon monoxide vol. percent	Nitrogen oxides ppm	Hydrogen vol. percent	Carbon dioxide vol. percent	Water vol. percent
Idle	750	5.2	30	1.7	9.5	13.0
Cruise	300	0.8	1500	0.2	12.5	13.1
Acceleration	400	5.2	3000	1.2	10.2	13.2
Deceleration	4000	4.2	60	1.7	9.5	13.0

*Note results taken with a flame ionization detector (FID) are about 80 percent higher.

Source: Starkman, 1971.

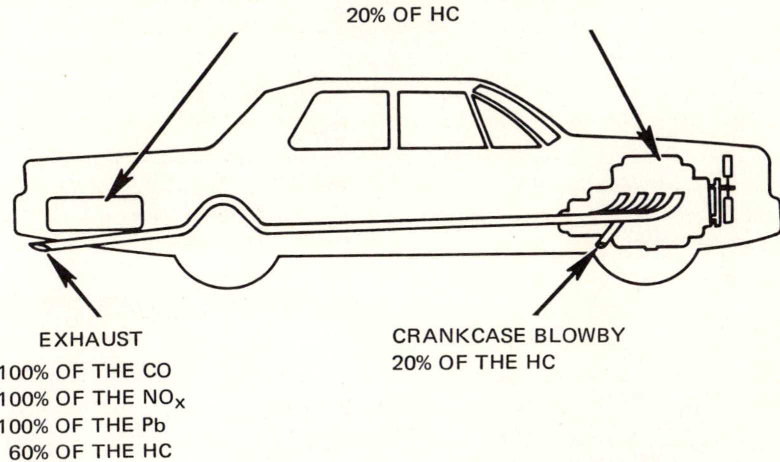

Fig. 6.9 Approximate distribution of automobile emissions by source. (*AP-66, 1970.*)

It is well at this point to recall some of the information of Chap. 4 on combustion. There we found that for the stoichiometric combustion of isooctane, C_8H_{18}, there should be no emissions of anything except H_2O and CO_2, that is,

$$C_8H_{18} + 12.5O_2 + 12.5(3.76)N_2 \rightarrow 9H_2O + 8CO_2 + 12.5(3.76)N_2 \quad (6.3)$$

Also the stoichiometric air/fuel ratio was 15 lb of air to every pound of gasoline. However, we know that the exhaust contains hydrocarbon compounds, NO_x and CO. Why? In Chap. 4 we also discussed the ideas of chemical equilibrium and reaction kinetics. We found that both CO and NO_x (even given time to reach chemical equilibrium) would exist in auto exhaust if it were at high temperatures. Since the exhaust is, in fact, not above 1500°K (2700°R) we would expect little or no CO and NO_x. Furthermore, the equilibrium calculations predicted no unburned hydrocarbons.

Conclusions from the chemical equilibrium calculations of Chap. 4 are reprinted here.

1. The effect of equivalence ratio, ϕ, or alternatively the air/fuel ratio, is very important. Air-rich mixtures, $\phi < 1$ will tend to produce minimal CO but will produce NO.

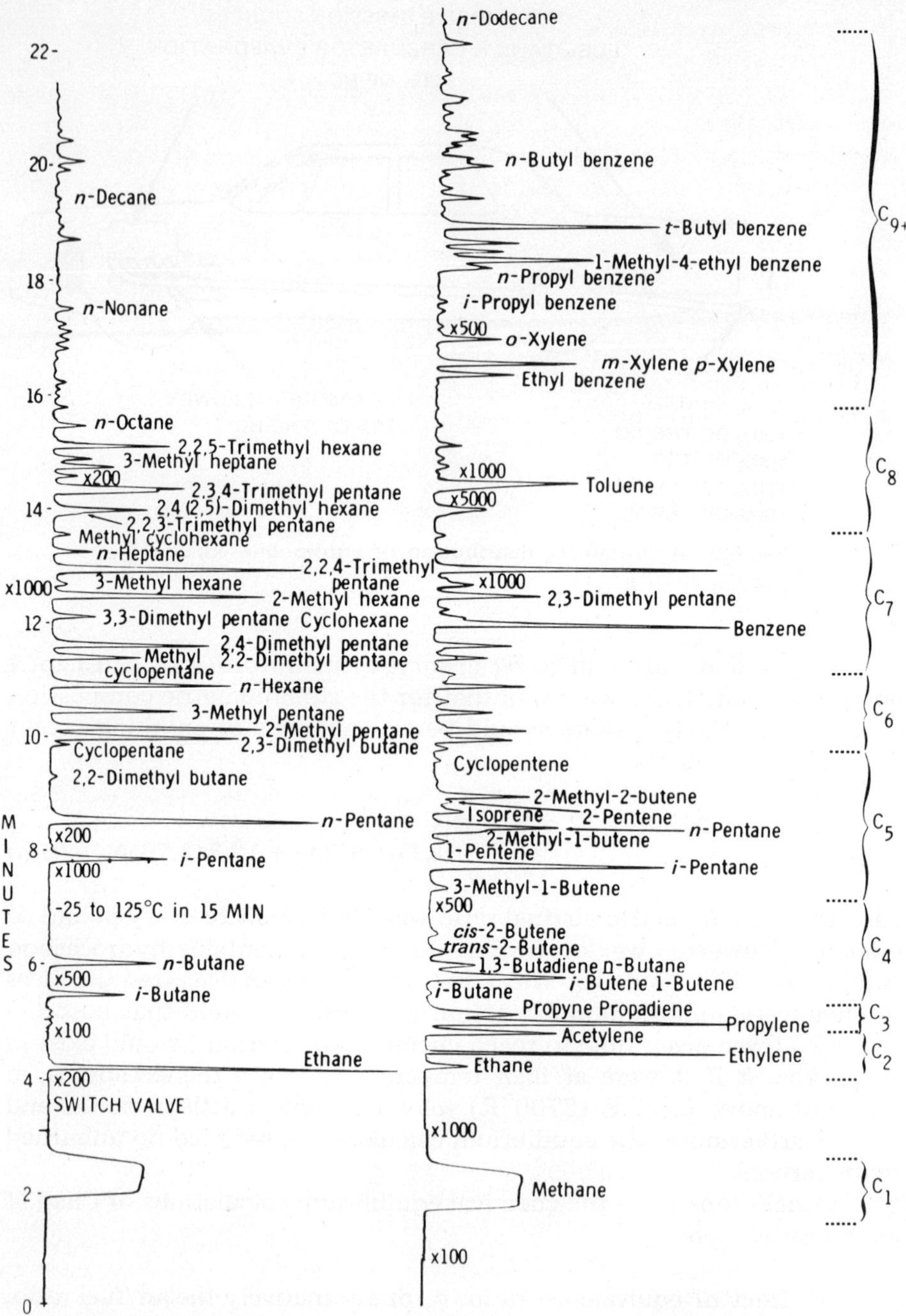

Fig. 6.10 Gas chromatograph showing products of combustion in automobile engine. (*Hafstad, 1969.*)

2. Keeping combustion temperatures down tends to reduce CO concentrations and NO formation.
3. No unburned hydrocarbons are predicted from equilibrium calculations.
4. Below 1500°K (2700°R) equilibrium calculations indicate little pollutant formation (but we know this is not true).

From (3) and (4) we were led to look at the effects of chemical kinetics and we found that the combustion process is usually kinetically limited and does not reach an equilibrium state. For example at exhaust conditions some concentration is reached which is more representative of a previous temperature and pressure and not the exhaust condition. Simplified equilibrium calculations for CO have been presented in Chap. 4. Similar calculations for NO are presented in Chap. 12. The formation of NO is enhanced by high temperatures and the availability of O_2, that is, larger than stoichiometric AFR.

CO

Carbon monoxide remains in the exhaust if the oxidation of CO to CO_2 is not complete. Generally this is due to a lack of sufficient oxygen. Our previous calculations indicated that increasing the AFR should decrease the CO concentration and this is correct. Figure 6.11 presents the results of calculations for carbon monoxide concentration for stoichiometric air resulting from the combustion of isooctane (Newhall, 1969). During the initial stages of expansion we see that the carbon monoxide concentration is not kinetically limited and can follow the

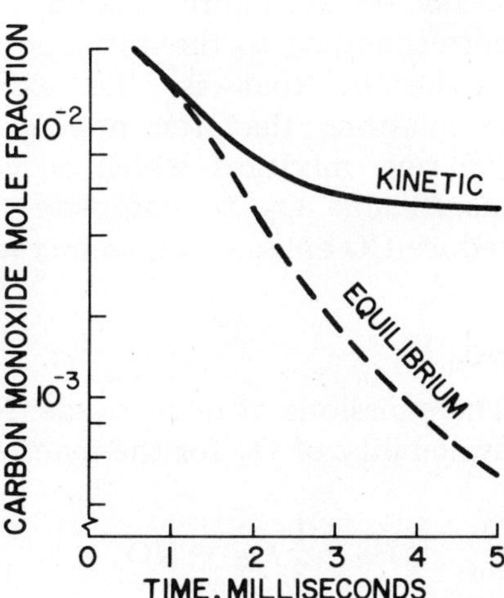

Fig. 6.11 Carbon monoxide concentration during expansion; fuel composition, C_8H_{18}; equivalence ratio, 1.0; engine speed, 4,000 rpm; compression ratio, 9:1. *(From Newhall, 1969.)*

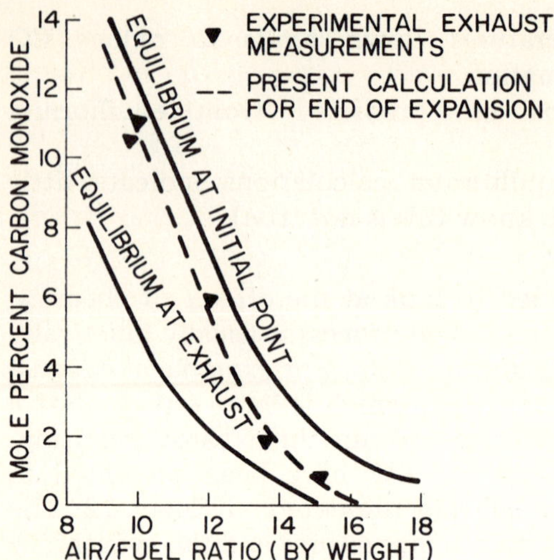

Fig. 6.12 Comparison of measured exhaust carbon monoxide concentrations with concentrations computed for end of expansion; fuel composition, C_8H_{18}. (*From Newhall, 1969.*)

equilibrium curve. However, the calculations show that as the expansion, noted here as time, continues, the process becomes kinetically limited. At the end of the expansion the CO concentration is much higher than the equilibrium value.

Figure 6.12 shows a comparison of measured and computed CO concentration. These are for the end of the expansion process and agreement between the kinetic calculations and data is good. Also shown on the figure as solid lines are chemical equilibrium calculations corresponding to the initial point of expansion and to the final point of expansion, that is, the exhaust. We see from the equilibrium calculations that the nonequilibrium behavior is most evident for fuel-rich mixtures which is as expected since at low AFR there is insufficient O_2 to complete the reaction. Thus an obvious way to reduce CO emissions is to increase the air/fuel ratio.

NO_x

The emissions of nitric oxide, NO, are governed by two parameters: the availability of O_2 for the reaction

$$\frac{1}{2}N_2 + \frac{1}{2}O_2 \rightleftharpoons NO \tag{6.4}$$

(really $O + N_2 \rightleftharpoons NO + N$ and $N + O_2 \rightleftharpoons NO + O$), and high enough temperatures to promote this reaction. Once the NO is formed equilibrium calculations indicate that it should decompose. Kinetic limitations again occur, however, as with CO, so that in fact NO remains in the exhaust and does not decompose. Figure 6.11 for CO is also a qualitative representation of what happens to the nitric oxide after formation. The concentration of NO is governed by the kinetic calculation and not the equilibrium calculation.

With NO the process is more complicated since kinetics also limit the formation of NO. The time for the NO to form is longer than the time available so that equilibrium at peak temperature conditions is not obtained. Thus the exhaust NO concentration is neither that corresponding to peak conditions nor to exhaust conditions.

Figure 6.13 (Myers et al., 1971) summarizes this, noting the equilibrium concentration compared to the actual concentrations. During the compression process the reaction rates are too slow to allow the NO concentration to reach peak equilibrium values. During expansion the temperature quickly drops to values low enough so that the slow decomposition reactions stop and the NO level is "frozen" at a higher than equilibrium value.

We would expect that higher temperature would promote the formation of NO by speeding the formation reactions. Ample O_2 supplies should also increase the formation of NO. We would then expect NO levels to be low in fuel-rich operation, that is, AFR < 15, since there is little O_2 left over to react with the N_2 after the HC's have had an opportunity to react. The maximum NO levels are formed with AFR about 10 percent above stoichiometric. More air than this reduces

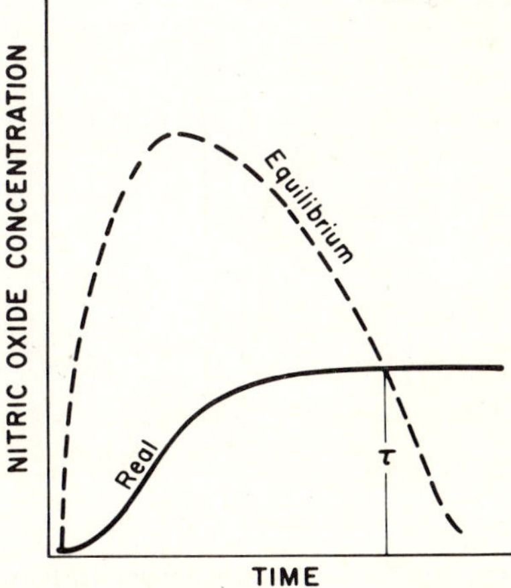

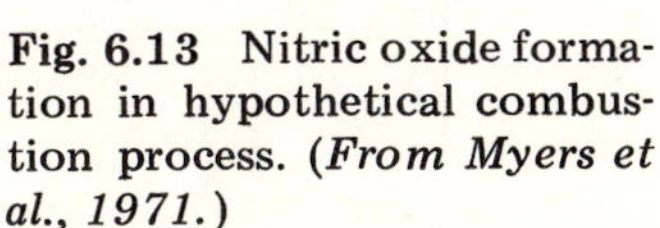
Fig. 6.13 Nitric oxide formation in hypothetical combustion process. (*From Myers et al., 1971.*)

the peak temperatures, since the excess air must be heated from energy released during combustion, and the NO concentrations fall off even with the additional excess oxygen.

Measurements taken of NO concentrations at the exhaust valve indicate that the concentration rises to a peak and then falls as the combustion gases exhaust from the cylinder. This is consistent with the idea that NO is formed in the bulk gases, since the first gas exhausted is that near the exhaust valve, followed by the bulk gases. The last gases out should be those from near the cylinder wall and should exhibit both lower temperatures and lower NO concentration.

HYDROCARBONS

As we have seen the HC emissions come from the crankcase, the fuel system, and the exhaust. Our equilibrium calculations predicted no unburned hydrocarbons and kinetic calculations indicate that the hydrocarbon reactions are fast and not kinetically limited. Thus we must seek another explanation for the fact of incomplete combustion.

Figure 6.14*a* indicates a typical temperature profile across the cylinder and through the cylinder wall. Because of the cooling there must be a cold zone next to the cooled combustion chamber walls. This region is called the quench zone, and because the temperatures are low the fuel-air mixture fails to ignite and remains unburned. If this model is correct the exhaust gases should show a marked variation in HC during the exhaust stroke. Figure 6.14*b* indicates why. The first gas out the exhaust exits from near the valve and is relatively cool; the last gas out has been scrapped off the cool cylinder walls and should also be

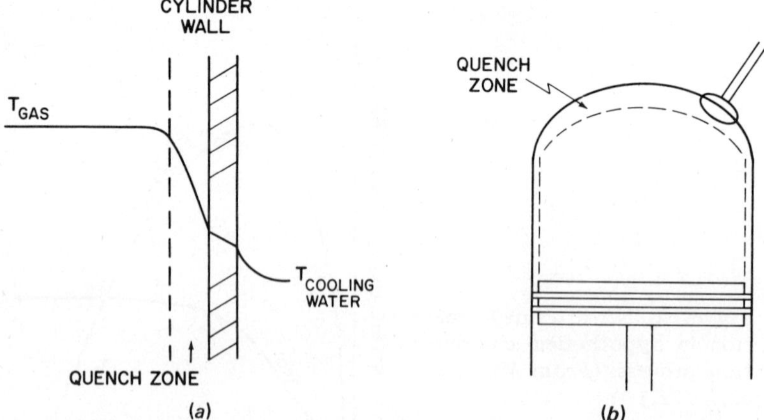

Fig. 6.14 The quench zone in the piston-cylinder engine, (*a*) and (*b*).

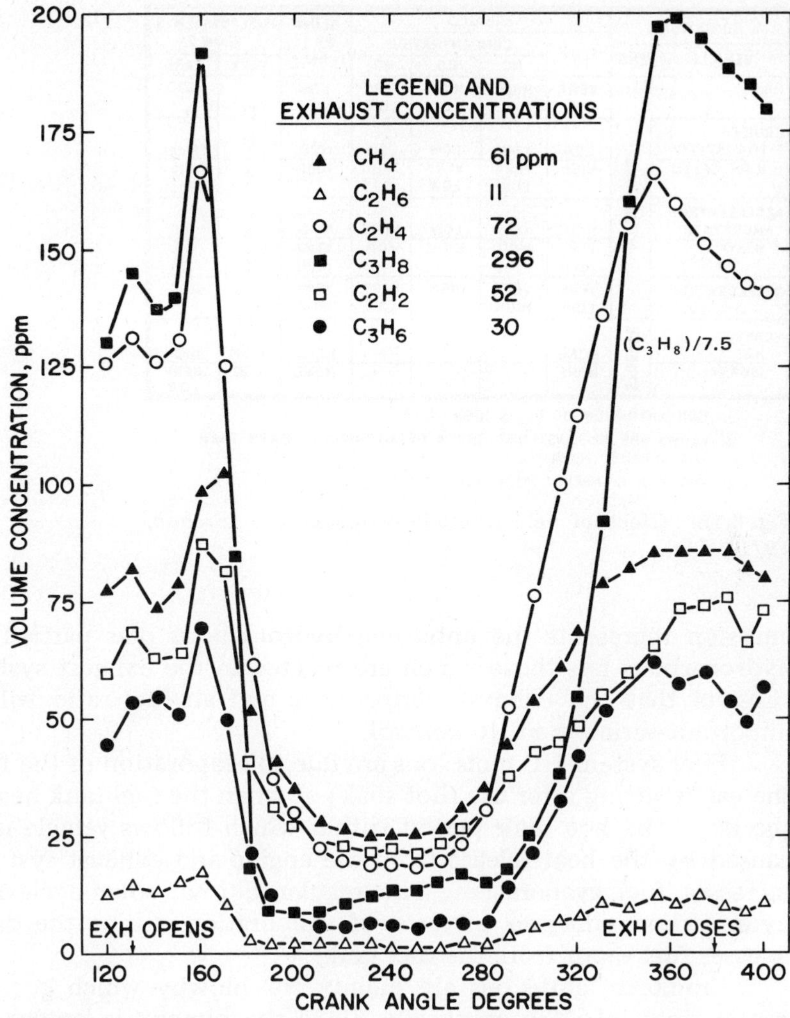

Fig. 6.15 Variation in HC concentration during exhaust stroke for an engine with propane as the fuel. (*Daniel, 1968.*)

relatively cool. During the exhaust stroke we would expect an initial high HC concentration, followed by a lower concentration as the bulk gases exhaust, and ending with a high concentration. Figure 6.15 from Daniel (1968) presents data which indicate the accuracy of this model. Thus control of hydrocarbon emission will partly involve minimizing the quench zone.

The hydrocarbons emitted in the exhaust are able to continue chemical reactions at the exhaust manifold. Therefore the exhaust

CONDITION		EXHAUST				BLOW-BY 1/ FLOW 2/	FUEL SYSTEM 3/ FLOW 2/	
			CONCENTRATION					
VEHICLE	ENG.	FLOW	HC	CO	NO$_x$		TANK	CARB.
IDLE	OPERATING	VERY LOW	HIGH	HIGH	VERY LOW	LOW	AVERAGE TO MODERATE	MOD.
CRUISE	OPERATING						AVERAGE TO MODERATE	
LOW SPEED	OPERATING	LOW	LOW	LOW	LOW	MOD.	AVERAGE TO MODERATE	SMALL
HIGH SPEED	OPERATING	HIGH	VERY LOW	VERY LOW	MOD.	HIGH	AVERAGE TO MODERATE	NIL
ACCELERATION	OPERATING						AVERAGE TO MODERATE	
MODERATE	OPERATING	HIGH	LOW	LOW	HIGH	MOD.	AVERAGE TO MODERATE	NIL
HEAVY	OPERATING	VERY HIGH	MOD.	HIGH	MOD.	VERY HIGH	AVERAGE TO MODERATE	NIL
DECELERATION	OPERATING	VERY LOW	VERY HIGH	HIGH	VERY LOW	VERY LOW	AVERAGE TO MODERATE	MOD.
SOAK	STOPPED							
HOT	STOPPED	NONE	—	—	—	NONE	HIGH	HIGH
DIURNAL	STOPPED	NONE	—	—	—	NONE	MOD.	VERY LOW

1/ CONCENTRATION OF HC IS HIGH

2/ FLOWS ARE AT LEAST ONE ORDER OF MAGNITUDE LOWER THAN THE EXHAUST FLOW.

3/ EMISSION IS NEARLY PURE HC

Fig. 6.16 Effect of vehicle mode on emissions. (*Brehob, 1971.*)

emission represents the unburned hydrocarbons plus partially burned hydrocarbons less those which are reacted in the exhaust system. This suggests that the exhaust temperature and air/fuel ratio will also be important variables in HC control.

Fuel system HC emissions are due to evaporation of the fuel when the car is sitting after use (hot soak) or when the fuel tank heats during the day. The hot soak period is that which follows vehicle use and is caused by the heat release from the engine and exhaust systems. This increases fuel evaporation. The regular diurnal solar cycle results in hydrocarbon emission because of evaporation during the day which releases fuel vapor from the fuel tank.

Crankcase emissions are mainly HC blowby which goes past the piston rings into the crankcase. Since the blowby is largely from the quench zone it is not surprising that it is high in hydrocarbon content.

Figure 6.16 from Brehob (1971) summarizes the emissions from cars as a function of operating conditions. Can you explain these trends?

CONTROLS

Hydrocarbon Controls

The earliest form of emission control was on crankcase emissions which form about 20 to 25 percent of the hydrocarbon emissions. California required some form of control in 1963 and all United States cars had mandatory equipment in 1968. Exhaust emission controls for HC and CO were required in California in 1966 and nationwide in 1968. Evaporative HC emission controls were required in California in the 1970 model and became nationwide in 1971.

Crankcase control has been achieved by closing off the vent to the atmosphere and recycling the blowby back into the intake of the engine. There are different systems for doing this but basically the intake manifold vacuum draws ventilation air and the HC's into the inlet mixture. The positive crankcase ventilation (PCV) valve regulates this process.

The evaporative loss control systems are designed to eliminate fuel vapors from the fuel tank and the carburetor. These account for about 20 percent of the HC emissions. There are different designs but Fig. 6.17 indicates the basic idea. The vapor is stored in an activated charcoal canister. The vapors can be purged through a control valve to the intake manifold and then burned in the engine. Crawford and Lindsay (1970) indicate that the evaporative loss control system results in slightly higher *exhaust* emission of HC (20 ppm extra) and CO (0.25 percent extra).

We now turn to the more difficult problem of reducing the HC exhaust emissions. The first controls were those required in California in 1966. Both CO and HC emissions can be reduced by increasing the air/fuel ratio and this is largely how the first California standards were met. Figure 6.18 indicates typical results for CO, HC, and NO_x emissions as a function of AFR. Although increasing the AFR did

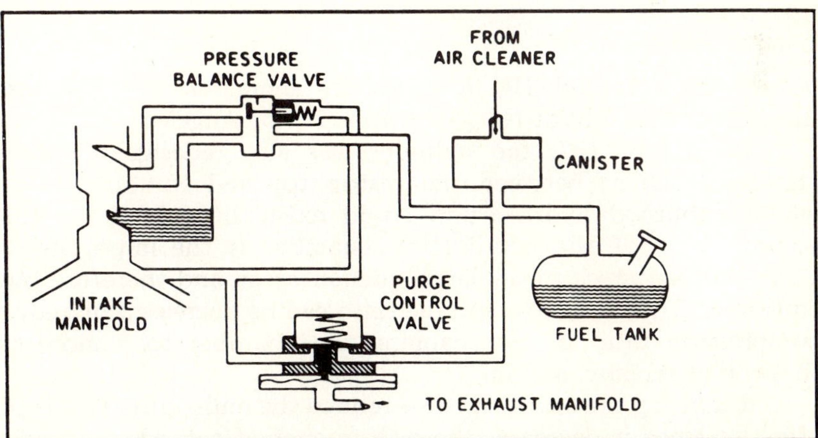

Fig. 6.17 Basic evaporative loss control system. (*From Crawford and Lindsay, 1970.*)

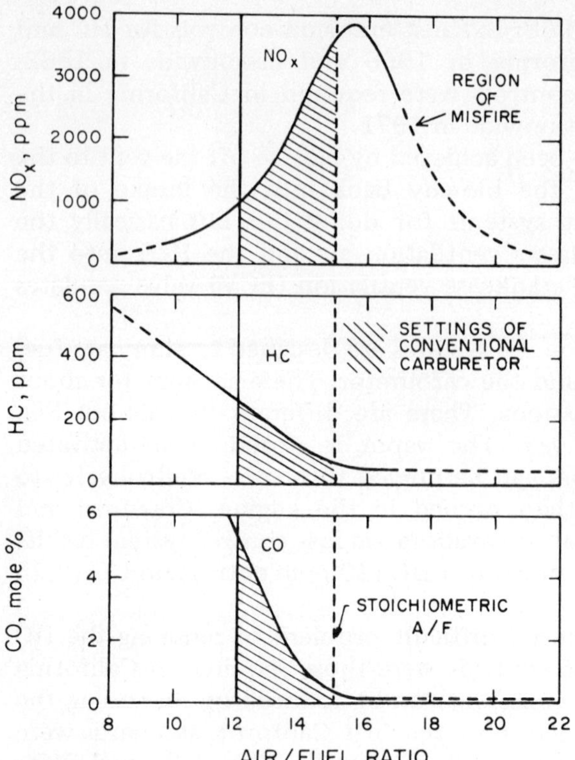

Fig. 6.18 Typical concentration of unburned hydrocarbons, carbon monoxide, and nitrogen oxides as a function of air/fuel ratio at 60 mph.

reduce the CO and HC it markedly increased the NO_x emissions (remember Fig. 5.5 for NO_x emissions in Los Angeles).

To reduce HC, the cylinder was also redesigned reducing the crevices (such as between the piston top and combustion chamber) where unburned hydrocarbon could exist. In general the surface to volume ratio of the combustion chamber is the important variable. Large surface areas mean large quench areas and therefore bigger HC emissions. The surface to volume ratio can be decreased by reducing the compression ratio or by changing the chamber to a more spherical shape, that is, short and fat.

Retarding the spark can also reduce the emissions. This is probably effective since it decreases the wall quench effect. Also the exhaust gas temperature is raised so that completion of the hydrocarbon reactions can occur in the exhaust.

Carbon Monoxide Controls

From Fig. 6.18 we see that an increase in the AFR will reduce CO emissions. For example, changing from an AFR of about 13 (4 percent CO) to 15 will reduce the CO to below 1 percent.

NO_x Controls

Unfortunately increasing the AFR to reduce HC and CO increases the production of NO_x. In fact, NO_x reaches a maximum at about the same AFR that minimizes HC and CO. Figure 6.19 indicates NO production as a function of AFR and retarding the spark. The data is from Campau and Neurman (1967). Note that retarding the spark does reduce the NO emission significantly. Reducing the compression ratio also will do this and these are two techniques which also help control HC emission. Figure 6.19 suggests several things. First, the effect of AFR is clear. Running rich or very lean reduces NO_x. Second, retarding the spark reduces peak temperatures and decreases the time available at high temperature for chemical reaction. Both of these reduce NO_x formation. But only a very few of these techniques are compatible with emission controls on the CO and HC.

Controlling NO_x is a difficult problem in itself but with the simultaneous required controls on CO and HC the problem becomes

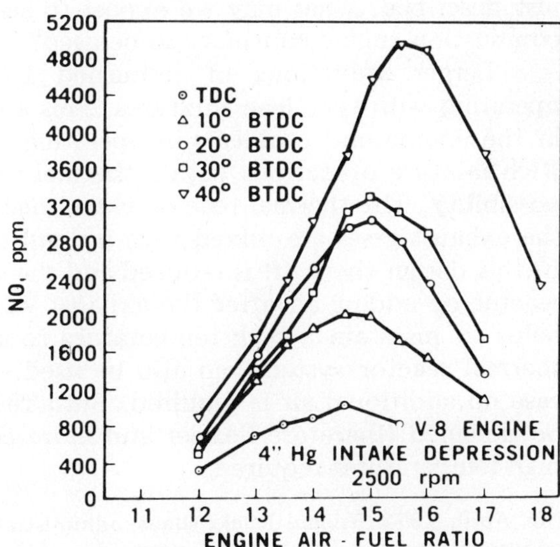

Fig. 6.19 Effects of fuel/air ratio and timing on NO_x. (*Campau and Neurman, 1967.*)

very severe. Initial NO_x controls were required in California with the 1971 model. Nationwide federal controls began with the 1973 model. Increasing controls on NO_x were required between 1972 and 1973 (initial controls) and between 1975 and 1976 (reduction from 3 g/mile to 0.4 g/mile).

FUTURE CONTROLS

Table 6.3 summarizes past and future emission limits for the typical automobile engine. Increasingly more stringent controls are required until 1976.[1] How far have we progressed? The 1970 model car, if it met requirements, emitted 23 g/mile of CO compared to 80 g/mile for an uncontrolled car. Thus CO had been reduced by about 70 percent. HC's were reduced from about 11 g/mile to 2.2 g/mile, a reduction of 80 percent. NO_x was uncontrolled nationally as of the 1970 models. The amended Clean Air Act of 1970 required further reductions for HC and CO of about 90 percent *of the 1970 levels*, which as we see were already substantially controlled, and reduction for NO_x of 90 percent of the 1971 level. So the new limits represent very stringent emission standards. The increase in CO and HC (noted in Table 6.3) between 1972 and 1973 is due to the change in test procedure whereby the driving cycle for testing emissions was changed from the California 7-mode test to the federal constant volume sample test.

Future controls will have to involve more than those techniques just described. What may we expect to see, assuming that the internal combustion engine continues to be used?

Larger reductions of unburned HC and CO may necessitate operating with very lean mixtures. This alternative causes power losses in the engine and problems in operating the engine without misfiring. Rich mixture operations with a thermal reactor in the exhaust is also a possibility. The thermal reactor is an enlarged exhaust manifold where the exhaust gases are mixed with air and the CO and HC are oxidized. In this design the AFR is reduced and the exhaust gas is oxidized in the reactor by adding air after the exhaust valve. The CO to CO_2 reaction helps to maintain a high temperature to aid in removing the HC. The thermal reactor system can also be used with lean operation in which case no additional air is required. But a reactor with less heat loss must be designed (therefore heavier and more complicated) to maintain the high temperatures required.

[1] In April, 1973 William Ruckelshaus, administrator of EPA, gave auto makers an additional year (as allowed under provisions of the 1970 Clean Air Act) to meet the *1975* standards for HC and CO. Interim standards for 1975 were set. An additional year for meeting the 1976 NO_x standards was granted later.

TABLE 6.3 New vehicle standards summary *(Light-duty vehicles under 6,000 lb)*

Year	Standard	Cold start test	Hydrocarbons	Carbon monoxide	Oxides of nitrogen
Prior to controls			850 ppm (11 g/mile)	3.4% (80 g/mile)	1000 ppm (4 g/mile)
1966–1967	State¶	7-mode*	275 ppm	1.5%	no std.
1968–1969	State & Federal	7-mode			
		50-100 CID	410 ppm	2.3%	no std.
		101-140 CID	350 ppm	2.0%	no std.
		over 140 CID	275 ppm	1.5%	no std.
			grams/mile		
1970	State & Federal	7-mode	2.2	23	no std.
1971	State &	7-mode	2.2	23	4
	Federal	7-mode	2.2	23	—
1972	State	7-mode or	1.5	23	3
		CVS-1†	3.2	39	§3.2
	Federal	CVS-1	3.4	39	—
1973	State	CVS-1	3.2	39	3
	Federal	CVS-1	3.4	39	3
1974	State	CVS-1	3.2	39	2
	Federal	CVS-1	3.4	39	3
1975	State	CVS-1	1	24	1.5
	Federal	CVS-2‡	0.41	3.4	3
1976	State	CVS-1	1	24	1.5
	Federal	CVS-2	0.41	3.4	0.4

The values in parentheses are approximately equivalent values.
*7-mode is a 137-second driving cycle test.
†CVS-1 is a constant volume sample cold start test.
‡CVS-2 is a constant volume sample cold start test average with a constant volume sample hot start test, both with the federal 22-minute driving cycle.
§Hot 7-mode.
¶State refers to California.

Catalytic oxidation is also under consideration. The catalytic reactor, or catalytic converter, can operate either on rich or lean mixtures and operates at lower temperatures than the thermal reactor. A catalytic device consists of the active catalyst deposited on a support system and placed in a can that looks about like a muffler.

Oxidizing reactors convert HC and CO to CO_2 and H_2O in an oxidizing atmosphere, that is, air rich. Selecting the catalyst is an art not a science and Starkman (1972) indicates that General Motors has evaluated over 800 materials as possible catalysts. Platinum and palladium are possibilities for the oxidizing catalyst.

Reducing reactors may be required in order to meet the 1976 standard for NO_x. A fuel-rich atmosphere is necessary to promote the reduction of NO to N_2. One possibility for the 1976 cars is a reducing reactor followed by air injection and a second reactor which would operate in an oxidizing atmosphere. The first reactor would remove NO_x, the second would remove HC and CO. Fuel-rich operation will, of course, lower fuel economy.

Unfortunately, while the aim of the reducing reactor is to reduce nitrogen oxides back to molecular nitrogen, in fact, most catalysts promote conversion of NO to ammonia, NH_3. The ammonia can be oxidized back to NO if there is an oxidizing reactor following! Ideally the ammonia should decompose to give N_2. It appears (Starkman, 1972) that the reducing reaction is

$$2NO + 5H_2 \rightarrow 2NH_3 + 2H_2O$$

and this is followed by the decomposition of ammonia as

$$2NH_3 \rightarrow N_2 + 3H_2$$

There are several other reactions which may be important (Starkman suggests that the reaction $2NH_3 + 3NO \rightarrow 3H_2O + (5/2)N_2$ is of interest in the ammonia removal process), but the essence of the problem follows from the two reactions above. The two-step reaction could be promoted by using two catalysts, for instance nickel for the second reaction and copper, platinum, or palladium for the first.

There are many problems in developing suitable converters, particularly the reducing reactor. Unless a way can be found to avoid oxidizing the ammonia back to NO the combination of a reducing reactor followed by an oxidizing reactor, obviously, will not work. Another difficulty is that lead, sulfur, and phosphorus act as poisons to the catalysts so that very low lead gasoline will be required if these devices are to operate successfully. In spite of these difficulties United States auto manufacturers are looking to catalytic converters as a necessary part of the system required to meet the 1975 and 1976 standards.

Future control techniques aimed at NO_x include exhaust gas recirculation, reducing catalytic reactors, and changed combustion sequences. In exhaust gas recirculation 10–20 percent of the exhaust is returned to the combustion chamber. Typically, NO_x reduction of 50 to 60 percent can be achieved with about 10 percent recirculation

(Brehob, 1971). This technique works by reducing peak temperatures, due to the thermal capacity of the recycled exhaust, without providing any additional oxygen. However, the use of exhaust gas recirculation (EGR) requires operating with a fuel-rich mixture; otherwise engine operation is poor. This results in increased CO and HC from the engine which must, in turn, be controlled. Both Brehob (1971) and Crawford and Lindsay (1970) indicate that recirculation increases HC emissions. However, they are in disagreement as to the amount of increase. Brehob suggests an increase of some 40 percent in HC with 10 percent recirculation at 40 mph. Crawford and Lindsay indicate a less than 10 percent increase.

The changed combustion sequence involves such ideas as initiating combustion with a fuel-rich mixture, to reduce NO_x formation, and then completing combustion, to remove HC and CO, after the expansion has reduced the chamber temperatures. A redesigned combustion chamber with a primary fuel-rich chamber and a secondary air-rich chamber has also been suggested (Newhall and El Messiri, 1970). For many years some diesel engine designs have incorporated such "precombustion chambers." This idea is used in the design of the Honda engine. Referred to as CVCC (compound vortex controlled combustion), the engine sequence starts with a fuel-rich mixture in a small chamber on top of the cylinder. In the cylinder an air-rich mixture is used.

Combustion in the small chamber occurs with no excess oxygen available for conversion to NO_x. The lean mixture in the cylinder burns at low enough temperature to minimize NO_x formation, yet it burns slowly enough to complete the oxidation of the CO and HC's. A two-barrel carburetor is used to produce the required rich and lean mixtures. Three valves are required, the conventional two and an additional one on the smaller chamber.

This design has reduced emissions after a 50,000-mile test to below the 1975 standards. However, the engine does not yet meet the 1976 NO_x standard. EPA tests rate the NO_x emissions from the 50,000 mile vehicle at about 1 g/mile whereas the standard is 0.40 g/mile.

The National Academy of Sciences report of January 1972 stated that the 1975 standards will require a car with improved carburetor performance, fast-acting and more accurate choke, inductive or electronic ignition system with modified timing, exhaust-gas recycle, secondary air pump with air injection into the exhaust ports or manifolds, and an oxidizing catalytic converter to complete the burnup of HC and CO. When its performance is optimized, this system can achieve very low HC and CO emissions once the catalyst attains its light-off temperature. However, unless additional controls are used, the catalytic converter warms up too slowly to control adequately emissions from a cold engine. The rich fuel-air mixture used while the engine is warming up results in high exhaust HC and CO emissions.

Several ideas are under study to achieve greater emission control during the engine warm-up phase. In one approach, the conventional exhaust manifold is replaced with a small-volume thermal reactor located upstream of the catalytic converter. The reactor is designed so that, very quickly after engine start, it will reach temperatures at which HC and CO will be oxidized. The combustion of some of the HC and CO raises the temperature of the gas leaving the reactor. Consequently, the catalyst bed heats up more rapidly. Once the catalyst reaches its operating temperature, the reactor and catalyst share the burnup duty, with the catalyst bed playing the major role.

FUTURE ENGINES

In order to appreciate the problem of producing an alternative engine to the present internal combustion engine we will look at the characteristics of several power sources. The Morse Report (1967) has suggested appropriate performance maps for such a comparison. We would first be interested in knowing the power per pound of engine, that is, hp/lb or watts/lb. Also the range in which we can operate an engine is important and this may be represented in units of (watts/lb) times the hours operated or watt-hr/lb. The watt/lb is the specific power and the watt-hr/lb is the specific energy.

As an example consider the internal combustion engine which might produce 200 hp and weigh 500 lb. Then the hp/lb is 0.4. This can be converted directly to watts/lb (746 watts = 1 hp) as 300 watts/lb of engine. What is the specific energy of this engine? The piston engine can run at 0.4 hp/lb for an indefinite period so it can produce 300 watts/lb for one hour or for several hours. If we plot this data as shown on Fig. 6.20 we find that the piston engine characteristics are almost a horizontal line in the upper right-hand side of the figure.

Now look at the velocity-range overlay on the figure. On a level road at a constant 50 mph, a 2,000-lb car requires a little less than 20 hp to overcome wind resistance and friction. This requirement increases greatly with acceleration or when climbing a grade. At 50 mph the required hp/lb of engine, assuming a 500-lb engine, is 20/500 or 0.04 hp/lb or 30 watts/lb. Further at 50 mph we can cover 100 miles in 2 hours, or the required watt-hr/lb is $30 \times 2 = 60$. Interpolating a speed of 50 mph for a range of 100 miles we see that, in fact, this point is at about 30 watts/lb and 60 watts-hr/lb.

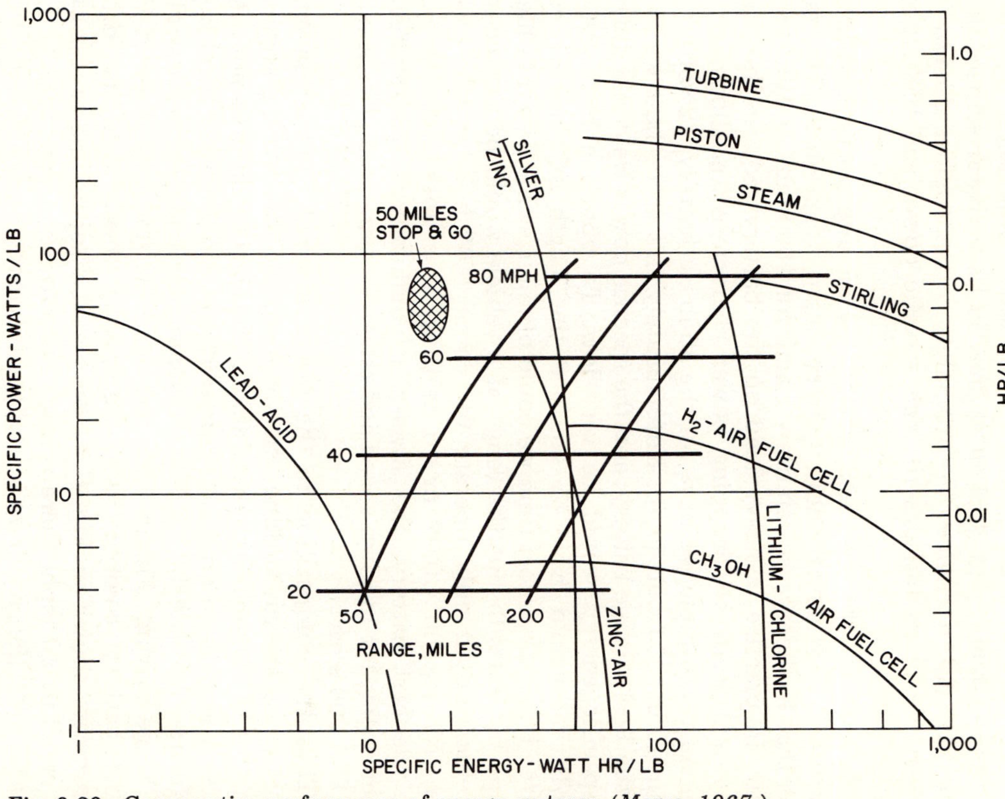

Fig. 6.20 Comparative performance of energy systems. (*Morse, 1967.*)

In this manner we can develop a performance map on which to compare alternative power sources. Figure 6.20 was developed for a 2,000-lb car, i.e., a compact, with a 500-lb engine. The ordinate and abscissa are for units of per lb engine. The piston engine can develop about 0.4 hp/lb engine. The continuous combustion cycles for the turbine, steam, and Stirling engines all have characteristics similar to the I.C. piston engine. Unfortunately, other engines do not have the desirable characteristics of high specific power and high specific energy.

How about the lead-acid battery as an alternative system? We know that a 100 ampere-hour battery might weigh 40 lb. This rating means that in theory, we can draw 1 amp for 100 hours or 100 amp for 1 hour. In practice, a heavy load on the battery reduces the voltage greatly and leads to dropping power characteristics, shown on the curve. As an example we have for the 40-lb car battery, 12 volts × 100 amp-hr/40 lb = 30 watt-hr/lb. Unfortunately this is the best we can do with the lead storage battery, and the specific power we could draw for this specific energy would be very low. To obtain a high value of watts/lb we would have to sacrifice the length of time we could operate. For a very short time we could obtain specific power as high as perhaps 50 watts/lb but the range of a vehicle would be limited since the vehicle could only operate for about 0.02 hour at this specific power (50 watt/lb × 0.02 hr = 1 watt-hr/lb as noted on Fig. 6.20). While the best battery, the silver-zinc combination, has satisfactory characteristics for in-town speed and range, its estimated price is a little high, $25,000 a car!

Thus the competition to the piston internal combustion engine appears to be the gas turbine, the steam engine, the Stirling engine, or alternative I.C. engines like the Wankel. The electric car does not appear to offer a realistic alternative to present cars. On the other hand we see that future light-weight, battery-powered cars could be successful for short range driving. Figure 6.21 indicates the status of some of these possible battery combinations. Comparing these with the driving requirements shown in Fig. 6.20 we see that for driving ranges of 50 miles at 30 to 40 mph, there is some hope for an electric vehicle if it is not necessarily the size or weight of today's car.

The gas turbine, steam, and Stirling engines are all *external* or *continuous combustion* engines which can operate with very high air/fuel ratios since the combustion process is steady state and not limited by the problems of operating in a piston internal combustion engine. Much of the air can be added in a secondary combustion zone after combustion is initiated in a near stoichiometric primary zone at the burner.

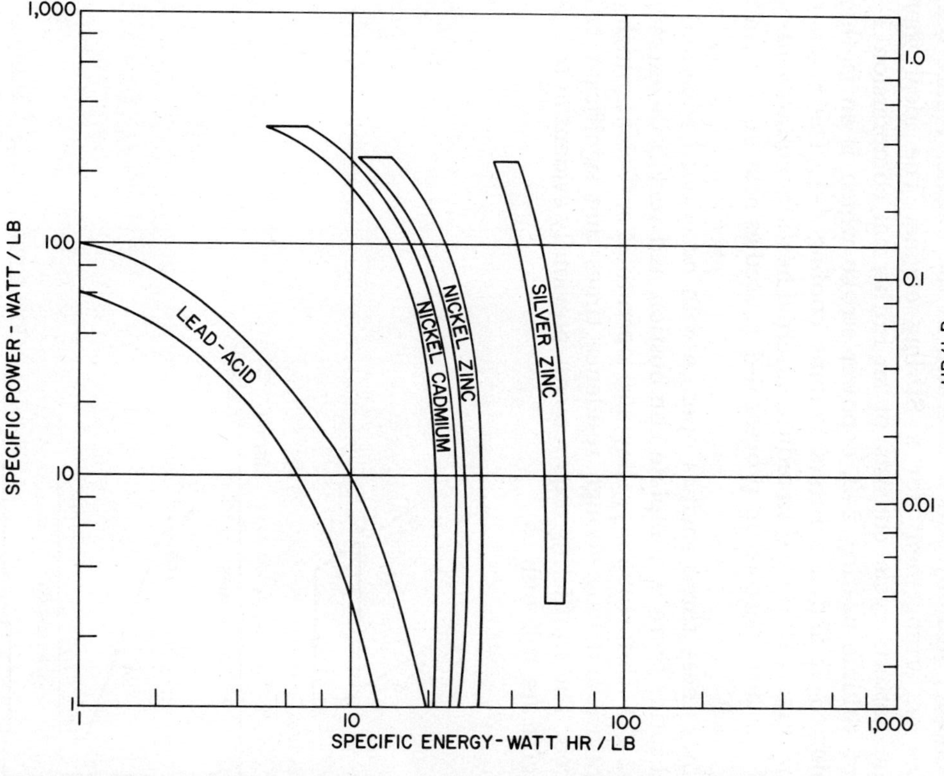

Fig. 6.21 Comparative performance of conventional batteries. (*Morse, 1967.*)

The gas-turbine cycle has already been described. The steam-car cycle is just the Rankine vapor cycle. Heat is added from an external source to vaporize the water and useful work is extracted by expanding the steam in a piston-cylinder arrangement. In the Stirling engine external combustion is used to provide energy for the working fluid (for example, helium) which is sealed in the engine. Figure 6.22*a,b* indicates an arrangement for a Stirling engine. The ideal thermodynamic pattern cycle consists of an isothermal compression 1-2, a constant volume heating 2-3, a power stroke which is an isothermal expansion 3-4, and a constant volume cooling 4-1. The regenerator noted in Fig. 6.22 is used to reduce external heating requirements since part of the heat rejected in process 4-1 is used as heat input in process 2-3.

Since these three engines operate with no quench zone and high air/fuel ratios there is complete combustion and very little emission of HC and CO. Unfortunately, they also operate at sufficiently high peak temperature and long enough residence times that significant NO_x is produced. The conflicting requirements previously evident in the piston I.C. engine are still with us.

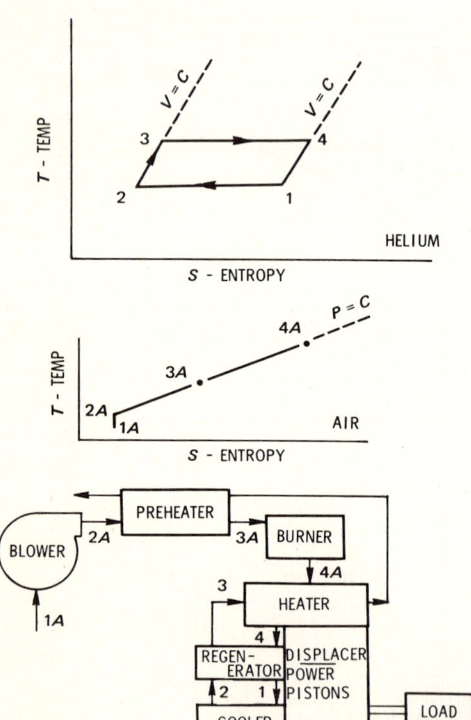

Fig. 6.22(*a*) Basic Stirling-cycle engine. (*Wade and Cornelius, 1971.*)

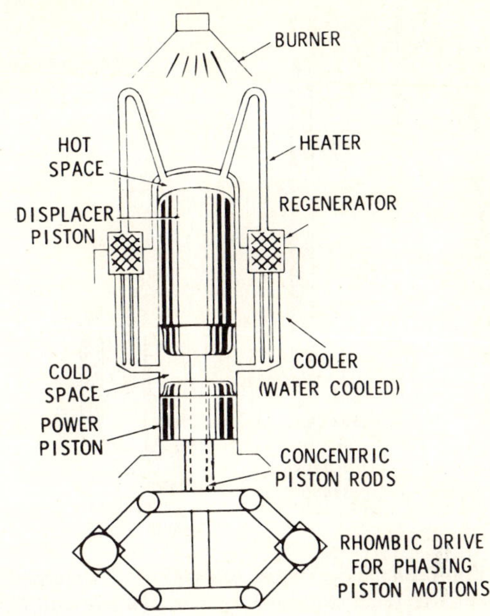

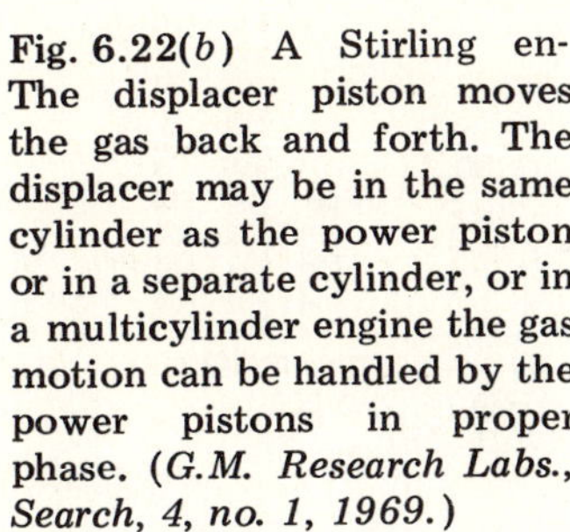
Fig. 6.22(*b*) A Stirling en- The displacer piston moves the gas back and forth. The displacer may be in the same cylinder as the power piston or in a separate cylinder, or in a multicylinder engine the gas motion can be handled by the power pistons in proper phase. (*G.M. Research Labs., Search, 4, no. 1, 1969.*)

How low are typical emissions from such engines? Figure 6.23 gives results for a General Motors turbine engine (Wade and Cornelius, 1971). The truck data were converted to a simulated emission for a 160 hp engine in a 4,500-lb car. Figure 6.24 gives results for steam engines and Fig. 6.25 for the Stirling engine, both from the above reference.

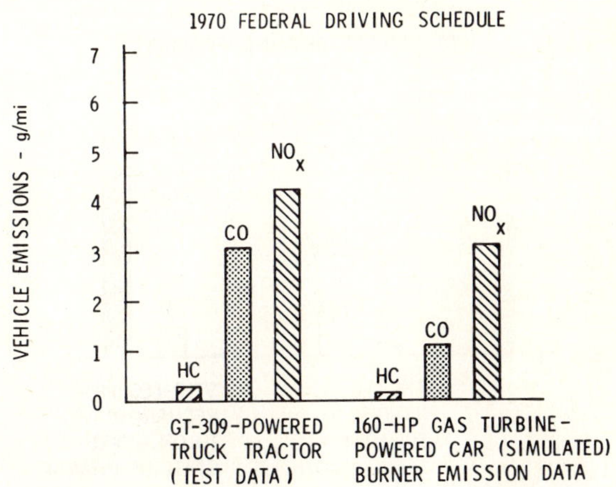

Fig. 6.23 True mass emissions of several vehicles powered by gas-turbine engines with standard burners. (*Wade and Cornelius, 1971.*)

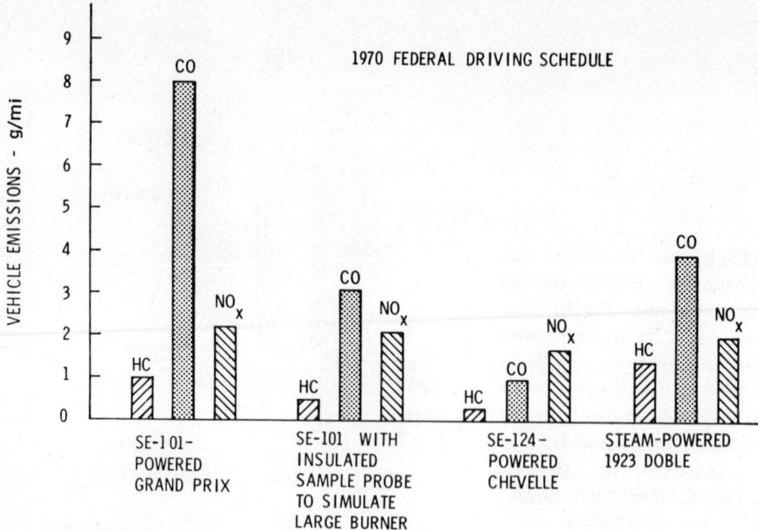

Fig. 6.24 True mass emissions of several vehicles powered by steam engines with standard burners. (*Wade and Cornelius, 1971.*)

The Stirling engine was used to drive an alternator which charged lead acid batteries and these were used as the power source for a D.C. motor to drive the car. Such a vehicle is known as a hybrid car. These emission results are noted both with and without recharge.

As expected, all three engine types are low in HC and CO. In particular the gas turbine and Stirling engines meet the 1976 federal CO

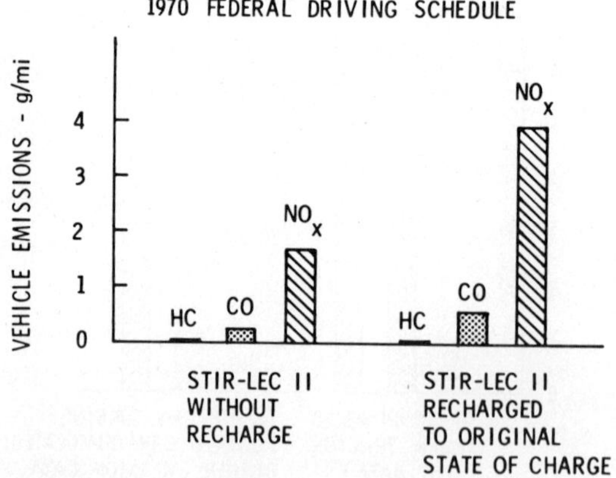

Fig. 6.25 True mass emissions of a vehicle powered by a Stirling engine with the standard burner. (*Wade and Cornelius, 1971.*)

and HC standards. In an effort to lower the NO_x emission several changes were made in the gas turbine engine. Figure 6.26 indicates the improvements. Control techniques are those already discussed, namely, ways to reduce peak temperatures and residence times at peak conditions.

It is clear from these results that the prime difficulty in meeting the 1976 federal standards, even for the external combustion engine, lies in the NO_x requirement. Whether the auto companies can meet the standard appears to be questionable.

There is one other engine of interest and this is the rotary or Wankel engine. In the Wankel engine compression, combustion, and expansion take place in a chamber with a rotor instead of a reciprocating piston (Fig. 6.27). The engine is an internal combustion engine but features far fewer parts, and is much more compact, leaving more room for clean-up devices. Because of its combustion system the engine has high unburned HC but relatively low NO_x and it is this latter feature that is of interest. The HC and CO can be controlled by some of the add-on devices that we have already described. General Motors has indicated that it may offer Wankel-powered cars with the 1975 model; a Japanese model is already selling in the United States.

Alternative fuels have also been suggested. In particular LPG (liquid petroleum gases, primarily propane and butane), LNG (liquid natural gas, primarily methane), and CNG (compressed natural gas) have been tried. The advantage to these fuels is that good mixture

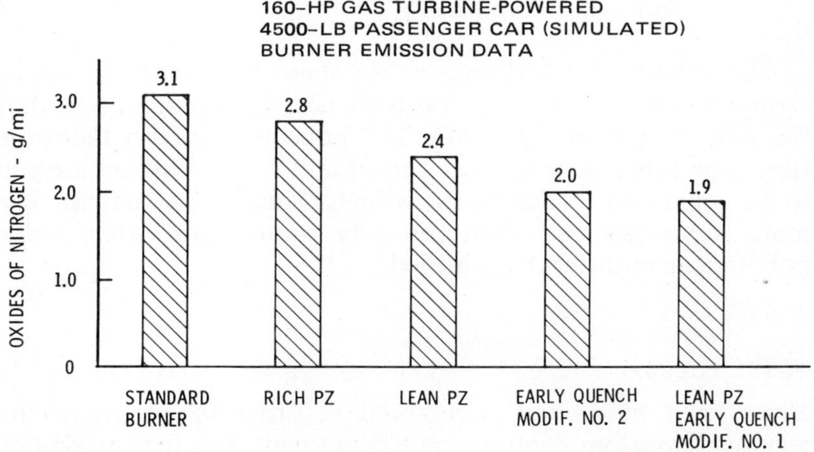

Fig. 6.26 True mass emissions of a simulated passenger car powered by a gas-turbine engine with modified GT-309 burners. (*Wade and Cornelius, 1971.*)

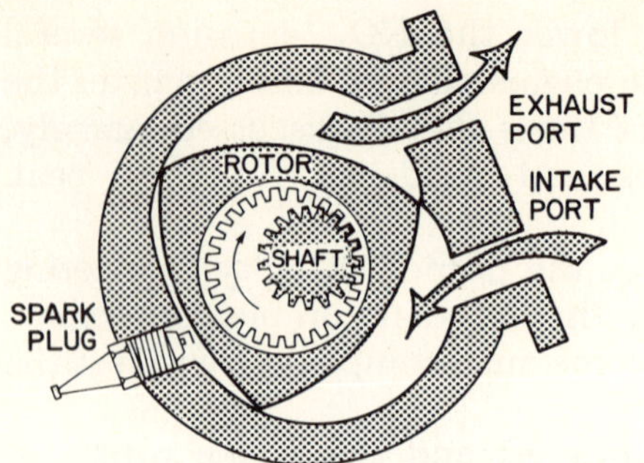

Fig. 6.27 The Wankel engine. The compact size and low NO_x emissions make this an attractive alternative to the conventional I.C. engine.

distribution is easy to obtain and so high AFRs can be used. This way HC and CO emissions are lowered. In addition the HC emitted are typically of low reactivity in the atmosphere; that is, they are saturated chain hydrocarbons. Since the AFR is large the NO_x is also reduced. EPA (1972) data indicate that "there are inconsistencies in test data but . . .

1. Data from eight 1968–69 model gasoline vehicles converted to LPG duel-fuel usage show the average emissions reduction is 25 percent for HC, 69 percent for CO and 13 percent for NO_x. Drivability impairment was noticeable but not critical.
2. Twenty 1970-model General Service Administration vehicles, converted to full-time LPG operation, approached 1975 federal emissions standards for new automobiles. The vehicles were tested before and after conversion. Emission reductions were 81 percent for HC, 86 percent for CO, and 64 percent for NO_x. Drivability effects ranged from barely noticeable to hazardous."

The State of California offers these fuels with no gas tax as an economic incentive to increase their use. Conversion costs about $300 for LPG and more for LNG. The hope in southern California is for fleet-type vehicles, which are fueled and maintained on a regular basis, to be converted to one of these fuels. Since these vehicles drive many more miles per day than privately owned cars, some reduction in pollutants could thus be achieved.

TEST CYCLES

Some kind of testing is required in order to determine accurately whether emission standards are being met. The first driving cycle was the California 7-mode cycle noted in Fig. 6.28. Cars to be tested were run through this cycle seven times, continuously, and selected cycles were chosen for analysis. The federal government adopted this driving

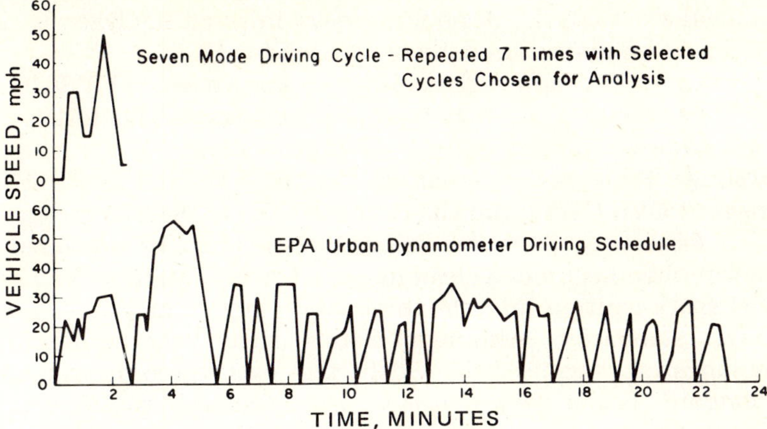

Fig. 6.28 Seven-mode test.

cycle for its testing in 1968 (HEW, 1968). In 1971 EPA announced a new driving cycle to start with the 1972 models. The cycle, a 23-minute one with a cold start, has higher maximum speeds and higher acceleration and decelerations. The sampling system has also been changed. The new constant volume sampler (CVS) is noted in Fig. 6.29. Basically, a constant volume pump draws exhaust and filtered air through a heat exchanger. A probe draws off a small sample at a constant rate and this sample is accumulated in a bag during the test for

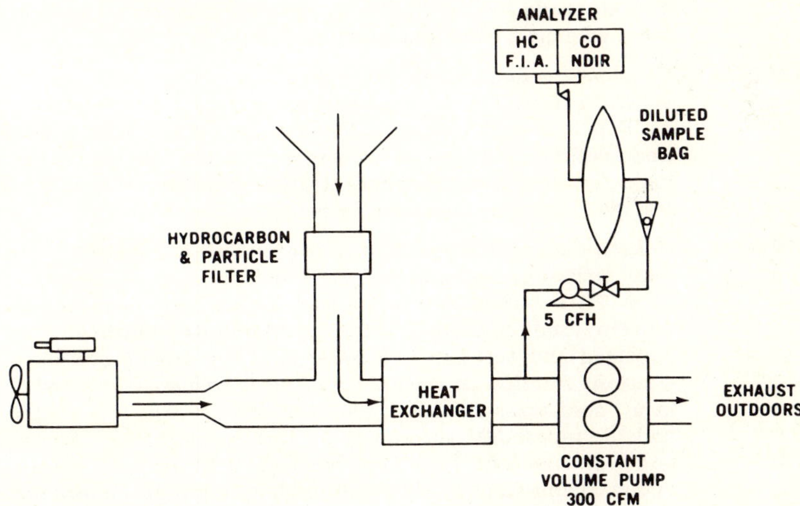

Fig. 6.29 Constant-volume sampler, basic concept. (*Brehob, 1971.*)

immediate analysis. A nondispersive infrared analyzer is used for CO and a flame ionization detector is used for HC.

In 1975 a new test is proposed known as the CVS-CH or constant volume sample test with cold and hot starts. This test is complicated and time consuming since a 12-hour wait at 70°F is required before testing. Table 6.4 summarizes the three tests: the old FTP test, the present CVS-C test, and the future (1975) CVS-CH test.

Obviously there are many questions to be answered before the automobile becomes a clean means of transportation. We have seen that the spark-ignition internal combustion engine may not be able to meet NO_x standards. Although claims have been made for external combustion engines it is evident that even these may exceed the NO_x standard. Some data (*Environ. Sci. Technol.* vol. 6, no. 6) indicate lower emissions for steam engines than the General Motors' results but one of these engines is for a bus. The conflict between meeting the HC–CO standards and meeting the NO_x standards is apparent and the dilemma is to reconcile the two.

A summary of past and possible future control hardware is given in Table 6.5. The cost estimates for this control, from the National

TABLE 6.4 Test procedures used to measure emissions

FTP	*Federal test procedure:* The driving cycle is the California 7-mode cycle repeated seven times. Pollutant concentrations in the exhaust are analyzed continuously throughout the 16-minute test. Concentrations in each mode are multiplied by weighting factors to give g/mile. This is not a true mass emissions measurement.
CVS–C	*Constant volume sampling procedure:* This is a cold start mass emissions test. The vehicle stands at constant temperature for 12 hr at 70°F before engine startup. The driving cycle is a 23-minute, 7.5-mile nonrepetitive pattern. A constant fraction of the exhaust flow is collected in a bag, and concentration measurements at the end of the test give true mass emissions in g/mile.
CVS–CH	*Constant volume sampling procedure:* This is a cold-hot start weighted mass emissions test. The vehicle stands at constant temperature for 12 hr at 70°F before engine startup. The driving cycle is the 23-minute pattern used in CVS-C. After a 10-minute shutdown, the engine is restarted and the first 505 seconds of the driving cycle is repeated. A constant fraction of the exhaust flow is collected: the first 505 seconds in a "cold transient bag"; the next 864 seconds in a "stabilized bag"; and the repeat 505 seconds in a "hot transient bag." Emissions in cold and hot transient bags are weighted 0.43 to 0.57 respectively, and added to emissions in the stabilized bag to give true mass emissions.

Source: National Academy of Sciences, 1972.

TABLE 6.5 Typical pattern of installation of automotive emission hardware

Model year added	Item
1966	PCV valve
1968	fuel-evaporation-control system
1970	(*a*) retarded ignition timing (*b*) decreased compression ratio (*c*) change of fuel/air ratio (*d*) transmission control system
1972	(*a*) antidieseling solenoid valve (*b*) thermostatic air valve (*c*) choke-heat bypass
1973*	(*a*) exhaust-gas recirculation (*b*) air-injection reactor (*c*) induction hardened valve seats (*d*) spark-advance control (*e*) air pump
1974*	precision cams, bores, and pistons
1975*	(*a*) proportional exhaust-gas recirculation (*b*) carburetor with altitude compensation (*c*) advanced air-injection control (*d*) air/fuel preheater (*e*) electric choke (*f*) electronic distributor (pointless) (*g*) improved timing control (*h*) catalytic (oxidizing) converter (*i*) catalyst pellet charge (*j*) cooling-system changes (*k*) improved underhood materials (*l*) body revisions

*Hardware listed for 1973, 1974, and 1975 model years is for a system deemed by the Panel to be promising.
Source: See Table 6.4.

Academy of Sciences (1972) is presented in Table 6.6. It is clear that the costs of control will not be cheap.

We have not considered in this chapter, such topics as pollution from diesel engines, alternative transportation means for commuters, that is, mass transit, or any detail on the economic consequences of the 1975–1976 Federal standards. Each could be a chapter in itself.

TABLE 6.6 Summary of cost for likely emission hardware based on panel estimates

Year	Yearly cost	Accumulated cost
1966	$ 3.00	$ 3.00
1968	15.00	18.00
1970	8.00	26.00
1971-1972	14.00	40.00
1973	60.00	100.00
1974	20.60	120.60
1975	193.40	314.00

Source: See Table 6.4.

REFERENCES

Brehob, W. M.: Mechanisms of Pollutant Formation and Control from Automotive Sources, Soc. Automotive Eng., Special Publication 365, June, 1971 (also SAE 710483).

Campau, R. M. and J. C. Neurman: Continuous Mass Spectrometric Determination of Nitric Oxide in Automotive Exhaust, *SAE Trans.*, vol. 75, 1967 (SAE 660116).

Control Techniques for Carbon Monoxide, Nitrogen Oxide, and Hydrocarbon Emissions from Mobile Sources, National Air Pollution Control Administration Publication AP-66, 1970.

Crawford, K. C. and R. Lindsay: Future of the Gasoline Engine—Environmental Conservation, Shell International Petroleum Company Unlimited, London, MOR 558F, April, 1970.

Daniel, W.: Engine Variable Effects on Exhaust Hydrocarbon Composition—A Single-Engine Study with Propane as the Fuel, *SAE Trans.*, vol. 76, section 1, pp. 774-795, 1968 (SAE 670124).

EPA: Control of Air Pollution from New Motor Vehicles and New Motor Vehicle Engines, *Federal Register*, vol. 36, no. 128, part II, July 2, 1971.

EPA: Environmental Facts, July, 1972.

Hafstad, L. R.: Automobiles and Air Pollution, in Argonne Conference on Universities, National Laboratories, and Man's Environment, July, 1969.

HEW: Control of Air Pollution from New Motor Vehicles and New Motor Vehicle Engines, *Federal Register*, vol. 33, no. 108, part II, June, 1968.

Morse Report, The Automobile and Air Pollution: A Program for Progress, U.S. Government Printing Office, October, 1967.

Myers, P. S., O. A. Uyehara, and H. K. Newhall: The ABCs of Engine Exhaust Emissions, Soc. Automotive Eng. Special Publication 365, June, 1971 (also SAE 710481).

National Academy of Sciences, Semiannual Report by the Committee on Motor Vehicle Emissions, January, 1972.

Newhall, H. K.: Kinetics of Engine-Generated Nitrogen Oxides and Carbon Monoxide, 12th International Combustion Symposium, pp. 603-613, August, 1969.

Newhall, H. K. and I. A. El Messiri: A Combustion Chamber Designed for Minimum Engine Exhaust Emissions, *SAE Trans.*, vol. 79, 1970 (also SAE 700491).

Starkman, E. S.: "Combustion-Generated Air Pollution," Plenum Press, New York, 1971.

——: Prospects for Attainment of the 1975-76 Vehicle Emission Standards, Am. Inst. Chem. Engr., Paper 26A, 65th annual meeting, November, 1972.

Wade, W. R. and E. Cornelius: Emission Characteristics of Continuous Combustion Systems of Vehicular Power Plants—Gas Turbine, Steam, Stirling, GMR-1135, September, 1971.

PROBLEMS

6.1 An idealized air-standard Diesel cycle has a compression ratio of 15. The energy input is idealized as a heat transfer of 700 Btu/lb. The inlet conditions are 70°F at 1 atm. Find the pressure and temperature at the end of each process in the cycle and determine the cycle efficiency.

6.2 The compression ratio on an ideal air-standard Otto cycle is 8.0. The inlet conditions are 1 atm at 70°F. The combustion process is idealized as a heat transfer of 700 Btu/lb. Find the pressure and temperature at the end of each process and determine the cycle efficiency.

6.3 Discuss the inadequacies in the models of Probs. 6.1 and 6.2. Which idealizations are likely to be the weakest, and what effects will they have on the predicted performance?

6.4 The thermal efficiency of the Brayton cycle was given as

$$\eta = 1 - \left(\frac{1}{P*}\right)^{(k-1)/k}$$

where $P* = P_2/P_1 = P_3/P_4$. Derive this relation.

6.5 The thermal efficiency of an Otto cycle is given as

$$\eta = 1 - \frac{1}{r^{k-1}}$$

where r is the compression ratio. Derive this relation.

6.6 Investigate the possibilities of a steam engine for automotive use. How does such an engine operate? What are typical operating conditions? In short write a brief paper on this topic.

6.7 The use of a modified combustion process using a primary and secondary chamber has been suggested by Newhall (SAE, vol. 79, 1970, 700491). Describe why this might help reduce pollution from the spark ignition I.C. engine.

6.8 Calculate the possible HC emissions due to filling an automobile fuel tank. Is this a significant source of HC if these emissions are not controlled?

6.9 Assume that one spark plug is misfiring every 10 firings in an 8-cylinder engine. If all of this HC is emitted in the exhaust, is this a significant source of HC?

6.10 Data in W. I. Doty, and L. J. Olejnik, (*Automotive Engineering*, *80* (1), pp. 68-72, Jan. 1972) indicates that if the air conditioner is operating or if the car is driven into a head wind the NO_x emissions go up. Can you offer an explanation for this?

6.11 Former emission standards were given in ppm (a volume measure) while present emissions are given in mass units of g/mile. Estimate for NO_x and CO the conversions between these two systems. This will require some assumptions about driving habits in terms of miles per gallon and miles driven per day.

6.12 Emission index is another measure of pollutant emissions. This is the grams pollutant per kg of fuel burned. Assuming a fuel specific gravity of 0.8 and the applicability of the 1976 Federal standards, set up a graph of allowable emissions index for CO, HC, and NO_x as a function of the fuel economy in miles per gallon.

6.13 Alternative power sources to the use of gasoline have been suggested for cars. Discuss the problems of the use of (*a*) electric cars, (*b*) LNG fueled cars. If LNG is used for cars, will this effect emissions from other sources?

6.14 The local auto expert says he can control auto emissions by just adjusting the air/fuel ratio to "burn out the carbon." Comment on this idea.

6.15 We can assume that cars have a lifetime of 10 years. Accounting then for past, present, and future controls, estimate the annual emissions of HC, CO, and NO_x from cars from 1940 to 1980 for the United States. Do the same problem for your own city.

6.16 Explain the results of Fig. 6.16 for emissions as a function of operating conditions. Are these results an argument for or against building freeways?

7
METEOROLOGY

> And not only do pollutants affect the weather in various other ways . . . as by nucleating clouds and producing acidic rainfall . . . but there are reciprocal effects. Weather conditions affect the transformation of automotive emissions into photochemical smog; and they can clamp the smog tightly upon the communities where it is produced. . . . When the surface air is cooler than a layer of air above it, and mixing cannot occur, the atmosphere is said to be stable . . . the condition is known as an inversion . . . and pollutants accumulate in the limited air space.
>
> *John Middleton, 1971*

Meteorology specifies what happens to a puff or plume of pollutant from the time it is emitted to the time it is detected at some other location. The motion of the air causes a dilution of air pollutant concentrations and we would like to be able to calculate how much dilution occurs as a function of the meteorology or atmospheric conditions. To put it another way, given a known emission we would like to calculate the resulting concentrations downwind of the source. To do this will require some knowledge of the elements of meteorology. That material will be presented in this chapter and then in the next two chapters we will investigate dispersion of effluent from stacks and the effect of buoyancy on the plume rise.

WIND PROFILES

We know that air motion requires a force and the wind is a result of an equilibrium produced by pressure, Coriolis, and friction forces. The pressure forces are caused directly by the existence of high and low

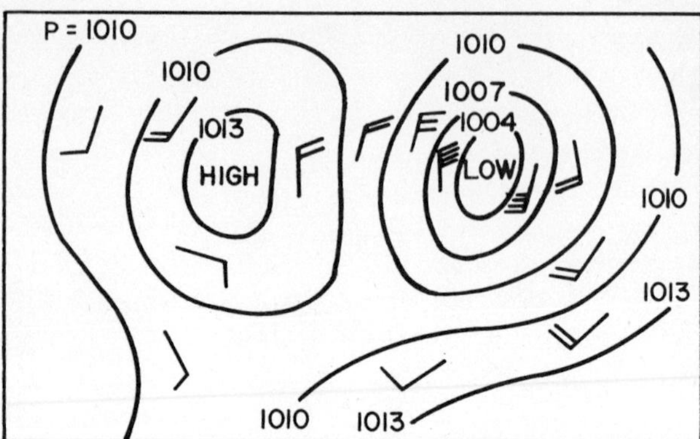

Fig. 7.1 Typical pressure patterns and the associated wind field. (*Reproduced by permission of Doubleday and Company, Inc. from "The Unclean Sky" by Louis J. Batten. Copyright © 1966 by Educational Services Incorporated.*)

pressure regions in the atmosphere, familiar to us from any weather map. In the Northern Hemisphere the air blows counterclockwise around low pressure centers while in the Southern Hemisphere the air blows clockwise. In the middle latitudes the low pressure centers, referred to as *cyclones*, tend to move northward and eastward.

Cyclone here refers to large-scale air masses and not to the violent but small tornados. High-pressure regions are called *anticyclones* and these are often the source of temperature inversions. An inversion limits the atmosphere available for dilution of pollutant emissions. In anticyclones the air moves clockwise in the Northern Hemisphere. Weather maps show these regions of high and low pressure and also denote wind direction and magnitude by means of vectors with small marks across them. The more marks the higher the wind speed. Figure 7.1 shows a typical piece of weather map. Note that in the Northern Hemisphere, shown here, winds blow so that low pressure areas are to the left-hand side, that is, counterclockwise, and high pressure areas are to the right hand side, that is, clockwise.

We might expect winds to blow across the *isobars*, lines of constant pressure, but the map shows the winds to be almost parallel to these. Why? The pressure force is the result of pressure differences between two points in space (see Fig. 7.2(*a*)). Pressure gradients alone would cause the air to flow, as suggested, across the isobars. However, as the air begins to move the effect of the earth turning beneath the flow becomes important. In the Northern Hemisphere the earth's rotation causes a turning of the flow to the right due to the Coriolis force. The result, shown in Fig. 7.2(*b*), is an equilibrium between the pressure and Coriolis forces with the wind blowing parallel to the isobars with low pressure on the left. The wind resulting from this force balance is called the *gradient* wind and, in fact, such winds are found in the upper layers of the atmosphere. Near the surface of the earth, however, the friction force comes into play. This force, just like that found on airfoils and the famous Blasius flat plate studied in fluid mechanics class, causes a change in wind velocity with height and a change in wind direction from the gradient case.

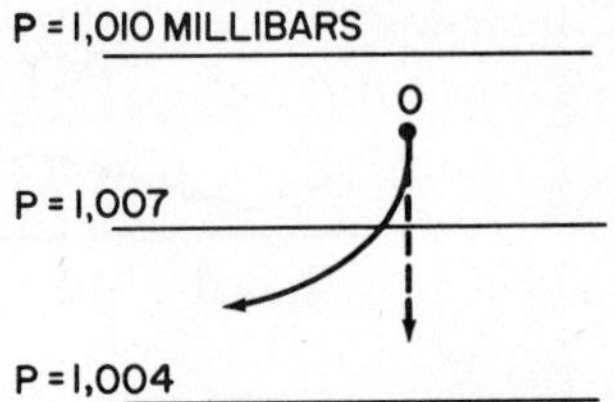

Fig. 7.2(*a*) The movement of air depends on the atmospheric pressure. (*Reproduced by permission of Doubleday and Company, Inc. from "The Unclean Sky" by Louis J. Batten. Copyright © 1966 by Educational Services Incorporated.*)

The wind speed must be zero at the earth's surface (just as the flow velocity must be zero on the surface of an airfoil) and then rise to the gradient value at a height which is usually a few hundred meters. The friction force causes the air to be turned slightly toward the low pressure. Because of friction the air does not blow exactly along the isobars but moves across them at a slight angle toward low pressure as noted in Fig. 7.1. In general, the air flow at the earth's surface is directed to the left of the gradient flow and the turning angle is about 15° by day over smooth surfaces, changing to as much as 50° by night over rough terrain. Above the surface the wind turns slowly in a clockwise direction so that this angle is reduced with height until at a few hundred meters the observed wind is equal to the gradient wind and runs parallel to the isobars. Figure 7.3 notes this change.

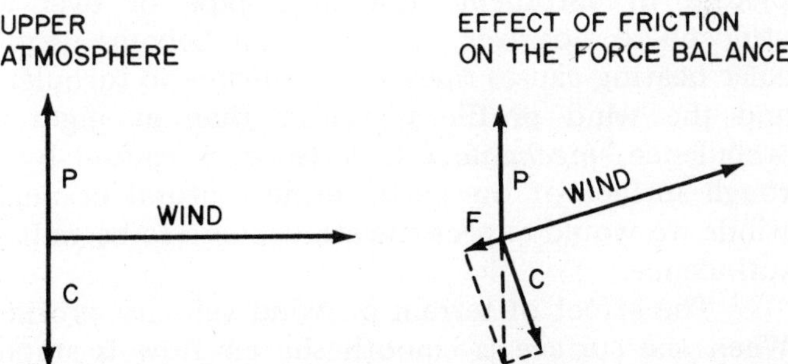

Fig. 7.2(*b*) The force balance between pressure, Coriolis, and friction forces. P = pressure gradient force; C = Coriolis force; F = friction. (*Smith, M., "Recommended Guide for the Prediction of the Dispersion of Airborne Effluents," Am. Soc. of Mech. Eng.*)

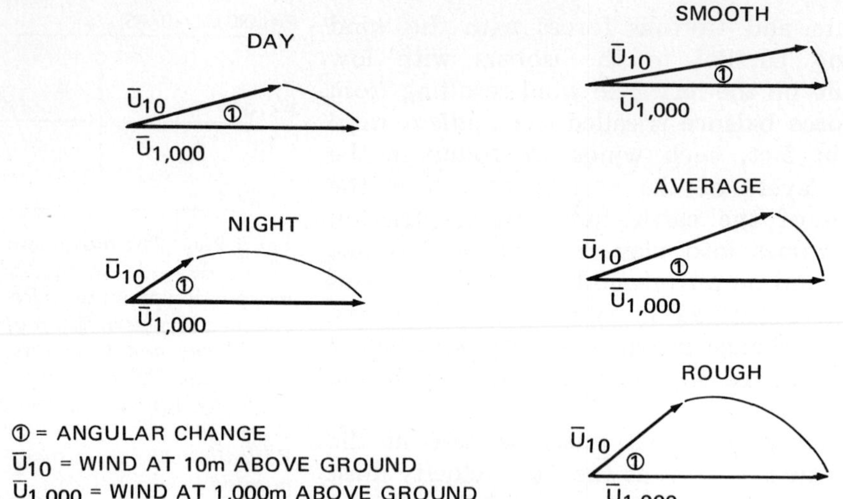

Fig. 7.3 Change of wind direction with height. Not only the wind speed, but the direction changes with height as a result of frictional forces. (*Smith, M., "Recommended Guide for the Prediction of the Dispersion of Airborne Effluents," Am. Soc. of Mech. Engrs.*)

TURBULENT DIFFUSION

The wind velocity profile is noted in Figs. 7.4 and 7.5. The change in velocity with height is a function both of the terrain and of the time of day. We model the air flow as turbulent flow with the turbulence represented by *eddy* motion. An eddy refers to a piece of air which moves randomly in a fluctuating manner just as do the eddies which we picture in turbulent flow in a pipe or over a flat plate. In the atmosphere, however, the eddies can become very large. During the day solar heating causes *thermal turbulence* so turbulent mixing is increased and the wind profile is flatter than at night. The other type of turbulence, *mechanical turbulence*, is caused by air motion over the rough surface of the earth, either natural or man made. During high winds we would expect the atmosphere to be well mixed by mechanical turbulence.

The effect of terrain on wind velocity profile is noted in Fig. 7.4. When the surface is smooth the air flow is smooth and the velocity profile becomes very steep near the ground. For rougher surfaces more mechanical turbulence is generated and the velocity profile becomes less steep and reaches deeper into the atmosphere.

The effect of eddy motion is very important in diluting concentrations of pollutants. If a piece of air is displaced from one level

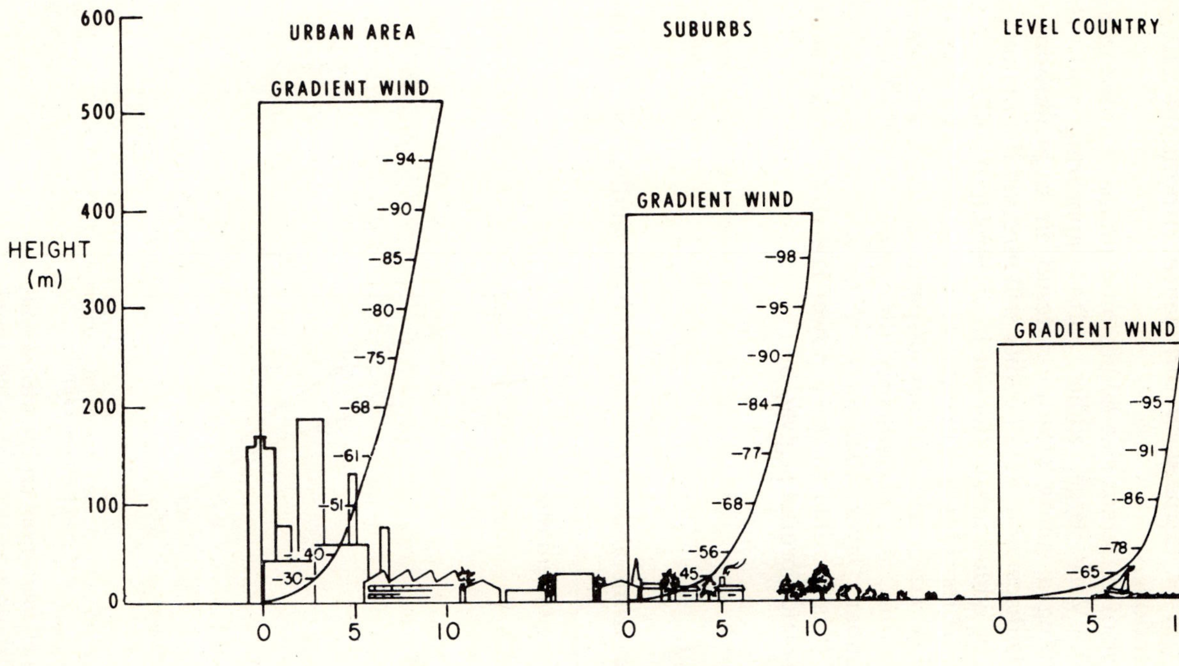

Fig. 7.4 Effect of terrain roughness on the wind speed profile. With decreasing roughness, the depth of the affected layer becomes shallower. (*Turner, 1970.*)

to another it can obviously carry momentum and thermal energy with it. It also carries whatever has been placed in it from pollution sources. Thus smoke will be diffused by turbulence in both the vertical and horizontal directions.

The effect of different size eddies on a plume is noted in Fig. 7.6 (Slade, 1968). Very large eddies, often appearing with thermal turbulence during a sunny afternoon result in large motions of the plume but little dispersion. Small eddies will increase the plume size but only slowly. The eddies most effective in dispersing the plume are those about the size of the plume. Fortunately the atmosphere usually has a variety of eddy sizes and plumes disperse as well as meander in the wind direction. An introductory discussion of turbulence in the planetary boundary layer, that is, $z = 0$ to $z = z_{\text{grad wind}}$, is given in Slade (1968). It is possible to use a power law to describe the velocity profile and one such profile is given by

$$\frac{\bar{u}_z}{\bar{u}_o} = \left(\frac{Z}{Z_o}\right)^n$$

where n goes from about 0.12 to 0.50 depending on atmospheric

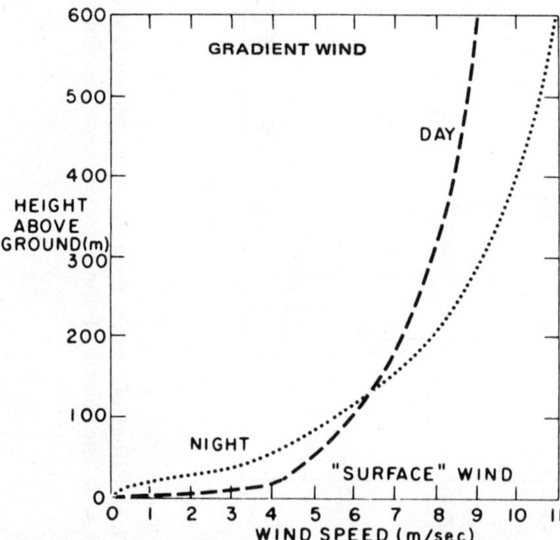

Fig. 7.5 Change of wind speed profile with stability. The frictional drag reduces the wind speed close to the ground below that found at the gradient level. (*Smith, M., "Recommended Guide for the Prediction of the Dispersion of Airborne Effluents," Am. Soc. of Mech. Engrs.*)

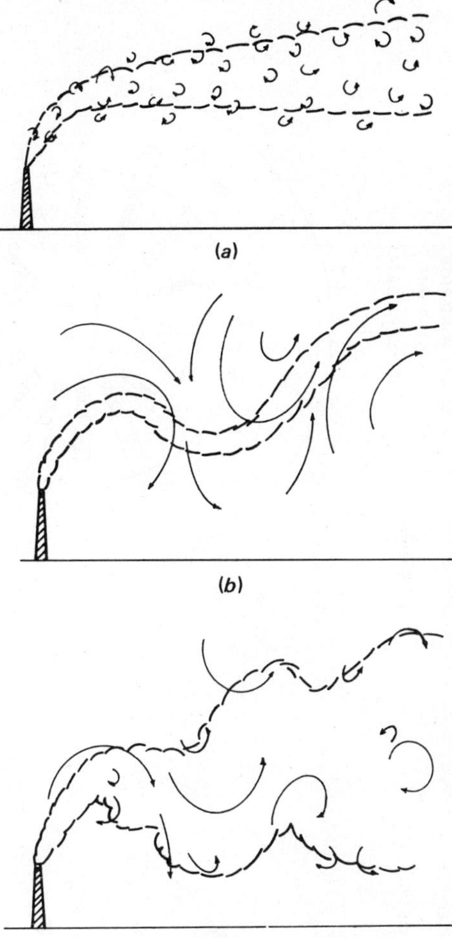

Fig. 7.6(*a*) Plume dispersing in a field of small eddies. A plume in a hypothetical field of small turbulent motions will move in a relatively straight line, with a gradual increase in its cross section. (*b*) Plume dispersing in a field of large eddies. If the eddies are all very large compared to the plume dimensions, the plume will grow very little in size, but will meander wildly. (*c*) Plume dispersing in a field of varied eddies. The typical daytime atmosphere has eddies of an infinite variety of sizes, and a dispersing plume both grows and meanders as it moves downwind.

conditions. The turbulent velocity profile for flow over a plate is sometimes taken as following a 1/7 power which would correspond to an $n = 0.14$. For a smooth surface with an adiabatic lapse rate in temperature (still to be defined), the 1/7 profile is a good approximation.

WIND ROSES

It is obviously important in predicting pollutant dispersion to know the direction of the wind and meteorologists use the *wind rose* to present wind direction and wind speed data. Figure 7.7 shows a wind rose from Cincinnati, Ohio, for January and July. Note that the wind direction is

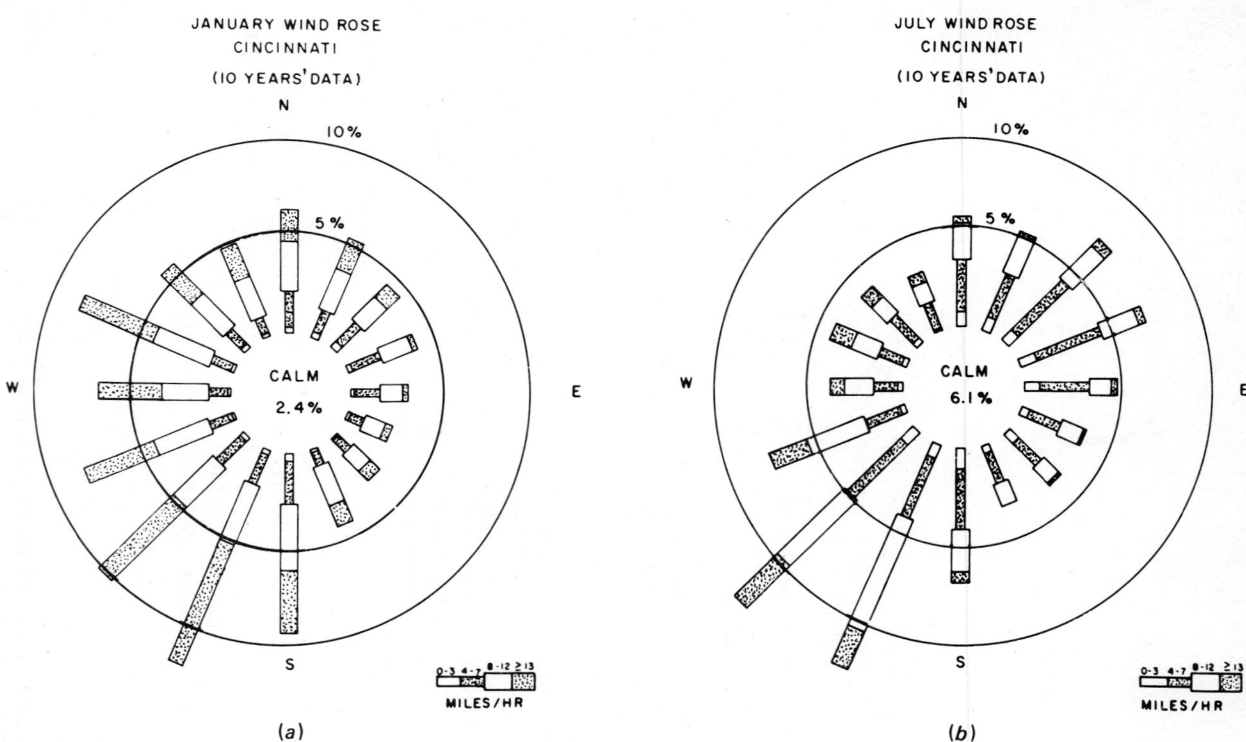

Fig. 7.7 January and July wind roses, Cincinnati. The monthly distributions of wind direction and wind speed are summarized on polar diagrams. The positions of the spokes show the direction from which the wind was blowing, the length of the segments indicate the percentage of the speeds in various groups. (*Smith, M., "Recommended Guide for the Prediction of the Dispersion of Airborne Effluents," Am. Soc. of Mech. Engrs.*)

always defined as that *from which the wind is blowing*, that is, a north wind comes from the north. In July the wind is generally from the southwest and occasionally is greater than 13 miles per hour (5.8 m/sec). In January the wind is somewhat more often from the west and is often greater than 13 miles per hour. The sum of all the vectors plus the percent calm must total 100 percent. Since wind roses indicate the direction from which the wind comes it follows that pollution is carried to the opposite quadrant, that is, a southwest wind deposits the pollution it carries to the northeast.

TOPOGRAPHICAL EFFECTS

Surface topography can affect the local wind, and under certain conditions this happens often enough to be of special interest. One such case is the sea-land or on-shore, off-shore breeze. In large bodies of water the thermal inertia of the water causes a slower temperature change than on the nearby land. For example, along an ocean coast line and during periods of high solar input, the daytime air temperature over the ocean is lower than over land. The relatively warm air over the land rises and is replaced by cooler ocean air. The system is usually limited to altitudes of several hundred meters which, of course, is where pollutants are emitted. The sea breeze or on-shore breeze develops during the day and is strongest in midafternoon. At night the opposite may occur, although usually not with such large velocities. At night the ocean is relatively warm and the breeze is from the cooler land to the warmer ocean. The on-shore breeze is most likely in the summer months; the off-shore land breeze may occur in the summer but is more likely in winter months. The action of such a wind-topographic system is shown in Fig. 7.8(a).

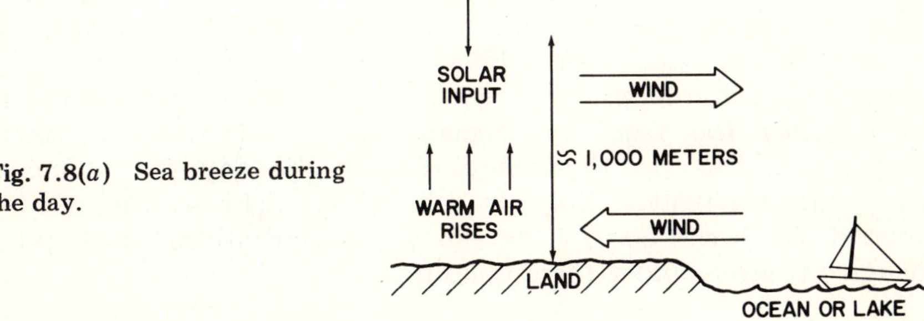

Fig. 7.8(a) Sea breeze during the day.

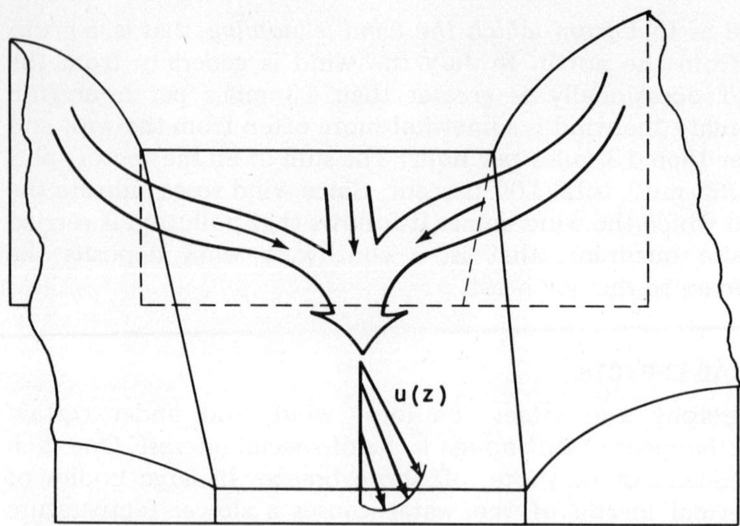

Fig. 7.8(*b*) Valley flow down the valley at night.

A second common wind system caused by topographic effects is the mountain-valley wind. In this case air tends to flow down the valley at night as noted in Fig. 7.8(*b*). Valleys are cooler at higher elevation and the driving force for the air flow results from the differential cooling. Similarly, cool air drains off the mountain at night and flows into the valley. During the daylight hours an opposite flow may occur as the heated air adjacent to the sun-warmed ground begins to rise and flows both up the valley and up the mountain slopes. However, thermal turbulence may mask the daytime up-slope flow so that it is not as strong as the nighttime down-slope flow.

Both the sea breeze and mountain valley wind are important in meteorology of air pollution. Large power stations are often located on ocean coasts or adjacent to large lakes. In this case the stack effluent will tend to drift over the land during the day and may be subject to fumigation (where a rapid downward mixing occurs) when the large eddies generated by thermal turbulence begin to reach the plume height. In the case of a pollutant source in a valley the regularly changing wind patterns may keep the emission trapped in the valley. During the day the plume would move up the valley only to return at night as the wind shifts toward the lower end of the valley. Concentrations can build up to dangerous levels under these conditions.

SEPARATED FLOWS

The heat island (previously discussed in Chap. 2) over large urban areas is another example of a topographic effect. Another man-made effect is the generation of mechanical turbulence caused by the nonuniform height of buildings in a city. Both the heat island and roughness effects tend to aid in the dispersion of pollutants. However, the *separation* of flow around buildings can cause high concentrations of pollutants in the downstream separated region. Figure 7.9 shows the pattern for flow around a cylinder. If there were no viscosity the stream lines would simply move around the cylinder and the flow would be symmetric about the body. We know that the flow pattern around a cylinder is not symmetric and the flow separates as noted in the figure. Close to the cylinder on the back side flow velocities are low, and large eddies are shed from the cylinder (forming the Karman vortex street at Reynolds numbers from about 60 to 5,000).

Separation behind a cylinder or chimney will also result in *downwash* of the effluent trapped in the separated region. Thus the plume may not rise above the stack level and may even be carried downward on the backside of the stack. This occurs when the stack velocity is about equal to, or less than, the ambient velocity. Figure 7.10(*a*) notes the downwash conditions. Other potentially serious cases of separated flow are noted in this figure.

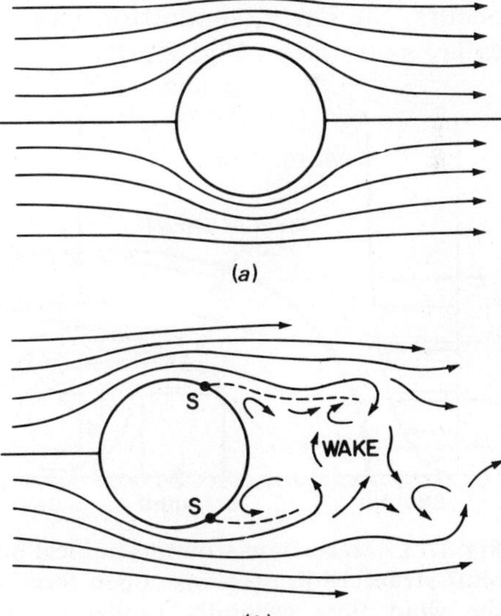

Fig. 7.9(*a*) The pattern of flow round a cylinder when there is no viscosity. (*b*) The effect of viscous forces is to bring the flow to a standstill at *S*, where separation takes place.

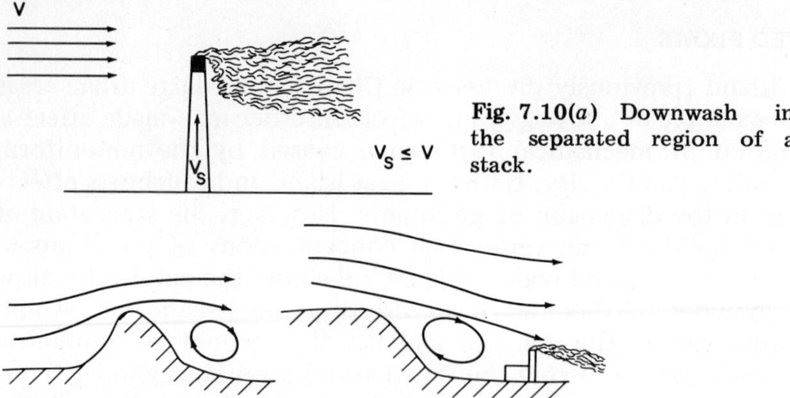

Fig. 7.10(*a*) Downwash in the separated region of a stack.

Fig. 7.10(*b*) Separated flow cases: Separation behind a hill (*left*); separation behind a cliff, where the plume can be carried to the ground when wind blows over a cliff (if the plume is emitted into the eddy, little mixing occurs).

A similar separation of flow occurs for flow over a building. Figure 7.11 shows the flow pattern around a building. The flow separates to form a large "cavity" behind the building. Backflow occurs within the cavity so that downwind sources are carried upwind if emitted in regions of separated flow. Pollution reaching this cavity tends to remain there since very poor mixing between the cavity and the main stream occurs. Similarly, separation can occur on the backside of a hill or in a valley as noted in Fig. 7.10.

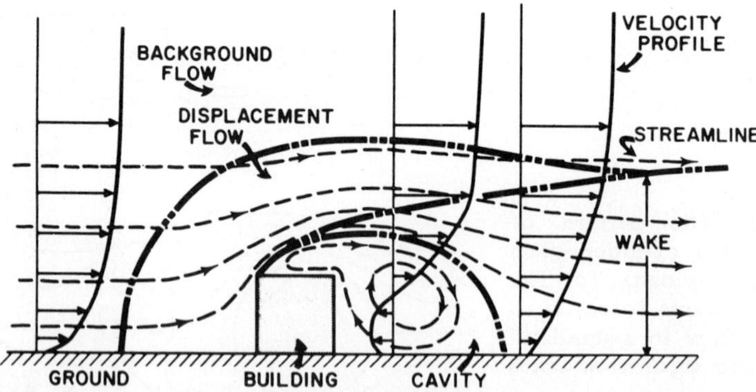

Fig. 7.11 Mean flow around a cubical building. The presence of a bluff structure in otherwise open terrain will produce changes in the wind flow generally similar to those shown. (*Smith, M., "Recommended Guide for the Prediction of the Dispersion of Airborne Effluents," Am. Soc. of Mech. Engrs.*)

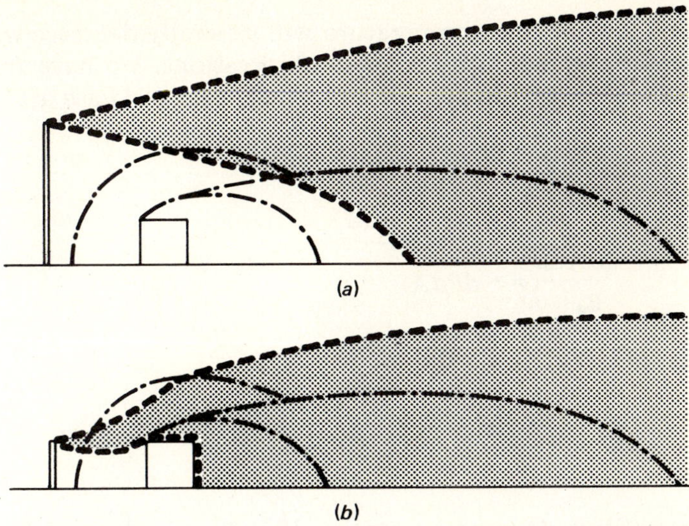

Fig. 7.12(*a*), (*b*) Separation effects on plume dispersion. *(Smith, M., "Recommended Guide for the Prediction of the Dispersion of Airborne Effluents," Am. Soc. of Mech. Engrs.)*

Figure 7.12 shows what happens to an upstream source when the separated cavity and wake behind a building interfere with the plume dispersal. Under conditions (*a*) the stack plume clears the cavity but enters the wake. Downward diffusion is thus increased by mixing occurring in the turbulent wake. In case (*b*) the plume enters the cavity from the front and high concentrations can then build up along the backside of the building. An industrial plant designer needs to be sure that stack plumes do not interact with the building in this manner. An empirical rule of thumb for stacks located on or near to a building is that $H_{\text{stack}} \geqslant 2.5\, H_{\text{building}}$ (Stern, 1968). Halitsky discusses gaseous diffusion near buildings and gives information on the flow pattern around structures (Halitsky, 1963, and in Slade 1968).

TEMPERATURE IN THE ATMOSPHERE

Change of temperature with altitude has great influence on the motion of air pollutants. For example, inversion conditions result in only limited vertical mixing. The amount of turbulence available to diffuse pollutants is also a function of the temperature profile. What do we know about the atmosphere? We expect that the pressure decreases with height from the normal sea level pressure of 1 atmosphere.[1] We

[1] 1 atmosphere = 14.7 lbf/in^2 = 29.92 in. Hg = 760 mm Hg = 1013 millibars. Millibars is a pressure unit used by meteorologists.

also expect that temperature will generally decrease with height. Let us develop a model from these observations. We have from a static force balance on an element of air of height dz and base area dA

$$\Sigma F = 0$$

or

$$\frac{\rho g dA dz}{g_c} = - dP dA$$

or

$$dP = -\frac{\rho g dz}{g_c} \tag{7.1}$$ [2]

A positive dz gives a negative dP since ρ is always positive. To integrate this relationship we need to know something about how ρ changes with P and/or z. We know that $P = \rho RT$ is a valid equation of state for air so that we have

$$dP = -\frac{Pg}{RT}\, dz$$

or

$$\int_{P_1}^{P_2} \frac{RT}{g}\frac{dP}{P} = -\int_{z_1}^{z_2} dz$$

$$z_2 - z_1 = \int_{P_2}^{P_1} \frac{RT}{g}\,\frac{dP}{P} \tag{7.2}$$

In order to solve we need to specify a model of how the temperature varies with height. For a first approximation let us assume that the atmosphere is isothermal, that is T = constant. Then we have

$$\ln \frac{P_1}{P_2} = (z_2 - z_1)\frac{g}{RT}$$

[2] The g_c will not be used in subsequent equations.

If $z_1 = 0$ at the earth's surface then,

$$P_2 = P_1 e^{-z_2 g/RT} \tag{7.3}$$

This would yield an exponentially decreasing pressure as a function of height since we have taken T to be constant. Also to get the pressure to zero in this model requires an infinite height which we know is not the case. Regardless, this model may be used by assuming a series of air layers each with a uniform but different temperature.

A better model for the atmospheric dependence of P,v, and T is the polytropic atmosphere which follows the equation

$$Pv^n = \text{constant} \tag{7.4}$$

Of course for $n = 1$ this yields the isothermal atmosphere. For air for $n = 1.4$ we have an isentropic atmosphere, that is one with a constant entropy as a function of height. If Pv^n = constant, how are P and T related, assuming perfect gas behavior? Since $Pv = RT$ where $v = 1/\rho$ we have

$$P_1{}^{1/n} v_1 = P_2{}^{1/n} v_2$$

$$P_1{}^{1/n} \frac{RT_1}{P_1} = P_2{}^{1/n} \frac{RT_2}{P_2}$$

or

$$T_2 = T_1 \left(\frac{P_1}{P_2}\right)^{\frac{1-n}{n}}$$

Inverting and generalizing T_2 to any T we have

$$T = T_1 \left(\frac{P}{P_1}\right)^{\frac{n-1}{n}} \tag{7.5}$$

Substituting this expression into Eq. (7.2)

$$\int_{P_2}^{P_1} RT_1 \left(\frac{P}{P_1}\right)^{\frac{n-1}{n}} \frac{1}{g} \frac{dP}{P} = z_2 - z_1$$

which upon integration becomes

$$z_2 - z_1 = \frac{n}{n-1} RT_1 \left(\frac{1}{g}\right) \left[1 - \left(\frac{P_2}{P_1}\right)^{\frac{n-1}{n}}\right] \tag{7.6}$$

Thus the pressure-height relationship is a parabola of order $n/(n-1)$. What is the temperature as a function of altitude for this model? Substituting from Eq. (7.5) we find

$$z_2 - z_1 = -\frac{nR}{(n-1)g}(T_2 - T_1)$$

Thus the temperature decreases linearly with height with a slope of $-(n-1)g/nR$. The decrease of temperature with altitude is known as the *lapse rate*. The normal or standard lapse rate, that is, the United States Standard Atmosphere, is $-3.5°$F/1,000 ft ($-0.65°$C/100 meters). What n does this correspond to?

$$\frac{-3.5}{1{,}000} = -\frac{(n-1)g}{nR}$$

for air $R = 53.34$ ft-lbf/lbm-°R (or $R = 1.986/M_i = 0.0686$ cal/g-°K) so

$$\left(\frac{n-1}{n}\right) = 0.186$$

$$n = 1.23$$

What height does the polytropic atmosphere reach with $n = 1.2$? We have from Eq. (7.6) that for $P_2 = 0$, $z_2 = nRT/(n-1)g$ or $z_2 = 31$ miles. We have already seen that the lower atmosphere ends at the tropopause which is at an altitude of about 12 kilometers in the mid-latitudes, higher at the equator, lower at the poles. In the troposphere $n = 1.2$ is a good *average* approximation. Above the troposphere and tropopause is the stratosphere which is a region of almost constant temperature, so $n = 1$.

The adiabatic case, really isentropic but usually referred to as adiabatic, is especially important and occurs when $n = 1.4$. We have

$$\frac{dT}{dz} = -\frac{(n-1)g}{nR} = -0.0054°\text{F/ft}$$

or

$$\frac{dT}{dz} = -5.4°\text{F}/1{,}000 \text{ ft}$$

(In metric units, $dT/dz = -0.0098°\text{C/meter}$, or about $-1°\text{C}/100$ meters.[3])

Consider now the motion of an air "particle" pictured as a little sphere of air shown in Fig. 7.13. If the piece of air is carried upward in the atmosphere it goes through a region of decreasing pressure and consequently expands. However, the expansion requires work against the surroundings and the "sphere" temperature drops. Since the process is usually reasonably rapid, it is a good approximation to assume that this occurs adiabatically, that is, without any heat transfer. *If* the atmospheric lapse rate were exactly the adiabatic lapse rate then the piece of air would reach its new position, z_2, at exactly the temperature of the surroundings. Then it would be at same pressure, temperature, and density of the surroundings and would see no buoyant force. We see, therefore, that an atmosphere with an adiabatic lapse rate is one of neutral stability; a displaced mass of air neither tends to return to its original position nor tends to continue its displacement.

Consider the second case in Fig. 7.13(*b*) however, where the piece of air moves upward in an adiabatic process but in an atmosphere with a subadiabatic lapse rate. The sphere follows a temperature change given by the adiabatic slope but when it arrives at point z_2 it is at a lower temperature than its surroundings but at the same pressure. Consequently, it is heavier than the surroundings ($P = \rho RT$) and tends to fall back to its original position. Such an atmospheric condition is

[3] For the isentropic case $n = k = c_p/c_v$ so

$$\frac{dT}{dz} = \frac{-g}{c_p g_c}$$

English units $$\frac{dT}{dz} = \frac{-(\text{ft/sec}^2)(\text{lbf-sec}^2/\text{lbm ft})}{(0.24 \text{ Btu/lbm}° \text{ R})(778 \text{ ft-lbf/Btu})}$$

$$\frac{dT}{dz} = -5.4°\text{F}/1{,}000 \text{ ft}$$

metric units, $g_c = 1$

$$\frac{dT}{dz} = -\frac{g}{c_p} = \frac{-(9.8 \text{ m/sec}^2)(28.97 \text{ g/g-mole})}{(6.95 \text{ cal/g-mole}° \text{ K})(4.186 \text{ kg m}^2/\text{cal sec}^2)} \times \frac{1}{(1{,}000 \text{ g/kg})}$$

$$\frac{dT}{dz} = -0.0098 \text{ °C/m} \cong -1°\text{C}/100 \text{ meters}$$

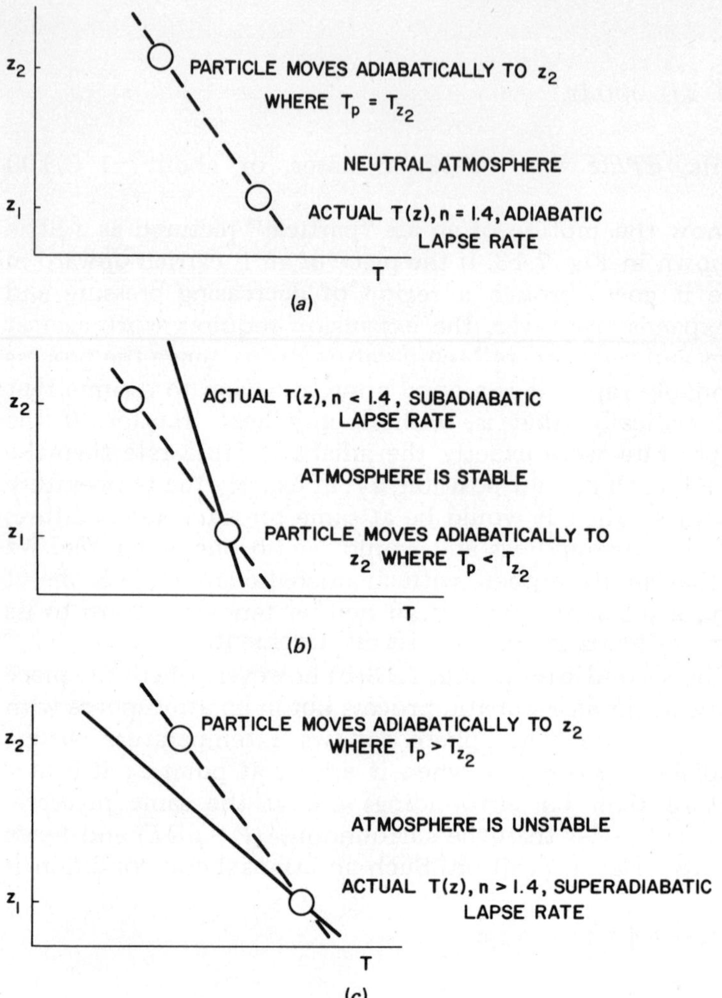

Fig. 7.13 Atmospheric stability.

called *stable*. In a stable atmosphere pollutants will only slowly disperse and turbulence is suppressed. Note that any atmospheric temperature change less than adiabatic is stable. An inversion noted in Fig. 7.14 is, of course, very stable but even the standard atmospheric lapse rate results in a stable atmosphere. In the conditions of Fig. 7.13(*c*) we see that the air motion results in a temperature, for the adiabatic process, which is greater than that of the surroundings. Consequently, the density of the piece of air is less than the surroundings and the air continues to rise. Such an atmosphere is *unstable*. Figure 7.14 summarizes the possible environmental lapse rates.

Meteorologists often use the term potential temperature, defined as the temperature which a parcel of dry air would acquire if brought *adiabatically* from its initial pressure to a standard pressure of 1,000 millibars. Consider the case of an atmosphere with an adiabatic lapse rate where the pressure falls according to Eq. (7.6) with $n = 1.4$. For this case, as we bring each parcel of air adiabatically from its particular pressure, and therefore height, to a pressure of 1,000 millibars, we find that the potential temperature is the same for each parcel of air. Thus the adiabatic lapse rate is one of constant potential temperature. In

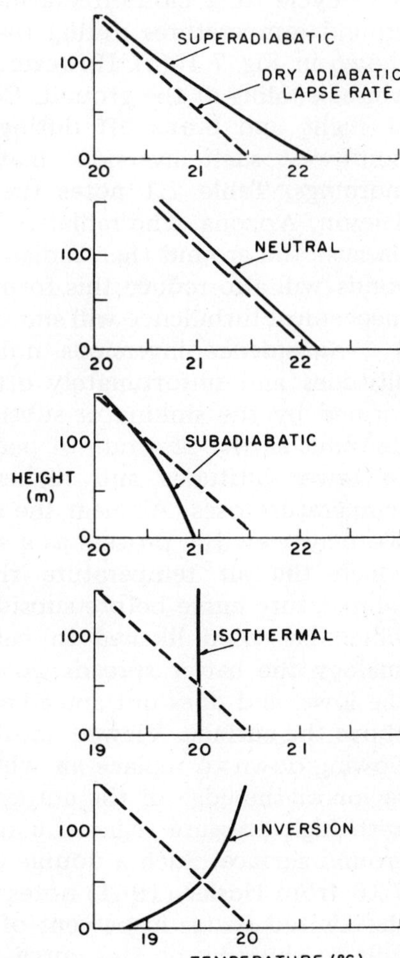

Fig. 7.14 Typical environmental lapse rates. Examples of vertical temperature profiles are shown in comparison with the dry adiabatic lapse rate (−1C/100 m). (*Smith, M., "Recommended Guide for the Prediction of the Dispersion of Airborne Effluents," Am. Soc. of Mech. Engrs.*)

a like manner superadiabatic lapse rates have a negative gradient of potential temperature and subadiabatic lapse rates have a positive gradient of potential temperature. Therefore positive gradients of potential temperature represent stable atmospheric conditions; a constant potential temperature represents a neutral atmosphere; and a negative potential temperature gradient gives an unstable atmosphere.

INVERSIONS

We have found that any atmosphere whose $dT/dz > (dT/dz)_{\text{adiabatic}}$ is a stable one. Inversion conditions cause a strongly stable atmosphere so we have a particular interest in this condition. Consider the diurnal solar cycle for a cloud-free atmosphere. Solar heating can result in high ground temperatures during the late morning and afternoon hours as shown in Fig. 7.15(a). However, at night radiation to the cold night sky causes cooling at the ground. Consequently a *radiative inversion* forms at night and burns off during the morning. In desert areas in the Southwest such inversions may form as often as 90 percent of the mornings. Table 7.1 notes the frequency of radiative inversions for Tucson, Arizona. The radiative inversion is suppressed with cloud cover because the ground then radiates to the relatively warm clouds. Strong winds will also reduce this form of inversion since the wind generated mechanical turbulence will smooth out the temperature gradients.

Subsidence inversions noted in Fig. 7.15(b) occur at modest altitudes and unfortunately often remain for several days. These are formed by the sinking or subsiding of the air in anticyclones. The air descends slowly at a rate of perhaps 1,000 meters/day. As the air sinks to lower altitudes and higher pressures it is compressed and its temperature rises. Air near the surface subsides less and is less affected. We may view the process as a series of adiabatic processes, $n = 1.4$, in which the air temperature rises to values higher than the local temperature curve before subsidence. Figure 7.15(b) notes the process which has been likened to pancake batter flowing in a pan. In this analogy the batter spreads slowly, from the high pressure region into the lows, and does not spread as much at the pan surface as it does just above the surface. Viewed another way the subsidence is caused by air flowing down to replace air which has flowed out of the high pressure region at the edge of the anticyclone. Since clear skies are often found with high pressure regions, a radiative inversion may also form at the ground surface. Such a double inversion is noted in Fig. 7.15(c). Figure 7.16 from Hosler (1961) notes the occurrence of *surface* inversions in the United States in percent of total hours. The Los Angeles problem fails to show up on this curve since it is a subsidence inversion (often

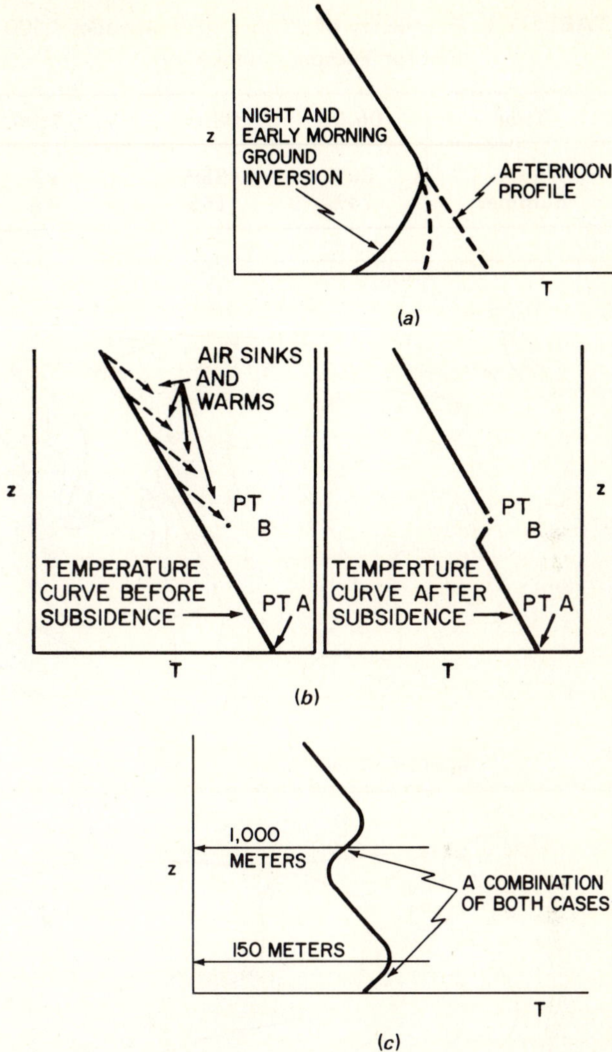

Fig. 7.15 Temperature inversions. *(a)* During the day. *(b)* The sinking of air leads to warming aloft the formation of inversions. *(c)* A combination of cases.

accompanied by a morning radiative inversion) at an altitude of perhaps 3,000 feet. The Los Angeles basin is walled in by mountains to the east and north and has a persistent anticyclone, the Pacific high, with which to contend.

Other types of inversions exist. The sea-breeze condition, discussed in this chapter, results in cool air near the ground and warmer air aloft.

TABLE 7.1 Frequency of $dT/dz > 0$ at altitudes <500 feet for Tucson, Arizona

Time	05:00	08:00	17:00	20:00
Winter	89%	83%	2%	65%
Summer	74%	15%	4%	19%

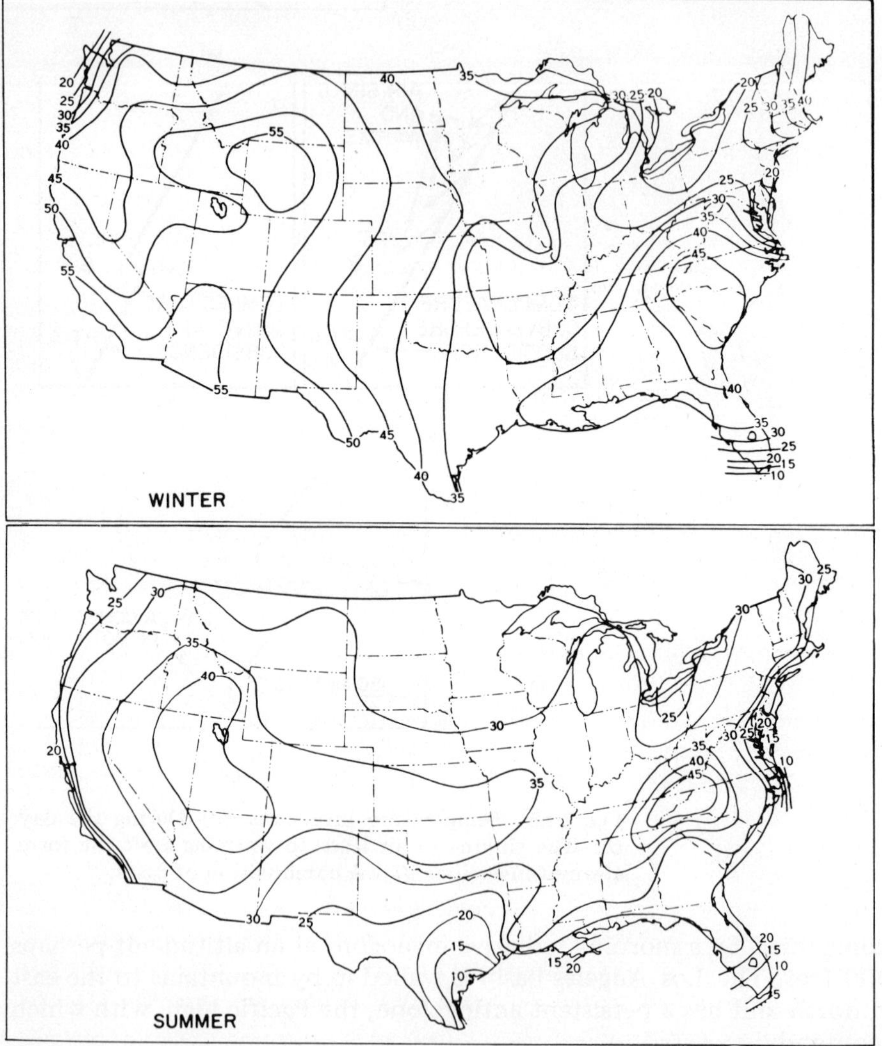

Fig. 7.16 Percentage frequency (percent of total hours) of the occurrence of inversions or isothermal conditions based below 500 ft during the winter and summer. (*Hosler, 1961.*)

The passage of a warm or cold front often results in an inversion at the front as noted in Fig. 7.17.

The *mixing height* is a measure of the vertical distance available for the mixing of pollutants and it accounts for the effect of inversions.

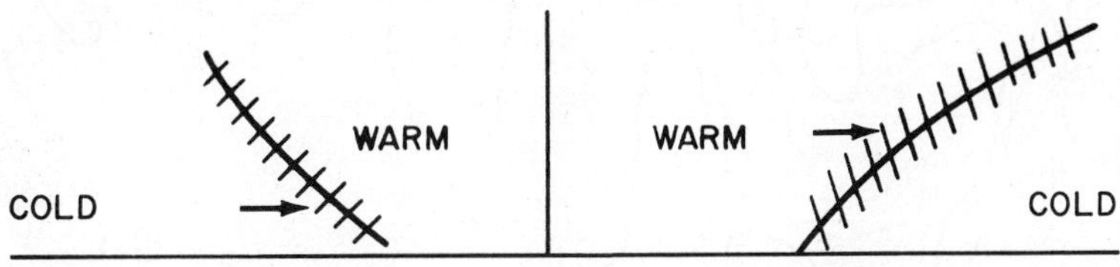

Fig. 7.17 An inversion may occur at the junction between cold air and warm air masses.

Figure 7.18 shows the results of Holzworth (1964) for *mean maximum* monthly mixing depth in January and July. These are the average of the highest mixing heights available. Surface inversions could trap pollutants close to the ground for many hours before the maximum mixing height occurs, typically in the midafternoon.

PLUME BEHAVIOR

Now that we have established this brief background in meteorology, let us turn to the classification of stack plumes as observed in the atmosphere. Figure 7.19 notes six classifications of plume behavior. The spread of the plume is directly related to the vertical temperature gradient as noted on the left-hand side of the figure. *Looping* occurs when the vertical temperature gradient is superadiabatic and the air is very turbulent. This occurs during warm seasons with clear skies where the solar insolation is high. The thermally generated turbulence causes large eddies which can carry the plume in its entirety down to the ground. The high degree of turbulence rapidly disperses the plume but high concentrations may occur close to the stack if the plume reaches the ground.

Coning occurs with a vertical temperature gradient which is subadiabatic but less than isothermal. The plume is shaped like a cone and dispersion is slower than for looping. However, the distance at which the bottom of the plume first reaches the ground is greater than for looping. Coning is likely to occur with overcast or night skies and moderate wind speeds.

Fanning occurs when the temperature gradient is positive, that is, in an inversion, both above and below the plume. The vertical plume spread is minimized since the air is very stable. The horizontal spread of

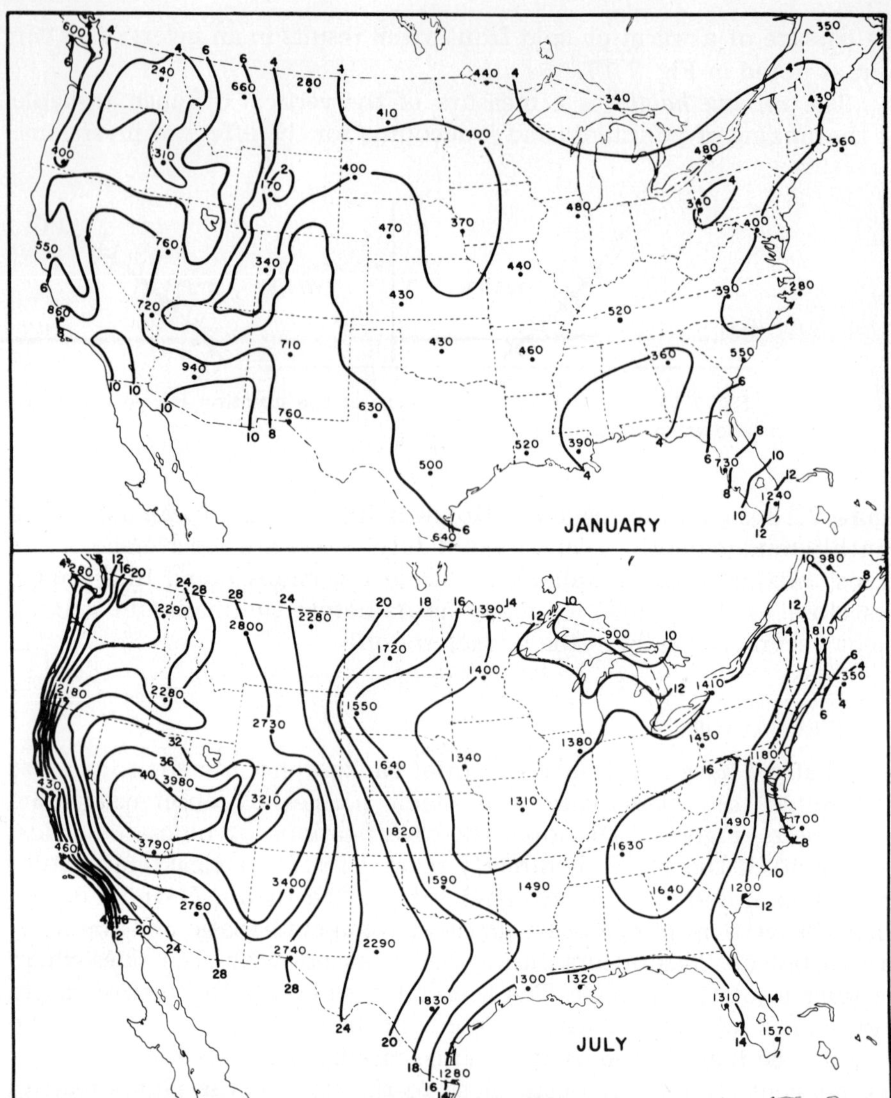

Fig. 7.18 The mean maximum mixing depth (m) for January and July. These data were computed from atmospheric temperature soundings obtained at 45 points in the United States. (*Holzworth, 1964.*)

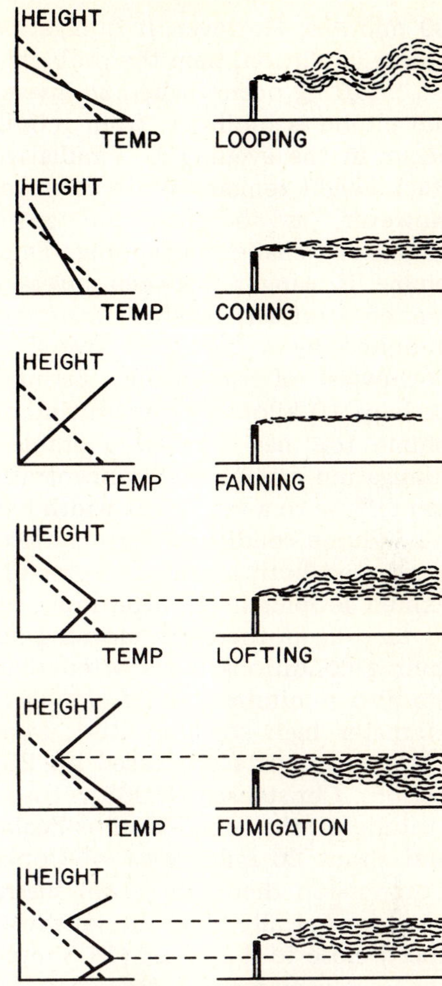

Fig. 7.19 Types of stack-smoke behavior.

the plume may also be quite small. The plume, however, can meander and substantially disperse in the horizontal direction. Therefore the prediction of concentrations is difficult for this case. Although fanning does not lead to high ground concentrations, it is often the predecessor to *fumigation* conditions. As the sun heats the ground the surface-level inversion is "burned-off" and mixing occurs in the region of negative temperature gradient. When the unstable air reaches the plume height, rapid downward mixing may occur and high ground-level concentrations result. Fumigation caused by the breakup of a ground-level inversion normally does not last for more than a period *on the order of*

30 minutes. However, if fumigation is caused by, say, an on-shore sea breeze condition, then the problem can exist for several hours.

Lofting occurs when an inversion exists only below the plume and the plume is inhibited from mixing downward. This condition might occur in the evening as a radiative ground inversion develops. If the stack height remains above the inversion then the condition will persist. However, as the inversion grows past the stack height, lofting conditions change to fanning. *Trapping* refers to conditions where the plume is caught between inversions and can only diffuse within a limited vertical height. Trapping also describes plume expansion inhibited by a high-level inversion which the plume first reaches far downwind of the stack. Estimates of ground-level concentrations perhaps 10 kilometers downwind must account for this condition if the plume top has reached a stable layer. The condition is similar to fumigation except that concentrations are much lower since the plume can diffuse to a very great width before it reaches the inversion.

Plume conditions for a particular stack depend on where the stack is situated both locally and globally. Stacks in warm, dry climates will exhibit looping in the afternoon, and depending on stack height, lofting or fanning in the early morning hours. Cloudy, wet climates will have coning conditions more often than looping. Strong ground inversions are also minimized under cloudy conditions. Data for a plume from a 60-meter high stack in Risö, Denmark, are shown in Figs. 7.20 and 7.21. These data are based on 25,000 observations over a three-year period, Christensen (1965). Risö lies between 55° and 56° north latitude on the large island of Zealand just north of the city of Roskilde and about 20 miles west of Copenhagen. The Danish Atomic Energy Commission there has a well-instrumented meteorological tower. The wind is typically from the southwest and at the stack height has a mean speed close to 6.5 m/sec (14.5 mph). Figure 7.20 shows the behavior of the stack plume for the year. During the summer months looping behavior is at a maximum and lofting is at a minimum. Can you explain why? The distribution of plume behavior by time of day is seen in Fig. 7.21 for January and July. In winter we see that lofting is common and may last all day since weak insolation is unable to remove any inversion that develops during clear nights. Virtually no looping or fumigation occurs. Furthermore, we note few great changes during the day. On the other hand, in July, we see that looping and coning are likely during daylight hours with lofting probable during the midnight hours. Large changes occur during the day. Fumigation, while not common, is much more likely in July than January and tends to occur during the day when the night inversion burns off. We see that for this particular stack, coning and lofting are most common and fumigation the least frequent.

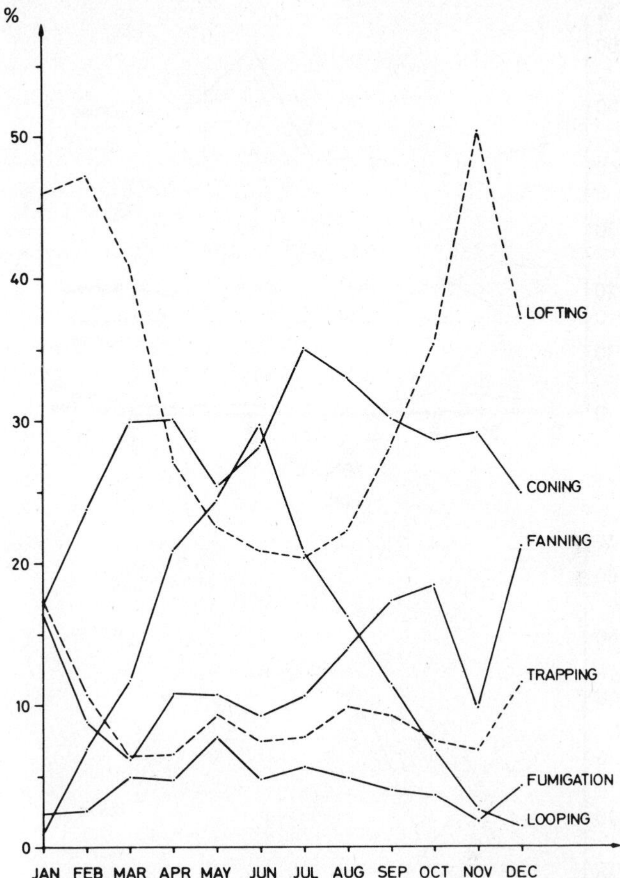

Fig. 7.20 Frequency of types of stack-smoke behavior (1962–1964) for a 60-meter stack. (*Christensen, 1965.*)

If the stack were emitting objectionable material, however, fumigation would represent a most critical situation.

Based on these results, when during the year would you expect temperature inversions below the stack height? When would you expect superadiabatic temperature profiles?

THANKSGIVING 1966 EPISODE

We conclude with a summary of the Thanksgiving 1966 air pollution episode in the eastern United States. The description is abstracted from AP-45, 1968 and uses much of the information that we have developed in this chapter.

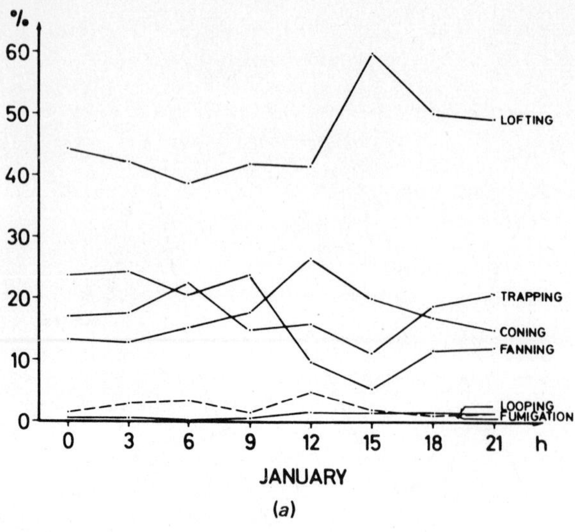

(a)

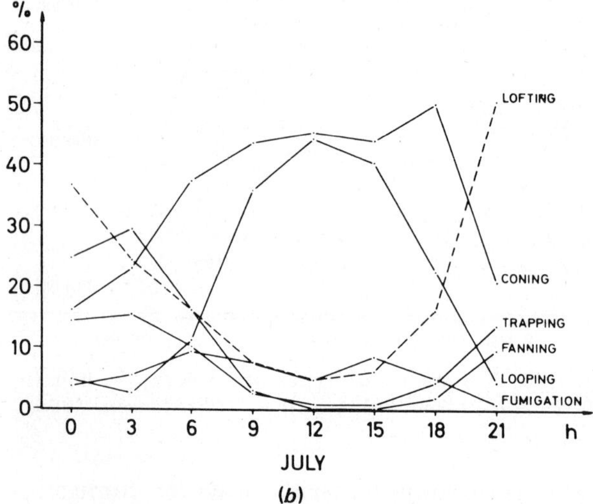

(b)

Fig. 7.21(a),(b) Daily variation of smoke behavior types (1962–1964).

Occasionally a high-pressure system becomes almost motionless over some part of the United States and tends to interrupt the usual cycle of ventilation. As a consequence, the usual daily afternoon dispersion and dilution are diminished, and pollutants may accumulate to high concentrations over a period of several days.

This section . . . describes one such stagnating high, which caused the Thanksgiving 1966 Episode. The development, progress, and breakup of this system are documented on a day-to-day basis.

On November 20, a surface high-pressure area, which had been moving steadily eastward across the United States, was centered over New York State. This high was classed as "cold" since, at upper levels the temperature was relatively low compared to temperatures of surrounding air. Over New York State the temperature at the 500-millibar level was -25°C. (The approximate height of this level is 18,000 feet). [A weather map] . . . shows that an intrusion of 10°C warmer air appeared over the northcentral states. The wind pattern carried this warmer air eastward. The replacement of the cold air over the surface high by warm air was largely responsible for the ensuing high-air-pollution-potential episode.

On November 21, the center of the surface high moved on into upper New England. This area of light winds, that is, poor horizontal ventilation, became elongated from northeast to southwest. In response to the sea-level isobaric pattern, surface winds were blowing clockwise around the center of the high with moderate northeasterly winds along the eastern seaboard and strong southwesterly winds over the upper Great Lakes region. At the 500-mb level the warmer air had moved eastward from the northcentral states to a position over the Great Lakes.

On November 22, the surface high remained in the same general location, although it continued to elongate to the northeast and southwest. It extended along the Atlantic seaboard from Newfoundland to Virginia. At the 500-mb level over New England and southward, the air temperature had warmed to above -20°C. The spread of this warm air aloft began to cause restriction of vertical dispersion.

On November 23, continued elongation of the surface high resulted in two high-pressure centers, which were connected by a ridge across New England. On the west side of the elongated high, moderate southwesterly winds occurred ahead of an advancing cold front. At the 500-mb level the warm temperatures continued to spread east and south. The warming of air at the upper levels had, by November 23, caused the reclassification of the high to "warm." Ventilation in the vertical direction was restricted by this upper-level warming. These meteorological conditions occur often, but are usually followed by the passage of a cold front with accompanying brisk winds and an influx of cleaner air. In the November, 1966, case the forecast indicated that the cold front approaching from the west would be delayed, and an advisory of high air-pollution potential was issued (for parts of the States of North Carolina, Virginia, West Virginia, Maryland, and southern Pennsylvania).

On November 24, Thanksgiving Day, the two surface highs, one east of Newfoundland and the other over northern Georgia and Alabama, were still joined by a ridge of high pressure across New England. The cold front moving across southern Canada had stalled along the St. Lawrence Valley, but a wave on this front was developing in the vicinity of Iowa. With the expected approach of the Iowa disturbance, it appeared that the western part of the forecast area would have better ventilation, and the advisory (for this area) was discontinued. However, because of the development of the high to the south, the advisory of high air-pollution potential was extended in that direction (to include parts of Alabama, Georgia, and South Carolina).

On November 25, the surface high over the Southeast had moved to near New Orleans. The frontal wave over Iowa developed rapidly and was moving over the Great Lakes. The ridge over New England was being

TABLE 7.2 Meteorological data for selected cities during the November 1966 air pollution episode (*La Guardia Field, New York, N.Y.; National Airport, Washington, D. C.; Logan International Airport, Boston, Mass.*)

Date	Average temp.	Cloud cover (tenths)	Afternoon mixing depths (meters)	Average wind speed (m/s) through mixing depth	Ventilation (m^3/s)	Resultant wind direction	Average surface wind speed (mph)
New York							
Nov. 13	43	0	1117	7.6	8490	N	14.4
14	41	1	911	7.1	6460	N	9.3
15	44	3	1409	7.7	10830	NNW	13.7
16	44	6	1065	2.8	2985	SSE	9.5
17	55	9	577	8.2	4730	SW	10.8
18	57	10	270	5.7	1540	WSW	12.2
19	48	5	1138	11.0	12520	N	18.3
20	36	0	774	5.0	3870	NE	10.8
21	41	0	1094	4.4	4820	NE	6.6
22	41	0	682	2.7	1840	NE	6.8
23	46	3	759	3.3	2505	SW	5.0
24	52	5	347	4.1	1420	NE	3.9
25	54	8	469	6.2	2910	SSE	7.3
26	52	5	164	5.2	853	N	8.6
27	50	7	671	2.3	1541	E	6.0
28	54	10	650	9.8	6370	SE	15.5
29	44	8	1437	10.6	15250	SSW	13.1
30	40	5	M	M	M	W	13.2
Washington, D.C.							
13	45	4	1291	5.2	6720	N	10.5
14	43	0	957	2.9	2775	NW	6.5
15	46	3	1047	11.3	11840	NNW	9.5
16	45	6	850	7.9	6710	S	7.9
17	55	7	499	2.1	1048	S	7.5
18	57	8	533	7.2	3835	SSW	6.9
19	50	6	1199	10.0	11990	NNW	13.4
20	38	1	1138	4.9	5580	NNE	7.0
21	41	5	1181	4.1	4850	NE	8.5
22	41	0	1261	0.9	1137	N	3.6
23	43	3	601	2.6	1560	S	4.2
24	51	7	1038	6.5	6750	SSW	3.7
25	58	10	191	M	M	S	5.2

TABLE 7.2 *(continued)*

Date	Average temp.	Cloud cover (tenths)	Afternoon mixing depths (meters)	Average wind speed (m/s) through mixing depth	Ventilation (m^3/s)	Resultant wind direction	Average surface wind speed (mph)
Washington, D.C. (*continued*)							
Nov. 26	52	1	1556	M	M	N	5.9
27	51	6	M	M	M	ESE	6.0
28	49	10	b	7.2	M	S	12.1
29	40	9	2296	6.8	15620	SSW	11.6
30	39	7	2056	M	M	W	11.5
Boston							
13	42	0	1200	11.0	13200	NNW	16.8
14	39	5	1290	6.0	7740	NE	12.9
15	39	1	900	9.5	8550	NNW	16.1
16	38	6	1100	6.5	7150	S	10.5
17	54	9	1610	10.5	16905	SW	14.7
18	56	10	100	7.0	700	SW	13.1
19	45	3	1100	15.0	16500	NW	19.1
20	34	0	1200	6.0	7200	NNW	9.2
21	38	0	1150	4.0	4600	NNE	9.5
22	37	0	850	7.0	5950	NNE	8.2
23	40	2	500	4.0	2000	NW	6.2
24	42	5	300	1.5	450	ENE	6.0
25	45	8	470	2.0	940	E	8.5
26	51	10	250	8.0	2000	NE	9.5
27	50	10	450	3.5	1575	NNE	11.5
28	49	10	200	5.0	1000	ENE	12.7
29	50	8	750	7.0	5250	S	11.4
30	43	5	M	M	M	W	14.4

b = frontal passage caused mixing depth discontinuity.
M = data missing.

displaced seaward. At 500 mb, colder air had returned to the Great Lakes region. With the projected eastward movement of the Great Lakes storm, and the movement of the clean air behind it to the east and south, the advisories for the remaining areas . . . were discontinued as of 7:00 PM E.S.T.

On November 26, the cold front, which had moved through the Great Lakes region the previous day, passed offshore into the Atlantic. The weather map again showed a high over the New England states extending southward, but the high could not be considered as stagnant because it had just formed and extensive storm systems to the west threatened its continued existence.

> The high that had been over New Orleans had been displaced offshore to the southeast so that it, too, presented no threat of high air-pollution potential.
>
> [Table 7.2] presents meteorological data . . . during the period from November 13 through November 30. The data shown include the average daily temperature, which is computed as the mean of the daily maximum and minimum temperatures. The cloud-cover column shows the average observed daytime cloud cover estimated to tenths. The afternoon mixing depth is an estimate of the height to which convective currents rise during the most active period in the afternoon. These estimates are made using upper air temperature data obtained by radiosonde observations . . . The average wind speed is the average of the speed of each thousand feet, including the surface, up the height of the afternoon mixing depth. The column labeled "Ventilation" is the product of the afternoon mixing depth and the average wind speed, and is considered as the flow through a column 1 meter wide. The resultant wind direction is the vector sum of eight surface wind observations spaced through a 24-hour period. It indicates the direction of displacement of the surface air in the vicinity of the designated city. The average surface wind speed is the average of eight hourly observations per day at 3-hour intervals."

Because of the stagnant high pressure region and the resulting low value of vertical mixing heights noted in the table, pollutants built up to very high concentrations. New York City on November 24 experienced an hourly peak average for SO_2 of 0.97 ppm. The daily mean value for November 24 was a bit over 0.50 ppm. Particulate levels were also high and reached almost $400\mu g/m^3$ in Philadelphia on the 25th of November. It is interesting to speculate on how high the pollution level might have gone had this not been a holiday period with minimal industrial activity and a relatively warm period with minimal demand for space heating.

REFERENCES

Christensen, J.: Meteorological Measurements at Risö, Denmark 1962-1964, Risö Report 121, 1965.

Halitsky, J.: Gas Diffusion Near Buildings, *ASHRE Trans.*, vol. 69, pp. 464-484, 1963.

Holzworth, G. C.: Estimates of Mean Maximum Mixing Depth in the Contiguous United States, *Monthly Weather Rev.*, vol. 92, pp. 235-242, 1964.

Hosler, C. R.: Low Level Inversion Frequency in the Contiguous United States, *Monthly Weather Rev.*, vol. 89, pp. 319-339, 1961.

Middleton, John: Planning Against Air Pollution, *Am. Scientist*, vol. 59, pp. 188-194, 1971.

Slade, D. H. (ed.): "Meteorology and Atomic Energy, 1968," TID 24190.

Smith, M. (ed.): Recommended Guide for the Prediction of the Dispersion of Airborne Effluents, Am. Soc. Mech. Eng., 1968.

Stern, A. C. (ed.): "Air Pollution," 2d ed., vol. 1, Academic Press, New York, 1968.

Thanksgiving 1966 Air Pollution Episode in the Eastern United States, National Air Pollution Control Administration Publication AP-45, 1968.

Turner, D. B.: "Workbook of Atmospheric Dispersion Estimates," U.S. Public Health Service Publication 999-AP-26, revised 1970.

PROBLEMS

7.1 The simplest diffusion model for calculating pollutant concentrations over an urban area is to model the problem as if the emissions went into a box of height equal to the mixing height, H. Calculate the concentration of a pollutant of source strength Q g/sec-m^2 if the average wind speed into the up-wind side of the box is U. Comment on the accuracy of this model.

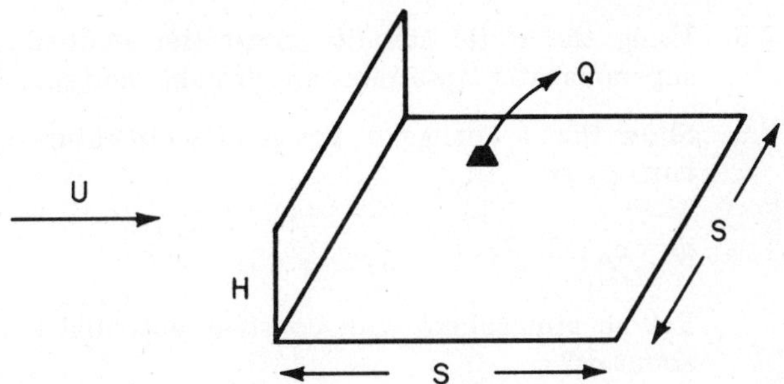

7.2 Now consider the model of Prob. 7.1 but including a transient buildup of the pollutant. Derive a differential equation which describes the concentration as a function of time. Show that the solution for long times approaches the solution in Prob. 7.1, that is,

$$\chi = \chi_0 + QS/UH$$

where χ_0 is the background concentration.

7.3 Find the mass of the atmosphere if the earth is a sphere of radius equal to 6,370 km.

7.4 The potential temperature is defined as

$$\theta = T \left(\frac{1000}{P}\right)^{(k-1)/k}$$

(*a*) Show that

$$\frac{1}{\theta}\frac{d\theta}{dz} = \frac{1}{T}\left[\frac{dT}{dz} + \frac{g}{c_p}\right]$$

(*b*) Show that for an adiabatic temperature distribution in the atmosphere

$$\frac{d\theta}{dz} = 0$$

(Therefore lines of constant potential temperature are also adiabats.)

(*c*) The static stability parameter, S_z, is defined as

$$S_z = \frac{\left(\frac{dT}{dz}\right)_{\text{environment}} - \left(\frac{dT}{dz}\right)_{\text{adiabatic}}}{T}$$

so that $S_z > 0$ is stable. Show that S_z can also be written as $(1/\theta)(d\theta/dz)$. What role does the potential temperature have in defining stability?

7.5 Using the static stability parameter defined in Prob. 7.4(*c*), show that superadiabatic lapse rates are unstable and subadiabatic lapse rates are stable.

7.6 Show that a change in potential temperature can be related to a change in entropy as

$$ds = c_p \, d(\ln \theta)$$

For an atmosphere with constant potential temperature, what can you tell about *ds*?

7.7 For the given atmospheric data, which of the layers are stable, unstable, or neutral? Which layer is the most stable? For a parcel of air rising initially from point *c*, where does the first deceleration occur?

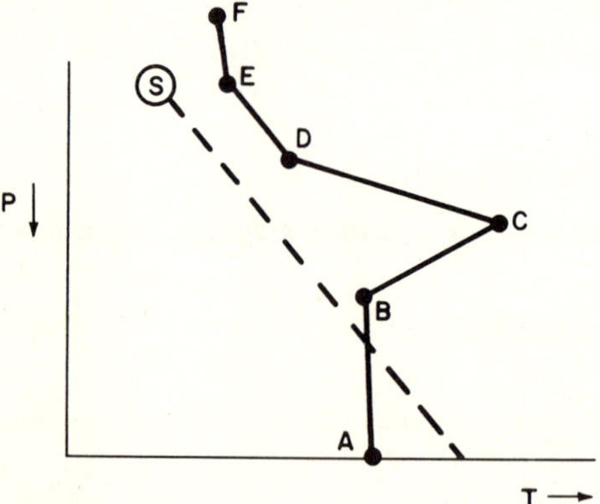

7.8 Mass deposition rates are given in the table for an aircraft flight defined as a take-off and a landing (an LTO). The deposition rates are in g/meter where the meter refers to a meter of vertical ascent *and descent* during an LTO. If we arbitrarily assume that the cross wind mixing disperses the pollutants to a width of 2 kilometers, calculate in $\mu g/m^3$ the future annual pollutant concentrations if there will be 10^6 LTO/yr and the average wind speed is 10 miles/hr (16 km/hr). (Reference is J. A. Fay, *M.I.T. Fluid Mechanics Lab. Report* 70-6, May, 1970.)

Mass deposition rates (g/m)
4-engine "smokeless" turbofan JT8D-7

particulates	12.8
carbon monoxide	17.6
oxides of nitrogen	15.0
hydrocarbons	1.0
sulfur dioxide	3.5

7.9 Measured concentrations are a function of the sampling time over which the measurement is averaged. Write a brief review of the effect of sampling time on measured concentrations.

Hino, M., Maximum Ground Level Concentration and Sample Time, *Atmos Environ*, **2**, 149-165, 1968.

McGuire, T., and Noll., K. E., Relationships between Concentrations of Atmospheric Pollutants and Averaging Time, *Atmos Environ*, **5**, 291-298, 1971.

You may also wish to consult Turner (1970) and Smith (1968).

7.10 The velocity profile for turbulent flow is often given in the form

$$U(z) = \sqrt{\frac{g_c \tau_o}{\rho}}\, K \ln z^*$$

where

$$z^* = \left(z \sqrt{\frac{g_c \tau_o}{\rho}} \Big/ \nu \right)$$

This can be developed after some simplifying assumptions which you can find in a fluid mechanics text book. Develop this equation. If $\sqrt{g_c \tau_o / \rho}$ has the value of 0.16 m/sec and K = 2.5 plot up $U(z)$ for air from $z = 1$ m to 200 m and compare your results to those given in the Figs. 7.4, 7.5.

7.11 Wind velocity results taken at Risö, Denmark are presented in the form

$$\frac{U}{U_1} = \left(\frac{z}{z_1}\right)^{n/n-2}$$

For June the values given in Christensen (1965) are $n = 0.25$ and for December $n = 0.45$. Compare these values to the 1/7 law and to the values recommended by Smith (see Chap. 8). Does this data represent stable or unstable conditions? Compare these results to the data shown in Figs. 7.4, 7.5.

7.12 The figure shows temperature data at an altitude of 2 m and 123 m as a function of time of day and month for Risö, Denmark. Using these results, describe why the plume behavior noted in Figs. 7.20 and 7.21 occurs. (Use June temperature results for the July plume.)

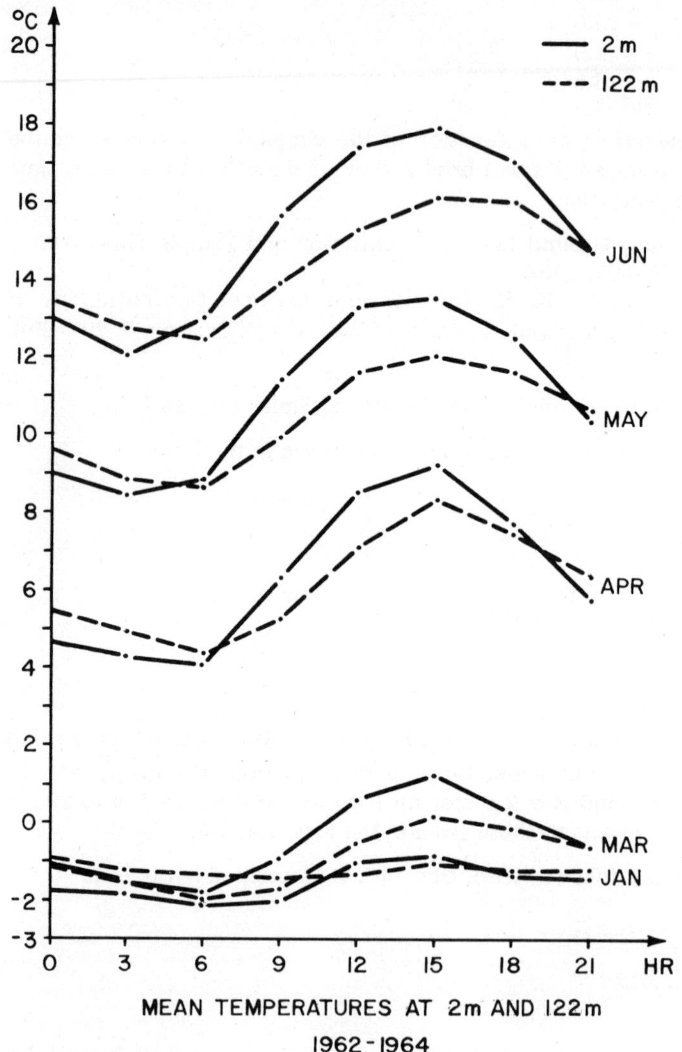

MEAN TEMPERATURES AT 2m AND 122m
1962-1964

7.13 One model for Jupiter's atmosphere indicates the following composition:

fraction by volume		
	H_2	0.86578
	He	0.13214
	CH_4	0.00062
	NH_3	0.00015
	H_2O	0.00102
	Ne	0.00013
	others	0.00016

Further, g has the value 2,500 cm/sec^2. Calculate for these atmospheric conditions the adiabatic lapse rate on Jupiter.

7.14 Show that the variation of g with altitude can be neglected in determining temperature variations in the atmosphere.

7.15 One model of the atmosphere assumes a constant density. Find the lapse rate for this atmosphere and indicate whether this is a stable or unstable case.

7.16 Consider air heavily polluted with particulates. Explain why this might cause a stabilizing effect on the atmosphere during the day and a destabilizing effect at night.

8
PLUME DISPERSION

> **Owing to the complexity of the problem it is highly improbable that one single, unifying model will ever be conceived which can account for all the combinations of meteorological, topographic, and source parameters. This is the case even if we study simple dispersion of an inactive gas from a point source, and exclude the complications such as those brought about by chemical reactions, fallout, washout, and absorption by vegetation.**
>
> *Niels Busch, 1971*

We will now discuss how to determine the atmospheric concentration of pollutants emitted from single sources either at ground level or from stacks. Such sources may emit for short times in "puffs" of pollution or may emit for long times as plumes. We shall spend most of our effort on plume dispersion calculations.

Consider Fig. 8.1 which shows a plume from a source at $x = 0$, $y = 0$. The plume moves downwind and as it does, it grows through action of the turbulent eddies. We have discussed in Chap. 7 how the very large eddies simply move the whole plume around and why diffusion is most effective with eddies on the order of the plume size. Note that the concentration of pollutant at a particular location under the plume depends on the sampling time. The "instantaneous" plume has very high concentrations over a narrow width. Over a ten-minute average the plume will touch a much broader area but concentrations will be lower. Likewise over a two-hour period the plume will reach further in the cross-wind direction (even with the same mean wind direction) and the peak concentration will be reduced even more. The concentration

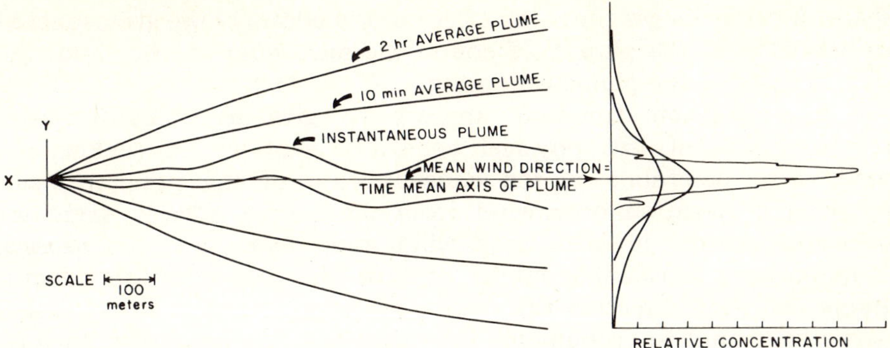

Fig. 8.1 The diagram on the left represents approximate outlines of a smoke plume observed instantaneously and of plumes averaged over 10 min and 2 hr. The diagram on the right shows corresponding cross-plume distribution patterns. (*Slade, 1968.*)

distribution perpendicular to the axis of the wind appears to be gaussian in this figure. The gaussian distribution will be our model for plume concentration calculations and the limits to this model will be discussed shortly.

The time dependence of plume width noted above also extends to calculations in the downwind direction. As the plume expands, larger and larger eddies are effective in dispersing the pollutants. Thus near the source, one range of eddy size is important and downwind another range is important. Since the size of the eddy can be related to a time scale (crudely speaking, it takes bigger eddies longer to go by) this means that we need to use different averaging times to study the plume as a function of downwind distance. Put another way, the rate at which the plume spreads is not only dependent on the intensity of the turbulence and the distribution of eddy sizes, but also on the dimension of the plume. The dimension of the plume, however, depends on the travel time from the source. Scorer (1968) suggests, as a rule of thumb, that the appropriate sampling time is equal to about the time for the wind to carry the plume from the source to the point of observation. Thus $\tau \cong x/U$. This point is important since the diffusion coefficients which we shall introduce are tabulated for a particular averaging time.

THE GAUSSIAN MODEL

There are several models available for predicting the concentrations downwind of a single source. We have just seen that the cross-wind area covered by a plume depends on the length of time that we observe the plume behavior. For very short times the plume has a narrow, sinuous shape but over longer times the effect of the eddies in the atmosphere is to broaden out the plume. Of course the maximum concentration still should be along the plume centerline.

The gaussian function appears to give us a mathematical representation of this physical behavior. What we are seeking is a method to determine χ, the concentration of the pollutant, as a function of position downwind from the source. Clearly there is a diffusion process in both cross-wind coordinates, and the gaussian function is a reasonable model of how the concentration behaves. Along the wind direction the convection is greater than the diffusion, so we have only to account for the effect of the wind in stretching the plume.

In what manner would the wind directly affect the dilution? Independent of the effects of wind on the generation of turbulence, the higher the wind speed the faster the pollutant is dispersed in the downwind direction. If we picture puffs of emission from a stack, then doubling the wind speed will double the distance between puffs or halve the concentration. Using χ for concentration we expect, therefore, that

$$\chi \propto \frac{1}{U}$$

We have already seen that plumes appear to have internal concentrations which are gaussian. Thus we might model the plume as a gaussian function in the vertical, z, and horizontal, y, coordinates and proportional to $1/U$, which is in effect along the x coordinate. Thus our solution for the plume behavior in the downwind direction should be directly proportional to the source strength, Q, usually given in g/sec, inversely proportional to U, and given by a gaussian function in y and z.

The mathematical form of the gaussian function in the y dimension is:

$$\chi \propto A \exp \left[-\frac{1}{2}\left(\frac{y}{\sigma_y}\right)^2\right] \tag{8.1}$$

where σ_y is the standard deviation. The function has the values

$y = 0$	$\chi = A$
$y = \sigma_y$	$\chi = A \exp(-0.5) = 0.607\,A$
$y = 2\sigma_y$	$\chi = A \exp(-2.0) = 0.136\,A$
$y = 2.15\sigma_y$	$\chi = A \exp(-2.31) = 0.100\,A$

Since we are dealing with diffusion in y and z we expect to have gaussians in both coordinates.

The gaussian function can be normalized so that the area under the curve has a unit value by picking A to be $(1/\sqrt{2\pi}\,\sigma_y)$, thus[1]

$$\chi \propto \frac{1}{\sqrt{2\pi}\,\sigma_y} \exp\left[-\frac{1}{2}\left(\frac{y}{\sigma_y}\right)^2\right] \tag{8.2}$$

This is important in our model since the integral of concentration over x, y, and z must equal the total amount of pollutant emitted.

We are now in a position to construct a solution for the concentration downwind of a point source of pollution.

Using the coordinate system shown in Fig. 8.2 the solution for the plume concentration takes the form

$$\chi(x,y,z,H) = \frac{Q}{2\pi\sigma_y\sigma_z U} \exp\left[-\frac{1}{2}\left(\frac{y}{\sigma_y}\right)^2\right]\left[\exp\left(-\frac{1}{2}\left(\frac{z-H}{\sigma_z}\right)^2\right) + \exp\left(-\frac{1}{2}\left(\frac{z+H}{\sigma_z}\right)^2\right)\right] \tag{8.3}$$

where χ, concentration in g/m^3 or μg/m^3
Q, source strength in g/sec
U, average wind speed in m/sec
σ_y, σ_z, diffusion coefficients in y and z directions in meters
H, effective height of source emission in meters.

[1]

$$\int_0^{\infty} e^{-h^2 x^2}\, dx = \frac{\sqrt{\pi}}{2h}$$

so

$$\int_{-\infty}^{\infty} e^{-\frac{1}{2}(y/\sigma_y)^2}\, dy = 2\frac{\sqrt{\pi}}{2}\left[\sqrt{2}\sigma_y\right] = \sqrt{2\pi}\sigma_y$$

thus

$$\frac{1}{\sqrt{2\pi}\sigma_y}\int_{-\infty}^{\infty} e^{-\frac{1}{2}(y/\sigma_y)^2}\, dy = 1$$

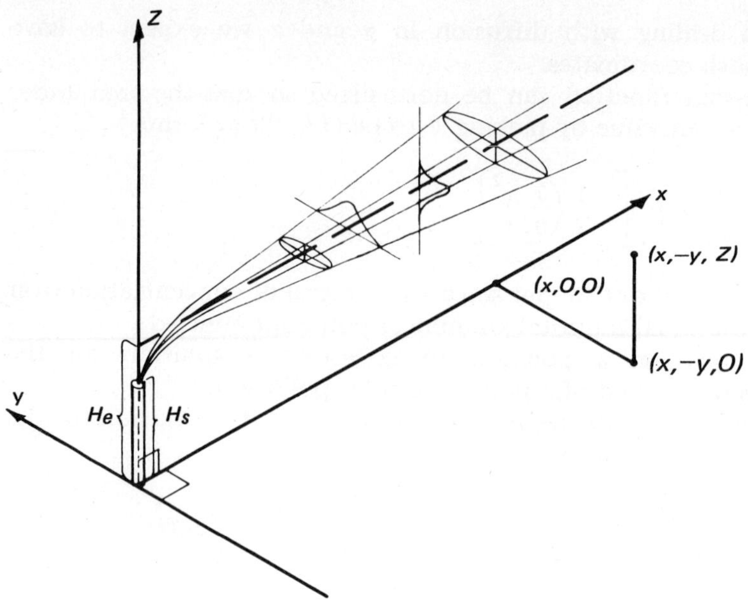

Fig. 8.2 Coordinate system showing gaussian distributions in the horizontal and vertical.

To repeat, this solution assumes no diffusion in the downwind, x, direction and thus is applicable to a plume and not a puff of pollutant.

The term $1/2\pi\sigma_y\sigma_z$ is to normalize the gaussian so that the integral over a volume $dxdydz$ of the plume will yield the amount of pollutant emitted, $Q\,dt$. It is clear that the term with y accounts for diffusion in the cross-wind, y, direction; but why the form of the term in the z direction? To explain this we must consider what happens to the plume if it reaches the ground. The two terms in z account for reflection of the plume when it touches the ground (see Fig. 8.3). If it is assumed that the earth's surface is a barrier to further diffusion and if there is no deposition or absorption, then we can account for this

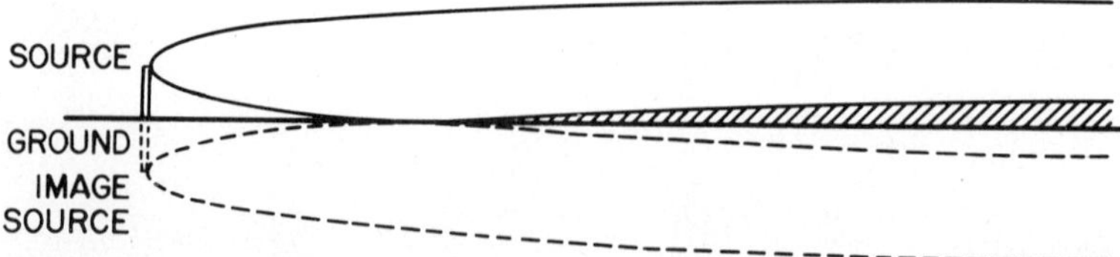

Fig. 8.3 Source and image source below ground, used to calculate the concentration owing to a plume. In the shaded region, the pollution is increased by the amount corresponding to the image source.

barrier by assuming an image source located symmetrically (with respect to the ground) to the actual source.

The term in $(z-H)$ accounts for the above ground real source and the term in $(z+H)$ accounts for the imaginary source below the ground as noted in Fig. 8.4. For a detector at ground level $z = 0$ and we find

$$\chi(x,y,0,H) = \frac{Q}{\pi \sigma_y \sigma_z U} \exp\left[-\left(\frac{y^2}{2\sigma_y{}^2} + \frac{H^2}{2\sigma_z{}^2}\right)\right] \tag{8.4}$$

The term in y^2 accounts for diffusion of the plume in the cross-wind direction. If one is interested in concentrations only along the plume centerline then $y = 0$ and the formula for ground level concentration reduces to

$$\chi(x,0,0,H) = \frac{Q}{\pi \sigma_y \sigma_z U} \exp\left[-\frac{H^2}{2\sigma_z{}^2}\right] \tag{8.5}$$

What U is used in this equation? Since U is a function of z, some mean value must be used. The appropriate value is the mean through the plume (Turner, 1970). However, the time-averaged wind speed at the stack height is commonly used. Unfortunately even this value may not be known, in which case an estimate must be made. This estimate could be based on an assumed power law velocity profile such as

$$U = U_1 \left(\frac{H}{z_1}\right)^n \tag{8.6}$$

Smith (1968) recommends $n = 0.25$ for unstable and $n = 0.50$ for stable conditions.

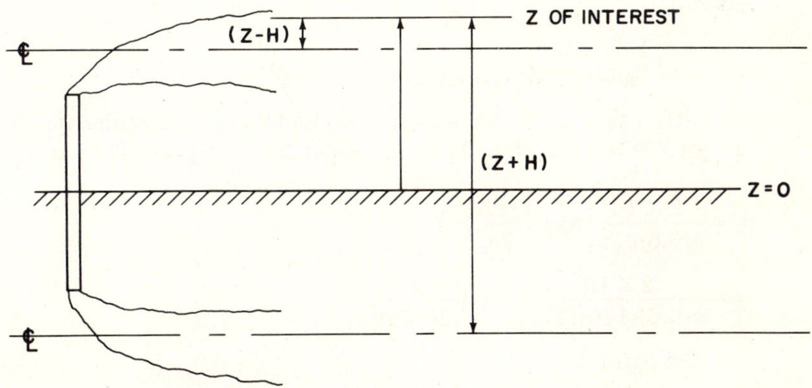

Fig. 8.4 Coordinate system for real source and imaginary source.

It is important to recognize that our model of gaussian behavior is only one of several possible models. It is, however, a useful tool for obtaining estimates of concentrations. The limitation to this model will be emphasized in the rest of this chapter.

Example 1. It is desired to estimate the concentration of SO_2 downwind of a 1,000-megawatt power plant burning 10,000 tons of 1 percent sulfur coal per day. The stack height is 250 meters. The wind speed has been measured on a clear, sunny day as 3 m/sec at the top of a 10-meter tower. Find χ at $x = 1$ km and at 5 km if the consulting meteorologist estimates the hourly mean value of σ_y and σ_z as

x	σ_y	σ_z
1 km	140 m	125 m
5 km	540 m	500 m

First note that since the diffusion coefficients are given as 1-hour averages, the concentration calculated will be an average concentration appropriate to 1-hour periods. Also in this example we shall use the height of the source emission as the stack height. We shall see in the next chapter how much *plume rise* occurs because of buoyancy forces acting on the hot plume. Our calculated result will be conservative since H would be actually larger than the stack height. Sunny, clear days are normally unstable so

$$U = U_1 \left(\frac{H}{z_1}\right)^n = 3\left(\frac{250}{10}\right)^{0.25} = 6.6 \text{ m/sec (14.7 mph)}$$

Let us now find Q as g/sec, assuming that all of the sulfur goes up the stack. In a coal-fired plant as much as 20 percent of the sulfur may remain in the ash and presumably is collected, at least in the new power plants.

$$\begin{aligned} \text{Sulfur} &= 10{,}000 \text{ tons/day} \times 2{,}000 \text{ lb/ton} \times 1/24 \text{ hr/day} \times 454 \text{ g/lb} \\ &\quad \times 1/3{,}600 \text{ hr/sec} \times 0.01 \text{ sulfur} \\ &= 1050 \text{ g/sec} \\ &= 1.050 \times 10^9 \text{ micrograms/sec} \end{aligned}$$

But SO_2 has a molecular weight of 64 and S has a molecular weight of 32, so we get 2 SO_2 for each S. Thus Q is about 2×10^3 g/sec. Then at 1 km

$$\chi = \frac{2 \times 10^9}{\pi(6.6)\sigma_y\sigma_z} \exp\left(\frac{-250^2}{2\sigma_z^{\;2}}\right)$$

$$\chi = \frac{2 \times 10^9}{\pi(6.6)(140)(125)} \exp\left(\frac{-250^2}{2(125)^2}\right)$$

$$\chi = 780\ \mu\text{g/m}^3$$

at 5 km

$$\chi = \frac{2 \times 10^9}{\pi(6.6)(540)(500)} \exp\left(\frac{-250^2}{2(500)^2}\right)$$

$$\chi = 315\ \mu g/m^3$$

Concentrations of this magnitude would be very unacceptable (24-hour United States limits are 365 $\mu g/m^3$ and annual averages are limited to 80 $\mu g/m^3$)! We shall find that the effect of buoyancy on the plume is to add enough height to reduce these concentrations significantly. Doubling H reduces χ by a factor of exp $(-H^2/2\sigma_z^2)$ to the fourth power which here at 1 km is a factor of $(0.136)^4$ or 3.4×10^{-4}.

Example 2. Returning to our diffusion model we see that for a ground-level source where $H = 0$, the ground concentration is given by

$$\chi(x,0,0,0) = \frac{Q}{\pi \sigma_y \sigma_z U}$$

Let us calculate the night-time concentration of oxides of nitrogen 1 km downwind of an open, burning dump if the dump emits 2 g/sec of NO_x. The wind speed is 4 m/sec at $z = 10$ meters. The 1-hour average diffusion coefficients under the neutral stability conditions have been estimated to be: $\sigma_y = 70$ meters and $\sigma_z = 50$ meters. If we assume the dump to be a point source and use the given wind speed as an appropriate average then

$$\chi(1,0,0,0) = \frac{2 \times 10^6}{\pi(70)(50)(4)} = 45\ \mu g/m^3$$

DIFFUSION COEFFICIENTS

How were these diffusion coefficients determined? The diffusion coefficients σ_y and σ_z can be related to the deviation in the wind direction given by σ_a in the azimuth angle (azimuth refers to the lateral or cross-wind direction) and σ_e in the elevation angle (elevation refers to the vertical, here z direction). The variation of these angles as a function of wind speed and stability condition is discussed in Slade (1968). The important point is that they can be measured simply by using a bivane.

Smith (1968) gives formulas relating σ_y to σ_a and σ_z to σ_e. As we might expect, the results of many experiments have been combined to establish σ_y and σ_z as a function of atmospheric conditions. Table 8.1 notes the atmospheric conditions and Figs. 8.5 and 8.6 give results for the diffusion coefficients as tabulated by Turner (AP-26, revised 1970). These values assume:

1. a sampling time of *about 10 minutes*

TABLE 8.1 Key to stability categories

Surface wind speed (at 10 m), m sec^{-1}	Day: Incoming solar radiation, Strong	Day: Incoming solar radiation, Moderate	Day: Incoming solar radiation, Slight	Night: Thinly overcast or ⩾4/8 low cloud	Night: ⩽3/8 Cloud
< 2	A	A-B	B		
2-3	A-B	B	C	E	F
3-5	B	B-C	C	D	E
5-6	C	C-D	D	D	D
> 6	C	D	D	D	D

The neutral class, D, should be assumed for overcast conditions during day or night.
Source: Turner, 1970.

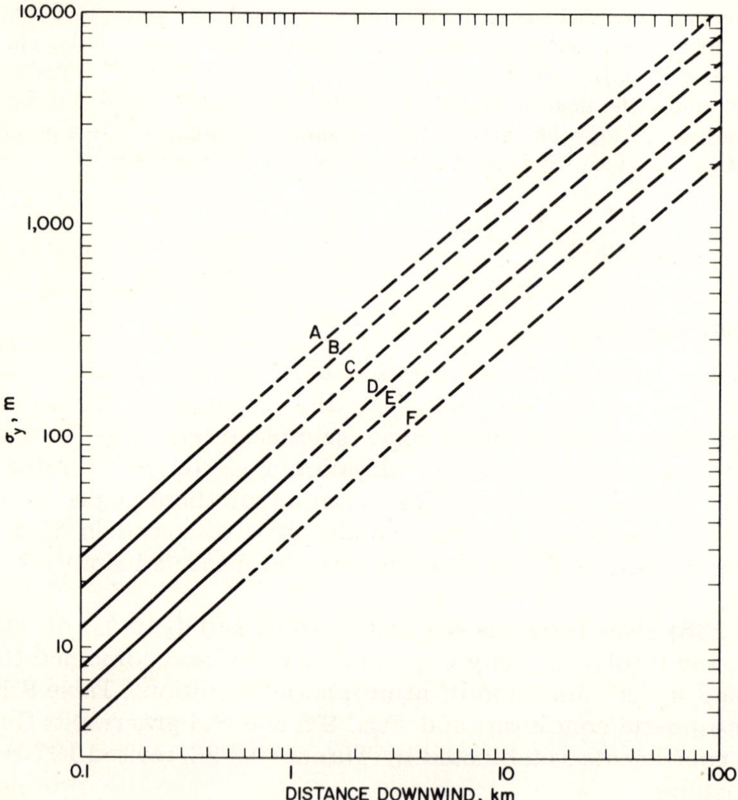

Fig. 8.5 Horizontal dispersion coefficient as a function of downwind distance from the source. (*Turner, 1970.*)

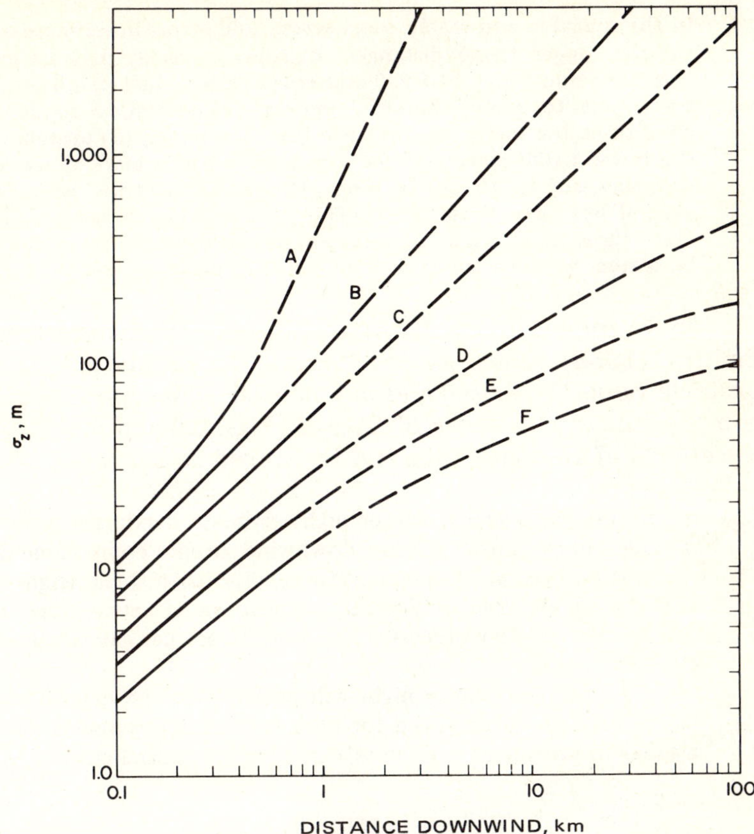

Fig. 8.6 Vertical dispersion coefficient as a function of downwind distance from the source. (*Turner, 1970.*)

2. the height values of interest to be in the lowest several hundred meters of the atmosphere since $\sigma = fn(z)$
3. a surface corresponding to conditions of open country.

For urban areas the diffusion coefficients are greater because of the heat island effect and the mechanical turbulence generated by surface roughness. So these results for diffusion coefficients would be conservative.

In Table 8.1 "strong" incoming solar radiation refers to solar altitude greater than 60° with clear skies; "slight" insolation refers to a solar altitude of 15 to 35° with clear skies. Cloudiness will reduce the solar radiation and an estimate of this must be included in picking the stability category.

How accurate are these predictions of σ_y and σ_z? Turner (1970) comments as follows:

> In the unstable and stable cases severalfold errors in estimate of σ_z can occur for the longer travel distances. In some cases σ_z may be expected to be correct within a factor of 2. These (cases) are: (1) all stabilities for distance of travel out to a few hundred meters; (2) neutral to moderately unstable conditions for distances out to a few kilometers; (3) unstable conditions in the lower 1,000 meters of the atmosphere with a marked inversion above for distances out to 10 km or more. Uncertainties in the estimate of σ_y are in general less than those of σ_z. The ground-level centerline concentration for these three cases (where σ_z can be expected to be within a factor of 2) should be correct within a factor of 3 including errors in σ_y and U.

In Examples 1 and 2 the diffusion coefficients were taken from Smith's (1968) tabulation of "hourly mean values." In Example 1 unstable values were used and in 2 neutral values were used. You should compare these with those of Turner (Figs. 8.5 and 8.6) to get a feel for the effects of averaging time and the spread in tabulated values.

Example 3. For an overcast winter night estimate maximum ground-level sulfur dioxide concentration 10 km downwind from a copper smelter if the wind speed is 3 m/sec at 10 meters. The smelter has a stack height of 100 meters and the plume rises sufficiently to give an effective stack height of 250 meters. The smelter processes 685 tons of ore per day which is chalcopyrite, $CuFeS_2$.

An overcast winter night will probably have a neutral atmosphere. Let us calculate the wind speed for both stable and unstable conditions and then average to provide us with an estimate for the neutral case.

$$\text{stable} \quad U = U_1\left(\frac{H}{z}\right)^{0.5} = 3\left(\frac{250}{10}\right)^{0.5} = 15 \text{ m/sec}$$

$$\text{unstable} \quad U = U_1\left(\frac{H}{z}\right)^{0.25} = 3\left(\frac{250}{10}\right)^{0.25} = 6.7 \text{ m/sec}$$

We can therefore estimate the wind speed for this example as about 11 m/sec at 250 meters.

The source strength can be estimated since every gram of ore will produce 64/184 = 0.35 g of sulfur. Since SO_2 has twice the molecular weight of sulfur, each gram of ore processed will result in about 2/3 g of SO_2 released. Or for this ore every unit of Cu produced results in 2 units of SO_2. Thus we have

$$SO_2 = \frac{685 \times 2{,}000 \times 0.67 \times 454}{24 \times 3{,}600} = 4{,}800 \text{ g/sec}$$

An overcast night corresponds to class D stability. So from Figs. 8.5 and 8.6 we find at 10 km

$\sigma_y = 550$ m $\quad \sigma_z = 135$ m

$$\chi = \frac{Q}{\pi \sigma_y \sigma_z U} \exp\left[-\frac{1}{2}\left(\frac{H}{\sigma_z}\right)^2\right] = \frac{4800}{\pi(550)(135)(11)} \exp(-1.71)$$

$\chi = 336\ \mu\text{g/m}^3$

These values represent short-time averages since our diffusion coefficients are for short times. Does our calculation meet Scorer's criteria that $\tau = x/U$? We have

$$\tau \cong \frac{10}{11}\left(\frac{1{,}000}{3{,}600}\right) \cong 0.25 \text{ hr}$$

Our σ are for a nominal 10-minute period so we are reasonably close to the proper averaging time.

What is the maximum concentration at ground level and where does it occur? Figure 8.7 from Turner (1970) gives results for the location of maximum concentration, x_{max}, as a function of $(\chi U/Q)_{\text{max}}$. This figure was calculated by plotting up Eq. (8.5) and finding the point of maximum concentration as a function of H. We find for our case with $H = 250$ and class D stability that

$$x_{\text{max}} = 13 \text{ km and } \left(\frac{\chi U}{Q}\right)_{\text{max}} = 8.4 \times 10^{-7} \ (1/\text{m}^2)$$

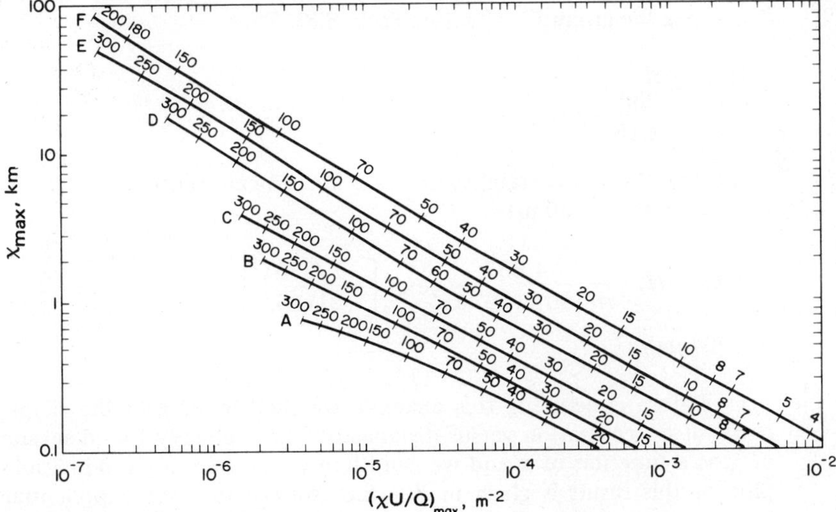

Fig. 8.7 Distance of maximum concentration and maximum $\chi U/Q$ as a function of stability and effective height (meters) of emission. (*Turner, 1970.*)

Thus at a distance of 13 km the concentration is

$$\chi = (8.4 \times 10^{-7}) \frac{4{,}800}{11} = 366\ \mu g/m^3$$

Let us check this answer directly from Eq. (8.5). At 13 km we have

$$\sigma_y = 700\ m \qquad \sigma_z = 160\ m$$

so

$$\chi = \frac{4{,}800}{\pi(700)(160)(11)} \exp\left[-\frac{1}{2}\left(\frac{250}{160}\right)^2\right] = 364\ \mu g/m^3$$

To continue the example further, we can ask what is the concentration at $x = 10$ km but 0.5 km off of the centerline? We have

$$\chi(x,y,0,H) = 336 \exp\left[-\frac{1}{2}\left(\frac{y}{\sigma_y}\right)^2\right] = 336(0.66) = 222\ \mu g/m^3$$

Where does the plume initially reach the ground and what is the concentration at that point? First we must determine the x position where the plume has just expanded sufficiently to reach the ground. Since the plume is assumed to be gaussian we must decide what constitutes "touching the ground." At a width of 2.15 σ the gaussian has a concentration of 10 percent of the value on the centerline and we shall arbitrarily use this as the "touching the ground" value (see Prob. 8.8). Thus

$$2.15\ \sigma_z = H$$

$$\sigma_z = \frac{250}{2.15} = 116\ m$$

From Fig. 8.6 for D stability, $x = 7.8$ km. The concentration at this position is given by ($\sigma_y = 450$ m)

$$\chi(7.8,0,0,H) = \frac{4{,}800}{\pi(450)(116)(11)} \exp\left[-\frac{1}{2}\left(\frac{250}{116}\right)^2\right]$$

$$\chi = 264\ \mu g/m$$

Before concluding this example we shall investigate the shape of the curve of concentration versus distance. We have already found several values of χ as a function of x and we can fill in with others as given in Table 8.2. A plot of this result is given in Fig. 8.8. We see that the exponential factor increases with x, approaching the value of 1 as x (and σ_z) becomes large. However, the preexponential factor with $1/\sigma_y\sigma_z$ steadily decreases with x approaching the value 0 as x becomes large. There is therefore a maximum value for χ as a function of x.

TABLE 8.2 Downwind concentrations

x	σ_y	σ_z	$Q/\pi\sigma_y\sigma_z U$	$-\frac{1}{2}(H/\sigma_z)^2$	exp	χ $\mu g/m^3$
3	188	64	0.0115	−7.6	0.0005	57
5	300	88	0.00526	−4.0	0.0185	97
7.8	450	116	0.00266	−2.32	0.099	264
10	550	135	0.00187	−1.71	0.180	336
13	700	160	0.00136	−1.22	0.294	366
20	1,000	202	0.00068	−0.77	0.465	316
50	2,200	328	0.00019	−0.29	0.748	142

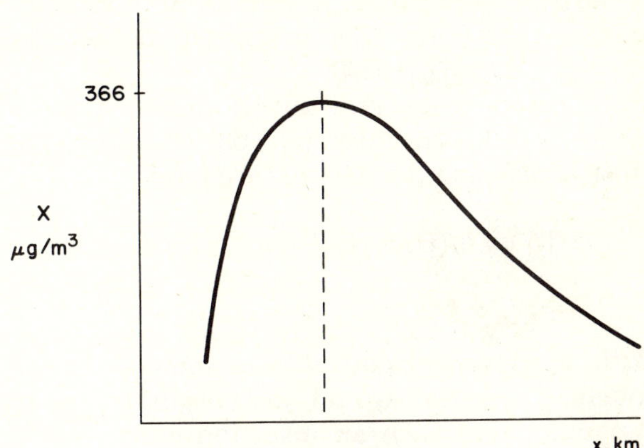

Fig. 8.8 Downwind concentrations.

INVERSION EFFECTS

Turner discusses in detail a procedure to handle diffusion when the plume expansion is limited by an upper-level inversion. The essence of the method is to calculate the concentrations as if these were distributed uniformly throughout the layer of height H_i where H_i is the distance from ground level to the inversion. The diffusion equation at ground level becomes for this case

$$\chi = \frac{Q}{(2\pi)^{1/2} U H_i \sigma_y} \exp\left[-\frac{1}{2}\left(\frac{y}{\sigma_y}\right)^2\right] \tag{8.7}$$

This can be applied well downwind of where the plume first reaches the H_i elevation, which we can estimate as where $2.15\sigma_z = (H_i - H)$ where

H is the effective stack height. Turner recommends using Eq. (8.7) for $x > 2X_i$ where X_i is the location where the plume reaches the inversion. Before X_i the regular diffusion result applies, and between X_i and $2X_i$ one must interpolate between the results at X_i and $2X_i$.

This same equation can be used to estimate concentrations under fumigation conditions. In this case H_i is the height to which the unstable air has risen, that is, the height to which the inversion has been eliminated. However, Q must be corrected then to the fraction of the plume which can be carried down to the ground. If the whole plume is just in unstable air, then $H_i \cong H + 2\sigma_z$ and Q is the total Q. Note that we use the stable classification to calculate σ_y and σ_z in predicting fumigation conditions. A correction to account for the plume spreading in the y direction as it is carried to the ground is given in Turner as $\sigma_{y,\text{fum.}} = \sigma_{y,\text{stable}} + H/8$.

Consider our previous example of the copper smelter. At what height would an inversion be necessary to affect the ground-level concentration at $x = 10$ km? We have

$$2.15\,\sigma_z = (H_i - H) \qquad \text{or} \qquad H_i = 2.15\,\sigma_z + H$$

$$= 2.15(135) + 250 = 540 \text{ m}$$

With an inversion lid at 540 meters the plume (using 2.15 as the touching criteria) would just begin to be influenced at $x = 10$ km. Assuming that such an inversion exists, calculate at ground level the centerline concentration at $x = 20$ km and compare to the previous result. We have

$$\chi = \frac{Q}{\sqrt{2\pi}\, UH_i\sigma_y} \exp\left[-\frac{1}{2}\left(\frac{y}{\sigma_y}\right)^2\right]$$

at $x = 20$ km, $\sigma_y = 1{,}000$ meters

so

$$\chi = \frac{4{,}800}{\sqrt{2\pi}(540)(1{,}000)(11)} = 322\ \mu\text{g/m}^3$$

This is bigger but not greatly different from our previous result of 316 μg/m^3.

LIMITS TO THE MODEL

We have already noted some limits to the model. For one thing the σ's are not very accurate. In addition, we have neglected the turning of the wind with height owing to friction effects (as discussed in Chap. 7). This tends to spread out the plume in the cross-wind direction. We have also neglected consideration of any absorption or deposition of pollutants when the plume reaches the ground. SO_2, for example, is known to interact with vegetation. We have left out any chemical reaction along the plume path. In the case of SO_2, a gradual change occurs to SO_3, and then to H_2SO_4. We have overlooked shifts in wind direction which occur over periods like the 0.25 hr required for our example plume to reach the 10-km position. Using diffusion coefficients developed for short-time averages to predict hourly or longer average concentration also causes error. Some information on this point is given in Turner (1970) and Smith (1968). Smith includes graphs for the diffusion coefficients for four classes of stability based on one-hour averages (see Prob. 8.11). One must be careful, in view of all these limitations, in using the simple gaussian plume model for more than first estimates of pollutant concentrations. Limits to the model are discussed in Scorer (1968). He comments: "So often have I met experts using formulae in situations to which they are not applicable that I may seem to give exaggerated emphasis to the complexities which make the formulae wrong and too little advice on what to do instead. But the complexities are the essence of it, and make almost every situation unique." The situation may not be quite this bad, but the message is clear—care is required in using plug-in formulas.

REFERENCES

Busch, Niels: Weather and Climate Factors in Industrial Site Evaluation with Respect to Air Pollution, Paper presented at Kem-Tek 2 Congress, Copenhagen, Nov. 4, 1971.

Scorer, R.: "Air Pollution," Pergamon Press, Oxford, 1968.

Slade, D. H. (ed.): "Meteorology and Atomic Energy, 1968" TID-24190, 1968.

Smith, M. (ed.): "Recommended Guide for the Prediction of the Dispersion of Airborne Effluents," Am. Soc. Mech. Eng., 1968.

Turner, D.B.: "Workbook of Atmospheric Dispersion Estimates," U.S. Public Health Service Publication 999-AP-26, revised 1970 edition.

PROBLEMS

8.1 "Because the atmosphere is usually enormously efficient in diluting material released into it, it will be, within limits, economically unwise to ignore completely or to fail to utilize properly this dispersive capacity" ("Meteorology and Atomic Energy, 1968," p. 4). Comment on this quote using material in this chapter and considering that an engineer has a responsibility to determine the least costly but most effective solution.

8.2 Show that the integral over $dy\,dz$ of the gaussian plume solution, Eq. (8.3), yields the results that the emissions equal Q/U. Now integrate over dx. Noting that $\Delta X = U\,\Delta t$ show that the total volume integral yields the result that the emissions are $Q\Delta t$, that is, the total pollutant emitted in time Δt.

8.3 Using Eq. (8.4) for the diffusion model and assuming that $\sigma_z = \sigma_y = ax$ find the position of ground-level maximum concentration and the value of maximum concentration. Is it realistic to assume this form for the diffusion coefficients?

8.4 Using Eq. (8.4) for the diffusion model and assuming that $\sigma_y = a\sigma_z$ find the position of maximum concentration and the value of maximum concentration.

8.5 The result of assuming that $\sigma_y = a\sigma_z$ is that the position of maximum concentration is that for which $\sigma_z = H/\sqrt{2}$ where $\chi = (2Q/\pi UeH^2)(\sigma_z/\sigma_y)$. Compare answers calculated from these results to those given in this chapter for the copper smelter example for the position of maximum concentration and the SO_2 value at this position.

8.6 Consider the case of a "puff" of pollution emitted from a stack. The time of emission τ is $< x/U$. Derive a gaussian diffusion solution for the concentration from this puff including diffusion in the x, downwind, direction.

8.7 In the previous problem you were asked to derive the concentration resulting from the diffusion of an instantaneous source, that is, one that puts out a puff of effluent. Any source having a plume time of emission $\tau < x/U$ has diffusion in the x direction and thus does not meet the conditions for Eq. (8.3). The diffusion solution under these puff conditions is

$$\chi = \frac{2Q_T}{(2\pi)^{3/2}\sigma_x\sigma_y\sigma_z} \exp\left\{-\frac{1}{2}\left[\left(\frac{x-Ut}{\sigma_x}\right)^2 + \left(\frac{y}{\sigma_y}\right)^2 + \left(\frac{z}{\sigma_z}\right)^2\right]\right\}$$

where Q_T is the total mass of release and the σ's are given in Turner as:

	$x = 100$ m		$x = 4$ km	
	σ_y	σ_z	σ_y	σ_z
Unstable	10	15	300	220
Neutral	4	3.8	120	50
Very stable	1.3	0.75	35	7

Calculate for neutral stability conditions the downwind concentrations 4 km from a source after a spill of "toxic" rocket fuel at ground level if the evaporation takes place for 1 minute and 3,500 g/sec are evaporated. The wind speed near the ground is 5 m/sec. Assume that $\sigma_x = \sigma_y$. If the toxic condition is such that 1,000 $\mu g/m^3$ can be tolerated for 1 minute is there any danger at 4 km? What if the toxic limit is 10,000 $\mu g/m^3$ for 1 minute?

8.8 In the example of the SO_2 concentration downwind of the copper smelter we calculated the x location where the plume "touched the ground" by using $2.15\ \sigma_z = H$. Considering the results of Fig. 8.8, comment on the validity of this criterion.

8.9 A stack is located 1 km upwind of a small hill of approximately the same altitude as the stack height. The Hellerup Is company proposes to put an ice cream stand on top of the hill to operate for an 8-hour day. Suggest a criterion for calculating the highest concentrations which might reach the ice cream stand, that is, the worst case.

8.10 In the July, 1971 issue of *Environmental Science and Technology* there is a paper by Stephens and McCaldin which discusses experimental measurements of the diffusion coefficient. Compare these results to those of Figs. 8.5 and 8.6.

8.11 Smith gives results for diffusion coefficients in the ASME guide. For σ_y the resulting equations which fit his curves are:

very unstable	$\sigma_y = 0.40\ x^{0.91}$
unstable	$\sigma_y = 0.36\ x^{0.86}$
neutral	$\sigma_y = 0.32\ x^{0.78}$
stable	$\sigma_y = 0.31\ x^{0.71}$

These are for only four stability classifications (unlike Turner's curves which are for six classifications) and they are for "hourly mean values." Compare these results for σ_y with those in Fig. 8.5.

8.12[1] It is estimated that a burning dump emits 3 g/sec of oxides of nitrogen. What is the concentration of oxides of nitrogen, averaged over approximately 10 minutes, from this source directly downwind at a distance of 3 km on an overcast night with wind speed of 7m/sec? Assume the dump to be a point source with no effective rise.

8.13 It is estimated that 80 g/sec of sulfur dioxide is being emitted from a petroleum refinery from an average height of 60 m. At 0800 on an overcast winter morning with the surface wind 6 m/sec, what is the ground-level concentration directly downwind from the refinery at a distance of 500 m?

8.14 Under the conditions of Prob. 8.13, what is the concentration at the same distance downwind but at a distance 50 m from the x axis? That is, $\chi(500,50,0,60) = ?$

8.15 A power plant burns 10 tons per hour of coal containing 3 percent sulfur; the effluent is released from a single stack. On a sunny afternoon the wind at 10 m above ground is 4 m/sec from the northeast. The morning radiosonde taken at a nearby Weather Bureau station has indicated that a frontal inversion aloft

[1] Probs. 8.12 to 8.22 are taken from the example problems in Turner (1970).

will limit the vertical mixing to 1,500 m. The 1,200-m wind is from 30° at 5 m/sec. The effective height of emission is 150 m. From Fig. 8.7, what is the distance to maximum ground level concentration and what is the concentration at this point? Does the inversion affect this result?

8.16 For the power plant in Prob. 8.15, at what distance does the maximum ground-level concentration occur, and what is this concentration on an overcast day with wind speed 4 m/sec?

8.17 For the conditions of Prob. 8.15 draw a graph of ground-level centerline sulfur dioxide concentration with distance from an x of 100 m to 100 km. Use log-log graph paper.

8.18 For the conditions given in Prob. 8.15 draw a graph of ground-level concentration versus cross-wind distance at a downwind distance of 1 km.

8.19 For the conditions given in Prob. 8.15, determine the profile of concentration with height from ground level to $z = 450$ m at $x = 1$ km, $y = 0$ m and draw a graph of concentration against height above ground.

8.20 For the condition given in Prob. 8.15 determine the ground-level centerline concentration and the centerline concentration at 150 m above ground for $x = 1.2$ kilometers. What do your results indicate?

8.21 For the power plant in Prob. 8.15 what will the maximum ground-level concentration be beneath the plume centerline and at what distance will it occur on a clear night with wind speed 4 m/sec?

8.22 For the situation in Prob. 8.21 what would the fumigation concentration be the next morning at this point ($x = 13$ km) when superadiabatic lapse rates extend to include most of the plume and it is assumed that wind speed and direction remain unchanged?

8.23 Using the data of Example 1 in this chapter, calculate the concentration at ground level which would result from a fumigation of the plume (initially in fanning behavior).

9

PLUME RISE

> Because the atmosphere is usually enormously efficient in diluting material released into it, it will be, within limits, economically unwise to ignore completely or to fail to utilize properly this dispersive capacity.
>
> *Meteorology and Atomic Energy, 1968*

An effluent plume rising from a stack does not immediately move horizontally. In a mild wind it may undergo a significant rise. This rise of the plume adds to the stack an additional height, ΔH, such that $H_s + \Delta H$ is the proper height to use in dispersion calculations. $H_s + \Delta H$ is called the *effective stack height*, H_e, and it is this height that is used in the gaussian plume calculations (see Fig. 9.1). Since the stack height enters the calculations for concentration as

$$\exp\left(-\frac{1}{2}\frac{H^2}{\sigma^2}\right)$$

the additional height due to plume rise can be very significant in reducing downwind ground-level concentrations.

We shall devote this chapter to discussing how to calculate ΔH. There are a bewildering variety of formulas available for calculating effective stack height or plume rise. Recently, however, some order has come from this array of formulas largely because of the efforts of

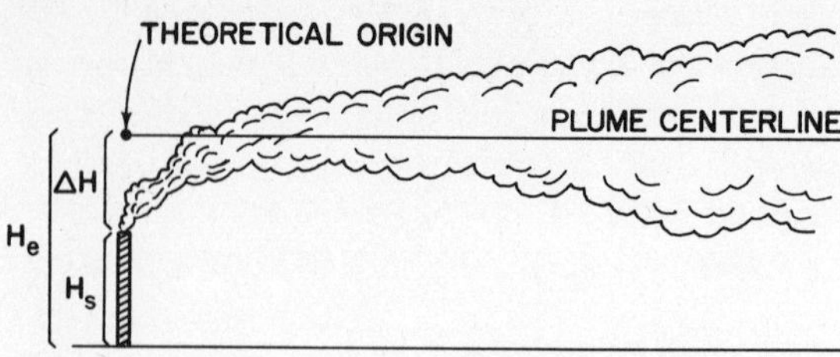

Fig. 9.1 Effective stack height. In the formulas used it is assumed that dispersion begins at a point directly above the stack, whose height (H) is the sum of the actual stack height (H_s) and the plume rise (ΔH).

Briggs (1969, 1971) and the M.I.T. group of Fay and Hoult (Fay, Escudier, and Hoult, 1970; Hoult, Fay, and Forney, 1969).

Consider the plume as it leaves the stack. It has an initial momentum which moves it in the vertical direction. In addition, if the gases are hot or of lower molecular weight than air, then the plume is less dense than the surrounding air. Consequently, there is a buoyant force which also causes the plume to rise. For most power plants the buoyant force on the plume is the dominant one and momentum effects are neglected.

As the plume rises it expands and if we assume no heat transfer between the plume and the ambient air, then we can model this process as adiabatic. Thus for a first model we can approximate the temperature change in the plume as following an isentropic process.

The plume rises at a velocity initially different from the ambient air so there is a shear force between the stack gases and the air. These, in turn, cause generation of mechanical turbulence and create mixing between the plume and the air. Note that this is a turbulence generated by the plume and not atmospheric turbulence. The plume thus entrains ambient air via this mixing action and the entrained air reduces the plume's upward momentum. Depending on the potential temperature gradient the plume may or may not lose its buoyancy through the entrainment mechanism. If the atmosphere is neutral then the buoyancy can be a conserved quantity, assuming that heat transfer between the plume and the ambient air can be neglected. If the atmosphere is unstable the buoyancy can even increase with height.

In a cross wind there is force on the plume which turns it and produces some horizontal momentum. The difference in the plume horizontal velocity and ambient velocity also produces shear forces, turbulence, and entrainment. The "bent over" plume can still have a vertical velocity component. If the atmosphere is unstable then the plume can continue to rise and may do so indefinitely. In a stable atmosphere the plume loses buoyancy by entraining the air, which is cooler in the lower atmosphere, and carrying it up adiabatically to the relatively warmer air above. In this case there must be a finite plume rise. Figure 9.2 indicates qualitatively how this behavior occurs.

We should expect different results for plume rise in the following cases:

1. plume rise with a cross wind in near neutral or stable conditions where the rise is limited
2. plume rise with a cross wind in unstable conditions where the rise may not be limited (but where we still need to locate the position and value of maximum ground level concentration)
3. plume rise with very light or no cross winds under conditions where the plume rises and spreads out without bending over.

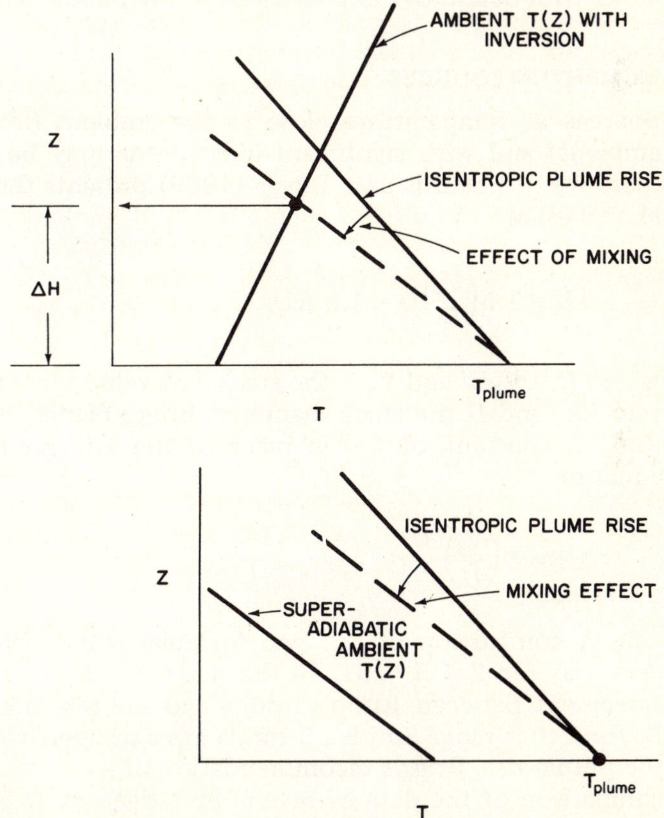

Fig. 9.2 Plume rise in a stable atmosphere (*top*); plume rise in an unstable atmosphere (*bottom*).

The above cases are for plumes dominated by buoyant forces. For plumes which are emitted at or near ambient temperatures another relation is used.

Note that even though a plume may no longer be visible to the eye, this does not mean it has ceased to rise. One must be careful to distinguish between the point of maximum rise and the region where the plume has expanded and been diluted sufficiently to be invisible. Infrared photography has been used to track plumes beyond their visible limit. The plume continues to expand after the centerline has ceased to rise. The plume top is usually easier to see and one may incorrectly decide that the plume is still rising when, in fact, the top is merely expanding with the centerline remaining horizontal.

Analyses for plume rise are found in Hoult, Fay, and Forney (1969) and Briggs (1969, 1971). We shall present their results here along with qualitative explanations of the plume behavior.

MOMENTUM SOURCES

Sources at temperatures close to the ambient (less than 50°F above ambient) and with significant exit speeds may be treated as jets in a cross wind. For this case Briggs (1969) presents the results of Rupp et al. (1948) as

$$\Delta H = 1.5\left(\frac{V_s}{U}\right)D = 1.5\,RD \tag{9.1}$$

where $R = V_s/U$ and V_s is the stack exit velocity, U the average ambient velocity, and D the stack diameter. Briggs (1969) himself recommends using a constant of 3.0 in place of the 1.5. Smith (1968) gives the equation

$$\Delta H = D\left(\frac{V_s}{U}\right)^{1.4} \tag{9.2}$$

A comparison of the two formulas is given in Table 9.1. Rupp's data was for $2 < R < 31$. In the range $2 < R < 10$ there is reasonable agreement between Rupp's results and the recommendation of Smith. Beyond that range Smith's formula gives progressively higher results for the plume rise. Brigg's recommendation of a constant of 3.0 is based on comparison of the data of several investigators. It is more conservative to underestimate plume rise than to over-estimate it (and therefore underestimate ground-level concentrations) so Rupp's formula is recommended.

TABLE 9.1 Comparison of the equations for calculating plume rise of a momentum jet

$R = V_s/U$	Eq. (9.1), Rupp $\Delta H/D$	Eq. (9.2), Smith $\Delta H/D$	Eq. (9.1) with 3.0 $\Delta H/D$
1	1.5	1.0	3.0
2	3.0	2.8	6.0
3	4.5	4.6	9.0
4	6.0	7.0	12
5	7.5	9.5	15
10	15	25	30
20	30	66	60

The jet rise to its final height is a function of the downwind position x and takes an $(x/D)^{1/3}$ dependence. An expression for determining ΔH as a function of x is given by Briggs (1971) as

$$\frac{\Delta H}{D} = 1.89 \left(\frac{R}{1 + 3/R}\right)^{2/3} \left(\frac{x}{D}\right)^{1/3} \tag{9.3}$$

BUOYANT PLUMES

Briggs (1971) indicates that for hot plumes buoyancy becomes the dominant factor, over momentum forces, for a distance downwind on the order of 5 seconds times the wind speed. Thus we need consider only buoyant forces when dealing with plumes from power plants, smelters, or other large industrial sources.

What parameters should be important in determining plume rise when the plume is subject to buoyant forces? Experience and intuition indicate that in a light wind the plume will rise much more than in a strong wind. Consequently, we expect that the height of rise should be a function of the wind speed and in particular should be inversely proportional to the wind speed so that

$$\Delta H \propto \frac{1}{U}^{a} \tag{9.4}$$

where a is some power.

Also we expect the ambient temperature gradient to be an important parameter. The results of Fig. 9.2 indicate that if dT/dZ is positive then the hot plume eventually will reach a point where its temperature is the same as the ambient. In fact this is true for any stable atmosphere. Consequently, the important parameter in the case of a stable atmosphere is not just dT/dZ but rather must involve both dT/dZ and the adiabatic lapse rate. The stability parameter, s, is defined as

$$s \triangleq \frac{g}{T}\left(\frac{dT}{dZ} + \Gamma\right) \tag{9.5}$$

where Γ is defined as a positive number equal in magnitude to the adiabatic lapse rate

$$\Gamma \triangleq \left|\frac{dT}{dZ}\right|_{\text{adiabatic}}$$

From Chap. 7 we have

$$\Gamma = \left(\frac{k-1}{k}\right)\frac{g}{R} = \frac{g}{c_P} = 5.4°\text{R}/1{,}000\text{ ft}$$

Thus

$$s = \frac{g}{T}\left(\frac{dT}{dZ} + 5.4\right)\ (1/\text{sec}^2) \tag{9.6}$$

For an adiabatic ambient atmosphere $dT/dZ = -\Gamma$ and $s = 0$; for a stable atmosphere $s > 0$; and for an unstable atmosphere $s < 0$. Having now established that ΔH is a function of $(1/U)^a$ and $(s)^b$, where b is some power, we must ask if any other parameters are required. Clearly buoyancy is important and we now must seek a parameter involving the buoyancy.

The plume can be buoyant if made up of gases lighter than air or if it is hotter than the ambient air. For most applications we are interested in hot plumes; however, in the case of accidental spills of chemicals such as ammonia, the ΔH is caused by direct differences in density. Here we shall confine the discussion to hot plumes.

The buoyant force must be proportional to the amount of exhaust gas, to the difference in density (caused by temperature difference), and to the gravitational constant g. Consider a parcel of gas of mass m as noted in Fig. 9.3(a). The buoyant force on such a parcel is given by

$$f = (m_{\text{air}} - m)g \tag{9.7}$$

where m_{air} is the mass of air displaced by the exhaust gas. For exhaust gas emitted we have (see Fig. 9.3(b))

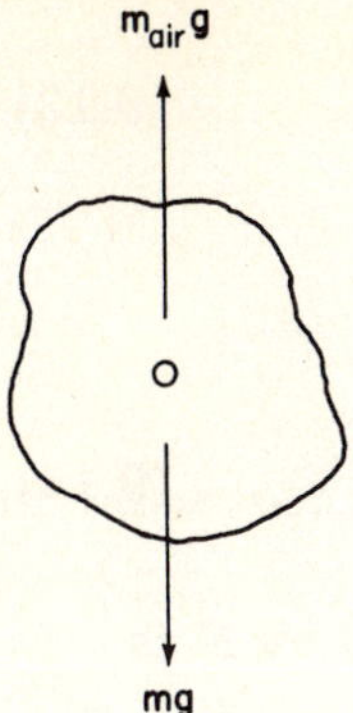

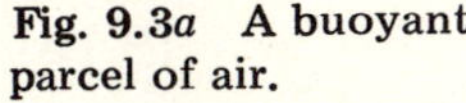

Fig. 9.3a A buoyant parcel of air.

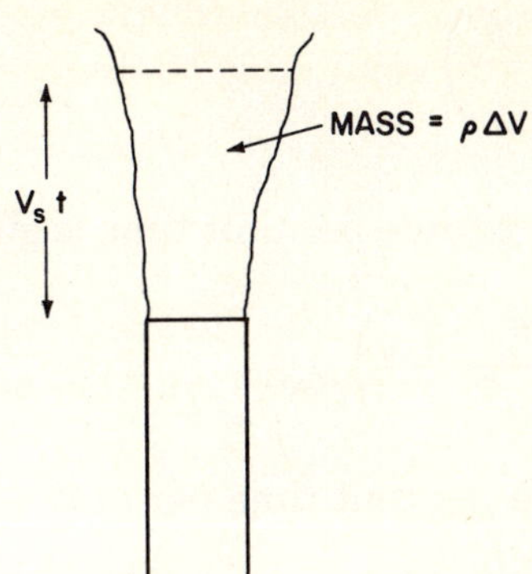

Fig. 9.3b A buoyant plume.

$$m = \rho_s(\text{Volume}) \propto \rho_s \left(\frac{D}{2}\right)^2 V_s t \tag{9.8}$$

where the volume must be proportional to the product of the stack exit velocity, V_s, the time of efflux and the stack area which is proportional to $(D/2)^2$. Thus the buoyant force is proportional to

$$f \propto (\rho_{air} - \rho_s) \left(\frac{D}{2}\right)^2 g V_s t \tag{9.9}$$

Now the difference in density can be related to the temperature difference either through the perfect gas equation or by noting that, in general

$$\rho = \rho_o + \left.\frac{\partial \rho}{\partial T}\right|_{0,P} \Delta T + \ldots \tag{9.10}$$

and

$$\left.\frac{\partial \rho}{\partial T}\right|_{0,P} = -\left.\frac{P}{RT^2}\right|_0 = -\left.\frac{\rho}{T}\right|_0 \tag{9.11}$$

so

$$\rho - \rho_0 = -\rho_0 \frac{\Delta T}{T_0} \tag{9.12}$$

for our case we have

$$\frac{\rho_s - \rho_\infty}{\rho_\infty} = \left(\frac{T_\infty - T_s}{T_\infty}\right) \tag{9.13}$$

Therefore the buoyant force per unit time is given by

$$\dot{f} \propto g\left(\frac{D}{2}\right)^2 V_s \left(\frac{T_s - T_\infty}{T_\infty}\right) \rho_\infty \tag{9.14}$$

and the buoyant force per unit time per unit mass, denoted by F, is

$$F \propto g\left(\frac{D}{2}\right)^2 V_s \left(\frac{T_s - T_\infty}{T_\infty}\right) \tag{9.15}$$

which has the units of length^4/sec^3.

Assuming we have included all important variables we can now put this together as

$$\Delta H \propto \left[\left(\frac{1}{U^a}\right)(s)^b (F)^c\right] \tag{9.16}$$

Since ΔH has the units of length, dimensional reasoning indicates that (see Prob. 9.16) $a = 1/3$, $b = -1/3$, $c = 1/3$. Therefore we have

$$\Delta H \propto \left(\frac{F}{Us}\right)^{1/3}$$

or, in equation form,

$$\Delta H = C\left(\frac{F}{Us}\right)^{1/3} \tag{9.17}$$

where C is an experimentally determined constant.

Stable or Near Neutral Conditions

For calculating plume rise we shall adopt the formula

$$\Delta H = 2.27\left(\frac{F}{Us}\right)^{1/3} \tag{9.18}$$

where U is the cross-wind speed at the stack, s is a stability parameter which can be calculated from

$$s = \frac{g}{T}\left(\frac{dT}{dZ} + \Gamma\right)$$

F is a buoyancy parameter given as

$$F = gV_s\left(\frac{D}{2}\right)^2 \left(\frac{T_s - T_\infty}{T_\infty}\right)$$

where T_s is the exit stack gas temperature. A consistent set of units must, of course, be used to determine F, s, and ΔH, for example °K, m, and sec or °R, ft, and sec. This form for calculating the plume rise is used by Smith (1968), Fay et al. (1970), and Briggs (1968, 1969, 1971). The constants recommended are 2.0 by Smith, 2.27 by Fay, 2.9 by Briggs (1969), and 2.6 by Briggs (1968). The three-place accuracy indicated in Eq. (9.18) is not warranted and a value of 2.3 can be used. Thus we have

$$\Delta H = 2.3\left(\frac{F}{Us}\right)^{1/3} \tag{9.19}$$

The buoyancy parameter can be related directly to the energy output from the stack, Q (megawatts or cal/sec), as shown in Fig. 9.4. Consider the definition of F given by Briggs (1969)

$$F = \frac{gQ}{\pi c_p \rho T} \tag{9.20}$$

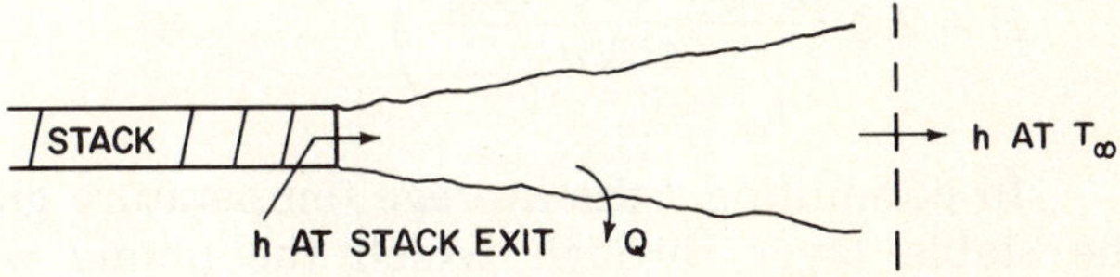

Fig. 9.4 An energy balance.

where c_p is the specific heat for air, Q is the "heat emission," and ρ and T are the environmental density and temperature. If we consider that the energy Q is eventually lost to the environment, then an energy balance yields, where h is now the enthalpy

$$Q = \dot{m}\,\Delta h = \left(\frac{\pi D^2}{4}\right)(V_s \rho_s c_p)(T_s - T_\infty) \tag{9.21}$$

where c_p is the specific heat of the effluent. Then substituting this expression into Eq. (9.20) we have

$$F = \frac{g\left(\frac{\pi D^2}{4}\right) V_s \rho_s c_p\,(T_s - T_\infty)}{\pi c_p \rho T_\infty} \tag{9.22}$$

If we can cancel out the two specific heats and ρ_s with ρ, then we find

$$F = g V_s \left(\frac{D}{2}\right)^2 \left(\frac{T_s - T_\infty}{T_\infty}\right) \tag{9.23}$$

Thus the form for F can be related directly to the energy release up the stack, and many plume rise formulas are given in terms of Q. For power plants the stack gases are largely nitrogen, water vapor, and carbon dioxide. Thus canceling out specific heats is not a bad assumption. The densities do differ since the stack gases are hotter than the ambient surroundings. Even these, though, are not greatly different (remember that absolute temperatures are used in the perfect gas equation and are typically about 750°R for the stack gas and 500°R for the ambient). The difference in constants used by the various authors may be at least partly because of slightly different definitions of F.

In any case the important thing to note is that ΔH = function (velocity, stability of atmosphere, and buoyancy) and we can calculate the plume rise as indicated in Eq. (9.19) repeated here as

$$H = 2.3\left(\frac{g V_s D^2 (T_s - T_\infty)}{4 U s T_\infty}\right)^{1/3} \tag{9.24}$$

In computing s the average temperature gradient is calculated for the stable layer through which the plume will pass. Although the expression is recommended for stably stratified and near neutral conditions, it is clear that for the neutral case where $s = 0$, the equation predicts that $\Delta H \to \infty$. For neutral or stable conditions we shall use a different expression. Equation (9.24) applies to "near neutral" stability conditions but only on the stable side of neutral.

To calculate plume path as a function of position x we shall use the relation

$$\Delta H = 1.6 \frac{F^{1/3} x^{2/3}}{U} \tag{9.25}$$

This form is used by many authors but again with a modestly different constant. Briggs (1971) uses 1.6, Fay et al. (1970) use 1.32.

The diffusion formulas given in Chap. 8 use the total plume rise so Eq. (9.24) is the one of interest for diffusion calculations of downwind concentrations a few stack heights away from the stack.

Unstable and Neutral Conditions

In this case the plume continues to rise since the buoyancy is not sufficiently diluted by entrainment to reduce plume temperature to the ambient value. However, in practice the plume eventually expands to reach the ground even though the plume centerline may continue to rise. Since the atmosphere is neutral or unstable the stability parameter is no longer an important variable, and we need only consider the effects of F and U. Plume rise may again be calculated as a function of x according to Eq. (9.25); the question is where do we cut off this rise. There is no agreement on the criteria for this position. Smith recommends

$$\Delta H = 150 \frac{F}{U^3} \tag{9.26}$$

Briggs (1965, 1968) suggests using 400 for the constant in this equation. As a second suggestion, Briggs recommends using Eq. (9.25) out to an $x = 10H_s$, then taking ΔH equal to a constant from that position on. He recommends this procedure for fossil-fuel plants with heat emissions of 20 megawatts or more. The data indicate (Briggs, 1969) that the constant of 150 is very conservative. However, it is difficult to determine plume rise in unstable conditions so we shall use a value of 150 for the constant.

Stable with Calm Conditions

Under atmospheric conditions of little or no wind there is no bending of the plume and it rises to some height where the buoyancy force is completely dissipated. Briggs (1968, 1969, 1971) recommends using a relation developed by Morton, Taylor, and Turner (1956) which is

$$\Delta H = \frac{5.0 F^{1/4}}{s^{3/8}} \tag{9.27}$$

One should calculate ΔH for the light wind case from this equation and Eq. (9.24) and use the smaller value.

Example 1. The Colbert plant of the TVA has four 200-megawatt units each with a 100-m stack. The stack diameters are 5 m, the flue gas exit velocity is 13.5 m/sec, and the gas temperature at exit is 145°C. Calculate the plume rise on a clear night with light winds when $dT/dZ = 0.5°\text{C}/100$ m. At the stack U is 4 m/sec and T is 15°C.

From Eq. (9.24) we have

$$\Delta H = 2.3\left(\frac{F}{Us}\right)^{1/3}$$

$$F = gV_s\left(\frac{D}{2}\right)^2\left(\frac{T_s - T_\infty}{T_\infty}\right)$$

$$F = (9.8)(13.5)\left(\frac{25}{4}\right)\left(\frac{418 - 288}{288}\right)\ \text{m}^4/\text{sec}^3$$

$$F = 375\ \text{m}^4/\text{sec}^3$$

$$s = \frac{g}{T}\left(\frac{dT}{dZ} + \Gamma\right)$$

$$s = \frac{9.8}{288}\left(\frac{0.5 + 1}{100}\right) = 0.00051\ (1/\text{sec}^2)$$

Then

$$\Delta H = 2.3\left(\frac{375}{4(5.1 \times 10^{-4})}\right)^{1/3}$$

$$\Delta H = 131 \text{ meters}$$

This is a very substantial addition to the actual stack height of 100 m.

Here we have neglected any effects of the interaction between stack plumes. If the plumes merge, then they may aid each other in a buoyant rise and the calculated ΔH will be too small. Let us now calculate ΔH under the above conditions but as a function of the ambient wind speed. We have

$$\Delta H = 2.3\left(\frac{375}{5.1 \times 10^{-4}}\right)^{1/3}\left(\frac{1}{U}\right)^{1/3}$$

from which results in Table 9.2 are calculated.

As expected, the greater the wind speed the less the plume rise since the plume is bent over more quickly. Note that in our diffusion equation result we have $\chi \propto 1/U$ so that the greater the wind speed the greater the diffusion. However, the effective stack height used in the gaussian formula exponential becomes less at larger wind speeds. Thus there is a maximum in ground-level

TABLE 9.2 Plume rise as a function of velocity for unstable conditions

U m/sec	$(1/U)^{1/3}$	ΔH, meters
1	1	207
2	0.794	164
4	0.630	131
6	0.550	114
8	0.500	103
10	0.464	96

concentration as a function of U. The wind speed which produces this maximum concentration is called the *critical wind speed*.

Assume now that the wind speed drops off to a very low value. Calculate the plume rise if the temperature gradient is the same as before. We have from Eq. (9.27)

$$\Delta H = 5 \frac{F^{1/4}}{s^{3/8}}$$

$$\Delta H = 5 \frac{(375)^{1/4}}{(0.00051)^{3/8}}$$

$$\Delta H = 374 \text{ meters}$$

The lower value calculated from Eq. (9.24) or (9.27) should be used. In this case it would require a very low wind speed before Eq. (9.27) would be used.

Example 2. Using the Colbert plant data, calculate plume rise if the air temperature changes at -1.2°C/100 m on a sunny afternoon with a wind speed of 6 m/sec.

This temperature gradient represents an unstable atmosphere so we have, using Eq. (9.26)

$$\Delta H = 150 \frac{F}{U^3}$$

$$\Delta H = 150 \frac{(375)}{(6)^3} = 261 \text{ meters}$$

Since this is a power plant with a rating of 200 megawatts electrical (therefore about 600 megawatts thermal) and if 5 percent of the total thermal energy goes up the stack, we are above the 20-megawatt limit suggested by Briggs. Problem 9.4 continues this example using the 10-stack height rule.

Example 3. A 750-megawatt coal-fired unit is to be built in a desert area using a tall stack to disperse the effluent. The design stack height is 250 m. The sulfur content of the coal is 1.6 percent; the stack diameter is 25 ft; the stack exit velocity is estimated to be 50 ft/sec and the design exit temperature is 275°F. Calculate the effective stack height for an ambient air temperature of 85°F on a sunny day with moderate wind speed at the stack altitude. Calculate the SO_2 concentration downwind of the plant at 5 km.

On a hot summer day the air will be unstable. A moderate wind speed is 5-7 m/sec so we shall use Eq. (9.24) with this velocity. We have

$$F = gV_s\left(\frac{D}{2}\right)^2\left(\frac{T_s - T_\infty}{T_\infty}\right)$$

Using English units to calculate F we have

$$F = (32.2)(50)\left(\frac{625}{4}\right)(735 - 545)\left(\frac{1}{545}\right)$$

$$F = 8.8 \times 10^4 \text{ ft}^4/\text{sec}^3$$

Then noting that 5-7 m/sec is about 20 ft/sec we find

$$\Delta H = 150\frac{F}{U^3} = 1650 \text{ ft} = 505 \text{ meters}$$

For a sunny day the insolation will be strong. The wind speed at 10 m is less than at the stack height of 250 m. We can estimate the 10-m wind speed as

$$\frac{U}{U_1} = \left(\frac{Z}{Z_1}\right)^{0.25}$$

$$U = 6\left(\frac{10}{250}\right)^{0.25} = 2.7 \text{ m/sec}$$

Thus, with strong insolation and this wind speed, class A-B stability is indicated. We shall use class B for a conservative answer.

We have for the ground-level concentration along the plume centerline

$$\chi = \frac{Q}{\pi\sigma_Z\sigma_y U}\exp\left[-\frac{1}{2}\left(\frac{H}{\sigma_Z}\right)^2\right]$$

and from Figs. 8.5 and 8.6 we find

x	σ_y	σ_Z	H_e
5 km	650 m	640 m	755 m

The source strength can be estimated from the rule of thumb that it takes about 1,000 tons of coal per 100 megawatt-day. We have

$$Q = \frac{750 \text{ Mw} \times 1{,}000 \text{ tons}}{100 \text{ Mw-day}} \times 2{,}000 \text{ lb/ton} \times 0.016 \text{ lb S/lb coal} \times 2 \frac{\text{SO}_2}{\text{S}}$$

$$\times\ 454 \text{ g/lb} \times \frac{1}{3{,}600} \text{ (hr/sec)} \times \frac{1}{24} \text{ (day/hr)}$$

$$Q = 2{,}520 \text{ g/sec}$$

Thus from the gaussian diffusion model we find

$$\chi = \frac{2{,}520 \times 10^6}{\pi(650)(640)(6)} \exp\left[-\frac{1}{2}\left(\frac{755}{640}\right)^2\right]$$

$$\chi = 320 \exp(-0.7) = 159\ \mu\text{g/m}^3$$

To appreciate the difficulty facing the power companies note that the federal standard annual average concentration for SO_2 is 80 μg/m^3. To achieve this standard for ambient air quality it may be necessary to set emissions standards that are very difficult to meet.

To gain an indication of the accuracy of the calculation consider the diffusion coefficients given in Slade (1968). These are the same as those of Figs. 8.5 and 8.6 for all classes of stability for σ_y and the same for σ_z for stability classes D, E, and F. However, they differ substantially for classes A and B and differ some for class C. For the above example one finds at 5 km that σ_z is 2,000 meters. This changes both the preexponential and the exponential factors and gives a concentration of 96 μg/m^3 instead of 159 μg/m^3.

Example 4. The location of maximum concentration can be estimated from the results of Prob. 8.5 since we are off scale in Fig. 8.7. We have χ_{max} occurring when $\sigma_z = H/\sqrt{2}$. For this example we find with $H = 755$ m, $\sigma_z = 535$ m. With B stability the x position is read from Fig. 8.6 as about 4.2 km. For this distance $\sigma_y = 560$ m and we find the peak concentration to be

$$\chi_{max} = \frac{2{,}520 \times 10^6}{(\pi)(6)(560)(535)} \exp\left[-\frac{1}{2}\left(\frac{755}{535}\right)^2\right]$$

$$\chi_{max} = 446 \exp(-1) = 165\ \mu\text{g/m}^3$$

SUMMARY

Briggs (1971) gives more elaborate suggestions for calculating the point of maximum plume rise and the value of the maximum rise under neutral or unstable conditions. More detailed analyses can also be found in the other references for this chapter. The caution expressed at the end of the previous chapter is also appropriate to this material, however, and care is required in using these formulas. Atmospheric conditions do not usually lend themselves to simple approaches.

The reader will encounter several points of interest in the problems that follow. These should be carefully reviewed.

REFERENCES

Briggs, G. A.: Plume Rise Model Compared with Observations, *J. Air Pollution Control Assoc.*, vol. 15, pp. 433–438, 1965.

——, in D. H. Slade (ed.): "Meteorology and Atomic Energy 1968," TID 24190, 1968.

——: "Plume Rise," A.E.C. Critical Review Series, TID 25075, 1969.

——: Some Recent Analyses of Plume Rise Observations, *Proc. 2d Intern. Clean Air Conf., Washington, D.C., 1970*, Academic Press, New York, 1971.

Fay, J. A., M. Escudier, and D. P. Hoult: A Correlation of Field Observations of Plume Rise, *J. Air Pollution Control Assoc.*, vol. 20, pp. 391–397, 1970.

Hoult, D. P., J. A. Fay, and L. J. Forney: A Theory of Plume Rise Compared with Field Observations, *J. Air Pollution Control Assoc.*, vol. 19, pp. 585–590, 1969.

Morton, B. R., G. I. Taylor, and J. S. Turner: Turbulent Gravitational Convection from Maintained and Instantaneous Sources, *Proc. Roy. Soc.* (London) Ser. A, vol. 234, pp. 1–23, 1956.

Rupp, A. F., S. E. Beall, L. P. Bornwasser, and D. F. Johnson: Dilution of Stack Gases in Cross Winds, *U.S. At. Energy Comm. Report*, AECD-1811 (CE-1620), Clinton Laboratories, 1948.

Smith, M. (ed.): Recommended Guide for the Prediction of the Dispersion of Air Effluents, *Am. Soc. Mech. Eng.*, New York, 1968.

Slade, D. H. (ed.): "Meteorology and Atomic Energy 1968," TID 24190, 1968.

PROBLEMS

9.1 Continue Example 3 in this chapter for the 750-megawatt coal-fired power station for the condition of class E stability with a wind speed of 4 m/sec. at the stack height and with an isothermal atmosphere. Calculate the ground-level concentration at 5 km from the stack and estimate the position of maximum ground-level concentration. The ambient temperature remains 85°F.

9.2 Continue Example 3 of this chapter for the 750-megawatt coal-fired power station for the condition of a winter day with an ambient air temperature of 35°F, wind speed at the stack of 6 m/sec and class C stability. Calculate the ground-level concentration 5 km downwind of the stack and estimate the position of maximum ground level concentration.

9.3 Using the results of Prob. 8.5 calculate the maximum concentration at ground level according to the conditions used in Example 4 of this chapter. Why is the agreement between the example and your calculation so poor?

9.4 For the data of Example 2 of this chapter calculate the plume rise assuming that Eq. (9.25) applies out to an $x = 10H_s$ and then ΔH remains constant. Compare your answer to that calculated in the example.

9.5 An ammonia plant has an emergency vent system such that in an accident ammonia will be vented at atmospheric temperature at a velocity of 85 ft/sec from a 50-ft stack of 3-ft exit diameter. Calculate the ground-level concentration 0.5 and 1 km downwind under partially cloudy night skies if the source strength is 1,000 g/sec and the wind speed at the stack height is estimated to be 7 m/sec. Estimate the position and value of maximum ground-level concentration. If the plant safety committee recommends doubling the stack height what would the new concentrations be? The World Health Organization has recommended that ammonia limits for an 8-hour working day be set at a maximum value of 37 mg/m^3. How do these concentrations compare to that figure?

9.6 Using Eqs. (9.1) and (9.3) of this chapter determine the jet rise as a function of x/D and determine the x position where we have assumed that the jet has become horizontal.

9.7 Plant data for several TVA power plants are given in the table taken from Briggs (1965). Find ΔH, the plume rise, at plant 1 for a neutral atmosphere with wind speeds of 10 mph and 24 mph. For plant 2, find ΔH for unstable conditions for a wind velocity of 15 mph. For plant 3, calculate the plume rise on a clear night when $dT/dz = 1.6°$F per 1,000 ft and the wind velocity is 12 mph. All ambient temperatures are 60°F. What do you conclude about the effect of wind speed on ΔH?

TABLE for Prob. 9.7

	Stack 1	Stack 2	Stack 3
Stack radius, ft	7	8.2	12.5
V_s, ft/sec	48	47	44
T_s, °F	290	290	290
H_s, ft	250	300	500

9.8 For an isothermal atmosphere, $dT/dz = 0$, calculate the x position of maximum plume rise as a function of wind speed U if the temperature is 300°K. If $U = 10$ m/sec what is the downwind location of the maximum plume rise?

9.9 Briggs suggests (1968, p. 28) that F can be calculated as $F = gQ/\pi c_p \rho T = 4.3 \times 10^{-3}$ ((ft^4/sec^3)/(cal/sec))Q_H or as $F = 3.7 \times 10^{-5}$ ((m^4/sec^3)/(cal/sec))Q_H. Verify either one of these expressions for a pressure of 1 atmosphere.

9.10 Briggs (1970, p. 27) gives the point of maximum plume rise under neutral conditions as $x = 119\ (F)^{2/5}$ for $F > 55$ m^4/sec^3. Using this expression and that in Eq. (9.25) find ΔH_{max} for several combinations of F and U and compare to Smith's recommendation of $H = 150 \ F/U^3$ and Briggs' recommendation of $H = 400 \ F/U^3$. What can you conclude about our ability to calculate ΔH under neutral or unstable conditions?

9.11 Concentrations downwind of an infinite line source can be calculated from the expression

$$\chi = \frac{2q}{\sqrt{2\pi} \ U\sigma_z} \exp\left[-\frac{1}{2}\left(\frac{H}{\sigma_z}\right)^2\right]$$

where q is the source strength per unit distance, that is, g/sec-m. On a clear autumn afternoon calculate the particulate concentration of small airborne particles downwind of a long line of burning agricultural waste if the source strength of such particles is 0.10 g/sec-m. Use a wind speed of 5 m/sec and an x position of 500 m.

9.12 The critical wind speed refers to the wind velocity where maximum ground-level concentration occurs because of the balance between dilution owing to wind speed and the decrease in effective stack height with increasing wind speed. Using the relation for unstable conditions that

$$H_e = H_s + 150 \frac{F}{U^3}$$

find the critical wind speed as a function of F and H_s. If a proposed stack height is 200 m high and F is 10^3 m^4/sec^3 what is the critical wind speed? If a proposed stack is 30 m high and F is 25 m^4/sec^3 what is the critical wind speed? (Hint: remember that $\chi_{max} \propto Q/UH_e^2$)

9.13 (*a*) A proposed paper plant will emit an estimated 4 tons of hydrogen sulfide per day. The limiting concentration is to be 1/10 of the smell threshold concentration, which is about 1,000 $\mu g/m^3$, at a distance 1 km from the stack. Estimate the required stack height for such a plant if plume rise is initially neglected and the design wind velocities are 3 m/sec and 10 m/sec. (Solution will require iteration, remember that $\chi_{max} \cong Q/\pi e U\sigma_z\sigma_y$ at the x position where $H/\sqrt{2} \cong \sigma_z$.)

(*b*) If the air conditions are near neutral or unstable calculate the plume rise from the paper plant stack if the parameters are

$V_s = 50$ ft/sec

$D = 6$ ft

$T_s = 250°F$; $T_{amb} = 70°F$

and the wind speed at the stack is an estimated 3 m/sec. Under these conditions would a stack even be needed? Suppose the wind speed increases to 10 m/sec and the atmospheric conditions are still near neutral but unstable. What is the plume rise and the required stack height?

(*c*) Calculate the critical wind speed for the conditions of this problem if the answer to Prob. 9.12 is $U_{crit} = 9.1\ (F/H_s)^{1/3}$.

9.14 For the stack conditions of Prob. 9.13*b*, find the plume rise for an isothermal atmosphere with a wind speed of 10 m/sec.

9.15 Typical data for a power plant stack might be a diameter of 10-20 ft, exit velocity of 45 ft/sec, and an exit temperature of 275-300°F. Calculate the exit Reynolds number based on stack diameter and show the stack gas flow is turbulent. If the load drops suddenly to half the full load does the flow remain turbulent?

9.16 Show that the formula given in Eq. (9.17) follows directly from Eq. (9.16).

9.17 It has been suggested that designing the stack exit as a nozzle will increase the buoyancy since the stack exit velocity will be increased. Is this a good suggestion?

10

PARTICULATES

Before placing our order for a precipitator, we surveyed the field in an attempt to find the best equipment available, which was over 4 years ago. We found many installations which were designed for 95, 96, and 97 percent efficiency but were actually being operated at 90 percent or less.

The precipitator we contracted for was designed at 98.4 percent at that time, and we have spent $600,000 to bring it up two-tenths of 1 percent to 98.6 percent at this time. Many contracts have been drawn on the basis of one-time test performance, but such contracts do not provide real protection to the environment.

Howard Allen, Vice President, Southern California Edison

The Mohave generating station was constructed without adequate consideration of the potential effects of its emissions on the air resources. Coal-burning plants generate huge quantities of particulate matter which, even if controlled by the most efficient collection devices, still emit considerable quantities to the atmosphere. Very small particles escape even the most efficient collectors and cause maximum reduction in atmospheric visibility. The Mohave plant is located in a national recreational area where the ability to clearly see distant mountains is an important asset. Whether any coal-burning power plant should have been allowed to be constructed in that area is a question that should have been asked before approval was given to build the plant.

Terry Stumph, Director of the Clark County Nevada District Health Department, Air Pollution Control Division[1]

Particulates form a major part of the emissions of air pollutants and come from such diverse sources as cars, trucks, steel mills, cement plants, and the local dump. Blowing dust is a problem in parts of the United States and even particulate fallout from forest fires can be occasionally troublesome. We have already seen in Chap. 2 that volcanos may be the largest single natural source of world particulate emission. Another significant source is ocean spray aerosols. As these tiny droplets move inland the water evaporates and leaves behind a particle of sea salt, such as sodium chloride. These sea salts are an important source of condensation nuclei for forming rain.

Chapter 2 described the sources of particulates and Chap. 3 listed the major sources in the United States. They were fuel combustion in stationary sources, "industrial processes," and forest fires. Table 10.1 shows the *suspended particulate* in $\mu g/m^3$ for particular United States'

[1] From "Problems of Electrical Power Production in the Southwest," Government Printing Office, 1972.

TABLE 10.1 Suspended particulate of city-center and rural areas in the United States, 1962–1966

Urban areas	Geometric mean $\mu g/m^3$
East Chicago, Ind.	183.7
Charleston, W. Va.	169.0
Phoenix, Ariz.	165.7
Philadelphia, Pa.	155.5
Omaha, Nebr.	107.3
Charlotte, N.C.	101.3
San Diego, Calif.	76.3
Minneapolis, Minn.	74.8
Honolulu, Hawaii	38.7
Cheyenne, Wyo.	33.7
Rural areas	
Shenandoah National Park, Va.	30.1
Grand Canyon, Ariz.	18.5
White Pine Co., Nev.	10.0

Source: Characteristics of Particulate Patterns, 1957–1966, Publication AP-61, 1970.

locations and Table 10.2 gives results by population class. Suspended particulate is measured by drawing a known volume of air through a filter and weighing the filter, both before and after. Chemical analysis is often made on the collected matter and the amount that is soluble in benzene is measured as an indication of the organic material in the air. Typically less than 10 percent of the suspended particulate is benzene soluble. The results of Table 10.1 are, of course, not complete but they indicate the wide variation possible in the suspended particulate of urban areas—a factor of 6 is indicated here. One must interpret any measurement of air pollution with some care. For example, the 1957–1961 geometric mean figure for suspended particulate in Phoenix, Arizona, was 206.5 $\mu g/m^3$, while the 1962–1966 results are given in AP-61 as 165.7 $\mu g/m^3$. The air may be cleaner in Phoenix but it may be that the city has grown so that the measuring station is now farther away from the dust production at the outskirts of the city.

Table 10.2 can be used to give a measure of the relative air quality with respect to suspended particulate in your city. First find out where and how your air quality is measured. Then compare your local results with those ranges given in the table.

TABLE 10.2 Distribution of selected cities by population class and particle concentration, 1957 to 1967
(*Avg. particle concentration μg/m³*)

Population class	<40	40 to 59	60 to 79	80 to 99	100 to 119	120 to 139	140 to 159	160 to 179	180 to 199	>200	Total cities in table	Total cities in U.S.
>3 million	—	—	—	—	—	—	1	—	1	—	2	2
1-3 million	—	—	—	—	—	—	2	1	—	—	3	3
0.7-1 million	—	—	1	—	2	—	4	—	—	—	7	7
400-700,000	—	—	—	4	5	6	1	1	1	—	18	19
100-400,000	—	3	7	30	24	17	12	3	2	1	99	100
50-100,000	—	2	20	28	16	12	6	5	1	3	93	180
25-50,000	—	5	24	12	12	10	2	1	2	3	71	—
10-25,000	—	7	18	19	9	5	2	3	1	—	64	*5,453
<10,000	1	5	7	15	11	2	1	2	—	—	44	—
Total urban	1	22	77	108	79	52	31	16	8	7	401	—

*Incorporated and unincorporated areas with population over 2,500.
Source: AP-49.

TABLE 10.2*a* Distribution of selected nonurban monitoring sites by category of urban proximity, 1957 to 1967

Category	Average particle concentrations, μg/m³				Total
	<20	20-39	40-59	60-79	
Near urban*	—	1	3	1	5
Intermediate†	—	5	6	—	11
Remote‡	4	5	—	—	9
Total nonurban	4	11	9	1	25

*Near urban—although located in unsettled areas, pollutant levels at these stations clearly indicate influence from nearby urban areas. All of these stations are located near the northeast coast "population corridor."
†Intermediate—distant from large urban centers, some agricultural activity, pollutant levels suggest that some influence from human activity is possible.
‡Remote—minimum of human activity, negligible agriculture, sites are frequently in state or national forest preserve or park areas.
Source: AP-49.

TERMINOLOGY

"Air Quality Criteria for Particulate Matter" (AP-49), defines a "particle" as any dispersed matter, solid or liquid, in which the individual aggregates are larger than single small molecules (about 0.0002μ in diameter) but smaller than about 500μ. Other definitions appropriate to aerosols or particulates have been given in Chap. 1. To review briefly, we consider aerosols formed by condensation of vapors or dispersion of liquids as tiny droplets to be *mists*. Mists are distinguished from smokes partly on the basis of particle size; *smokes* consist of finer particles than mists and are gaseous suspensions (often made up of what is not classified as a mist or dust). *Dusts* are aerosols made up of solid particles dispersed in a gaseous medium. In general, dusts consist of larger particles than smokes or mists. Practical experience is a good guide to classifying aerosols. Blowing dust is clearly a dust. Gaseous material from a chimney is a smoke and water droplets from a cooling tower would be a mist.

Figure 10.1 notes property parameters for particles. Atmospheric dust particles range over a wide spectrum of sizes from greater than 10μ to about 0.001μ ($1\mu = 10^{-6}$ m). Many particles are in the size range corresponding to visible light, that is about 0.5μ. Thus we would expect these to affect the transmission of light and form a haze. Note on the figure that the width of a human hair is given as about 50 to 200μ, which is 2 to 8 thousandths of an inch. The comparison of a particle to a human hair and to the wavelength of visible light may aid in gaining a feel for the size of particulates.

Figure 10.1 indicates that a variety of control techniques is necessary to cope with the enormous size range of the particles. The types of gas-cleaning equipment for removal of particulates is also noted on the figure and some of these will be discussed in this chapter. Figure 10.1 contains a great deal of information and should be studied carefully.

Particulates or aerosols are also classified according to size as indicated on Fig. 10.2. Aitken nuclei refers to those particles with radius less than 0.1μ. This size particle is important in providing condensation nuclei for rain and fog. In industrial areas the measurement of these particles provides an indication of the general level of pollution. Over the ocean typical concentrations are 800 to 5,000 particles per cm^3. In contrast, large city concentrations are 50,000 to 380,000 particles per cm^3 (Stern, Vol. 1, 1968).

Haze particles are those in the range of visible light, 0.38μ to 0.76μ, and these interfere with the passage of light through the atmosphere. The subject of visibility is a complex one and will not be

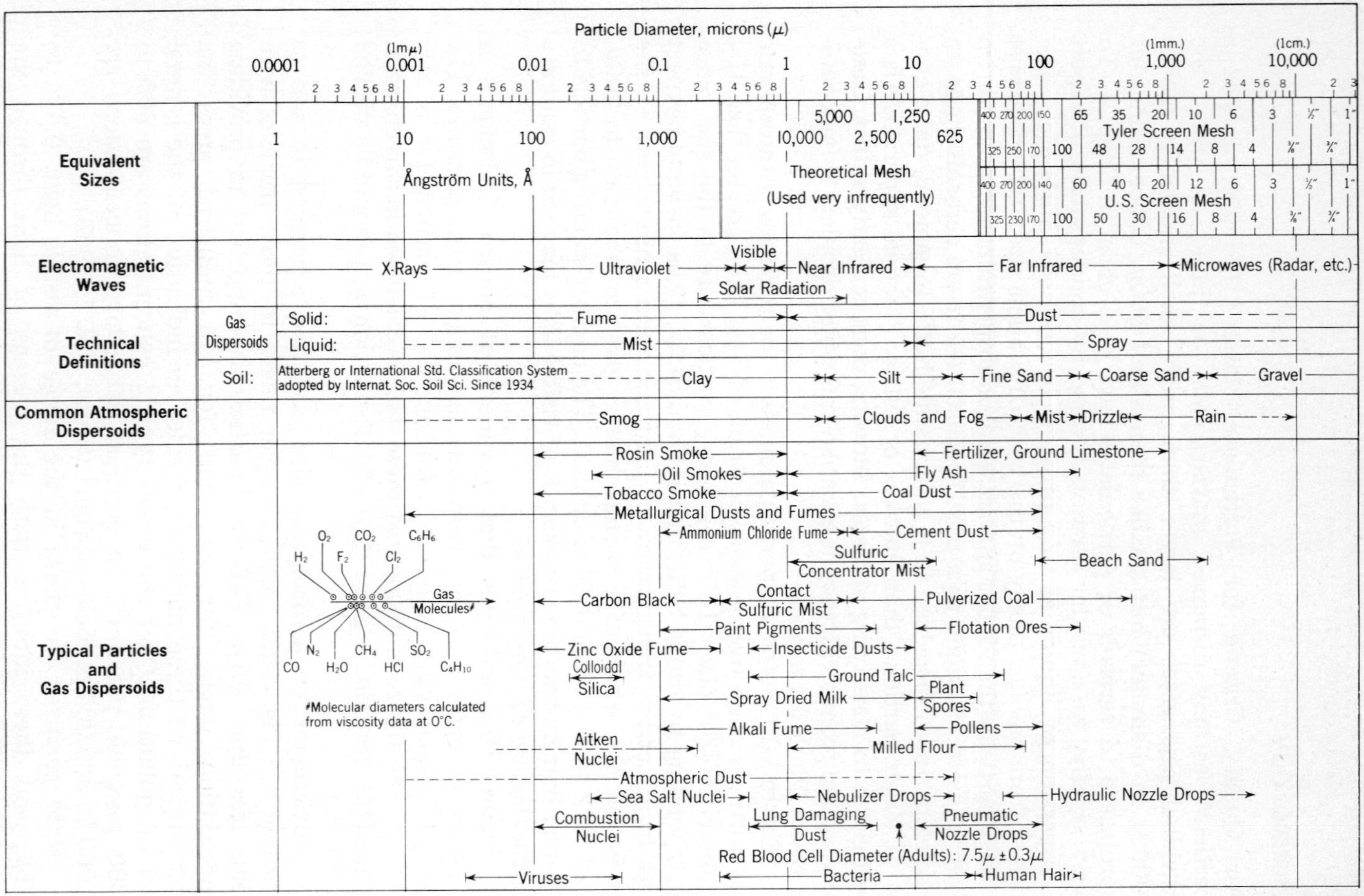

Particle Diameter, microns (μ)

0.0001 | 0.001 (1mμ) | 0.01 | 0.1 | 1 | 10 | 100 | 1,000 (1mm.) | 10,000 (1cm.)

2 3 4 5 6 8

Equivalent Sizes	Ångström Units, Å: 1, 10, 100, 1,000, 10,000 Theoretical Mesh (Used very infrequently): 10,000, 5,000, 2,500, 1,250, 625 Tyler Screen Mesh: 400, 270, 200, 150, 325, 250, 170, 100, 65, 48, 35, 28, 20, 14, 10, 8, 6, 4, 3, ½", ¼", 1", ¾" U.S. Screen Mesh: 400, 270, 200, 140, 325, 230, 170, 100, 60, 50, 40, 30, 20, 16, 12, 8, 6, 4, 3, ½", ¼", 1", ¾"
Electromagnetic Waves	X-Rays; Ultraviolet; Visible; Near Infrared; Far Infrared; Microwaves (Radar, etc.); Solar Radiation
Technical Definitions	Gas Dispersoids: Solid: Fume; Dust Liquid: Mist; Spray Soil: Atterberg or International Std. Classification System adopted by Internat. Soc. Soil Sci. Since 1934: Clay; Silt; Fine Sand; Coarse Sand; Gravel
Common Atmospheric Dispersoids	Smog; Clouds and Fog; Mist; Drizzle; Rain
Typical Particles and Gas Dispersoids	Gas Molecules[#]: H_2, O_2, F_2, CO_2, Cl_2, C_6H_6, CO, N_2, H_2O, CH_4, HCl, SO_2, C_4H_{10} [#]Molecular diameters calculated from viscosity data at 0°C. Rosin Smoke; Oil Smokes; Tobacco Smoke; Metallurgical Dusts and Fumes; Ammonium Chloride Fume; Carbon Black; Contact Sulfuric Mist; Paint Pigments; Zinc Oxide Fume; Colloidal Silica; Insecticide Dusts; Ground Talc; Spray Dried Milk; Alkali Fume; Aitken Nuclei; Atmospheric Dust; Sea Salt Nuclei; Combustion Nuclei; Viruses; Lung Damaging Dust; Nebulizer Drops; Bacteria Fertilizer, Ground Limestone; Fly Ash; Coal Dust; Cement Dust; Sulfuric Concentrator Mist; Beach Sand; Pulverized Coal; Flotation Ores; Plant Spores; Pollens; Milled Flour; Hydraulic Nozzle Drops; Pneumatic Nozzle Drops; Human Hair Red Blood Cell Diameter (Adults): $7.5\mu \pm 0.3\mu$

(continued)

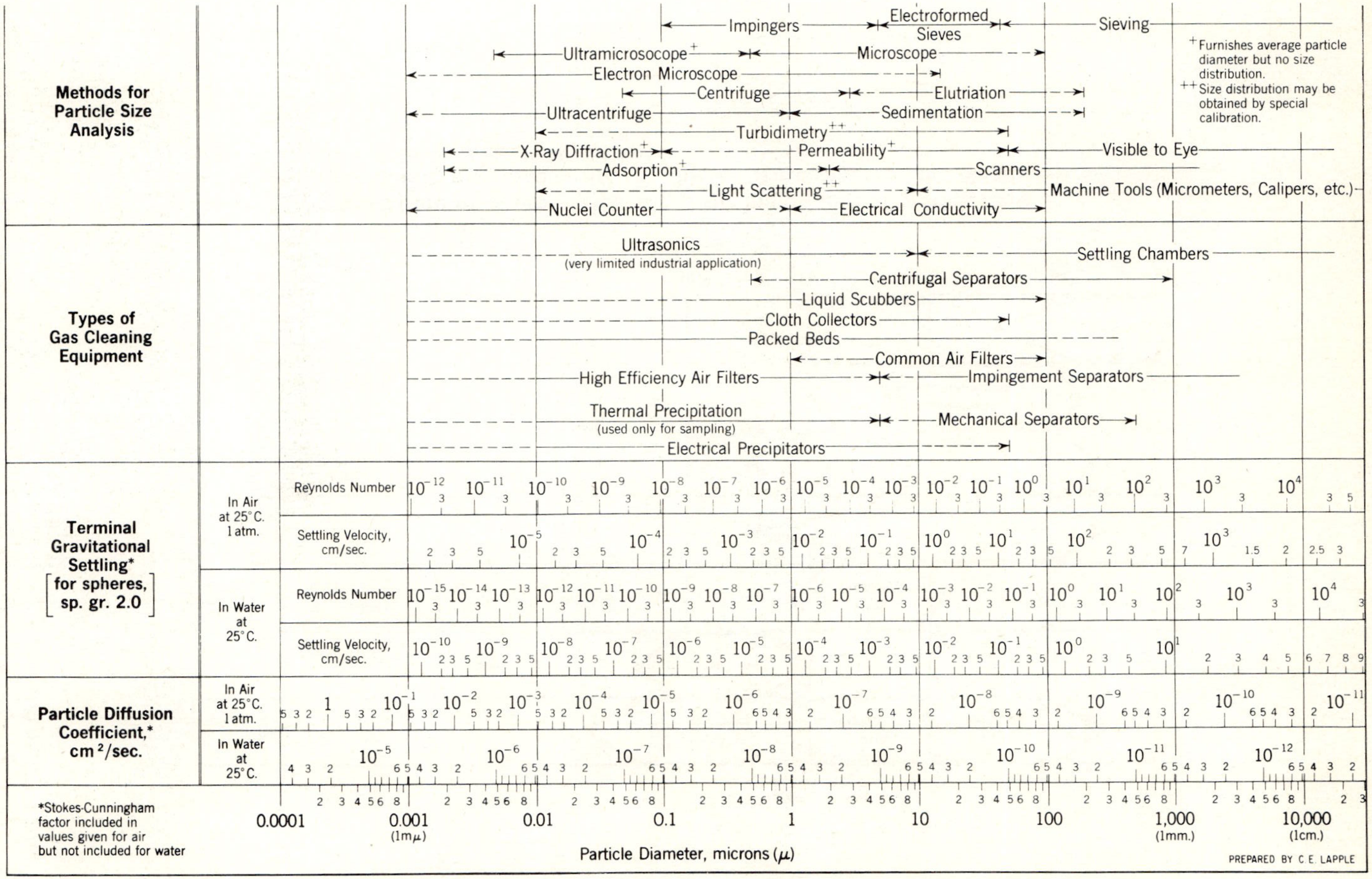

Methods for Particle Size Analysis

Impingers — Electroformed Sieves — Sieving

Ultramicroscope+ — Microscope

Electron Microscope

Centrifuge — Elutriation

Ultracentrifuge — Sedimentation

Turbidimetry++

X-Ray Diffraction+ — Permeability+ — Visible to Eye

Adsorption+ — Scanners

Light Scattering++ — Machine Tools (Micrometers, Calipers, etc.)

Nuclei Counter — Electrical Conductivity

+Furnishes average particle diameter but no size distribution.

++Size distribution may be obtained by special calibration.

Types of Gas Cleaning Equipment

Ultrasonics (very limited industrial application) — Settling Chambers

Centrifugal Separators

Liquid Scubbers

Cloth Collectors

Packed Beds

Common Air Filters

High Efficiency Air Filters — Impingement Separators

Thermal Precipitation (used only for sampling) — Mechanical Separators

Electrical Precipitators

Terminal Gravitational Settling* [for spheres, sp. gr. 2.0]

In Air at 25°C. 1 atm. — Reynolds Number: 10^{-12}, 10^{-11}, 10^{-10}, 10^{-9}, 10^{-8}, 10^{-7}, 10^{-6}, 10^{-5}, 10^{-4}, 10^{-3}, 10^{-2}, 10^{-1}, 10^{0}, 10^{1}, 10^{2}, 10^{3}, 10^{4}

In Air at 25°C. 1 atm. — Settling Velocity, cm/sec.: 10^{-5}, 10^{-4}, 10^{-3}, 10^{-2}, 10^{-1}, 10^{0}, 10^{1}, 10^{2}, 10^{3}

In Water at 25°C. — Reynolds Number: 10^{-15}, 10^{-14}, 10^{-13}, 10^{-12}, 10^{-11}, 10^{-10}, 10^{-9}, 10^{-8}, 10^{-7}, 10^{-6}, 10^{-5}, 10^{-4}, 10^{-3}, 10^{-2}, 10^{-1}, 10^{0}, 10^{1}, 10^{2}, 10^{3}, 10^{4}

In Water at 25°C. — Settling Velocity, cm/sec.: 10^{-10}, 10^{-9}, 10^{-8}, 10^{-7}, 10^{-6}, 10^{-5}, 10^{-4}, 10^{-3}, 10^{-2}, 10^{-1}, 10^{0}, 10^{1}

Particle Diffusion Coefficient,* cm²/sec.

In Air at 25°C. 1 atm.: 1, 10^{-1}, 10^{-2}, 10^{-3}, 10^{-4}, 10^{-5}, 10^{-6}, 10^{-7}, 10^{-8}, 10^{-9}, 10^{-10}, 10^{-11}

In Water at 25°C.: 10^{-5}, 10^{-6}, 10^{-7}, 10^{-8}, 10^{-9}, 10^{-10}, 10^{-11}, 10^{-12}

Particle Diameter, microns (μ): 0.0001, 0.001 (1mμ), 0.01, 0.1, 1, 10, 100, 1,000 (1mm.), 10,000 (1cm.)

*Stokes-Cunningham factor included in values given for air but not included for water

PREPARED BY C. E. LAPPLE

Fig. 10.1 Characteristics of particles and particle dispersoids. [*Reprinted by permission from C. E. Lapple, Stanford Research Institute Journal,* **5**, *94 (Third Quarter, 1961).*]

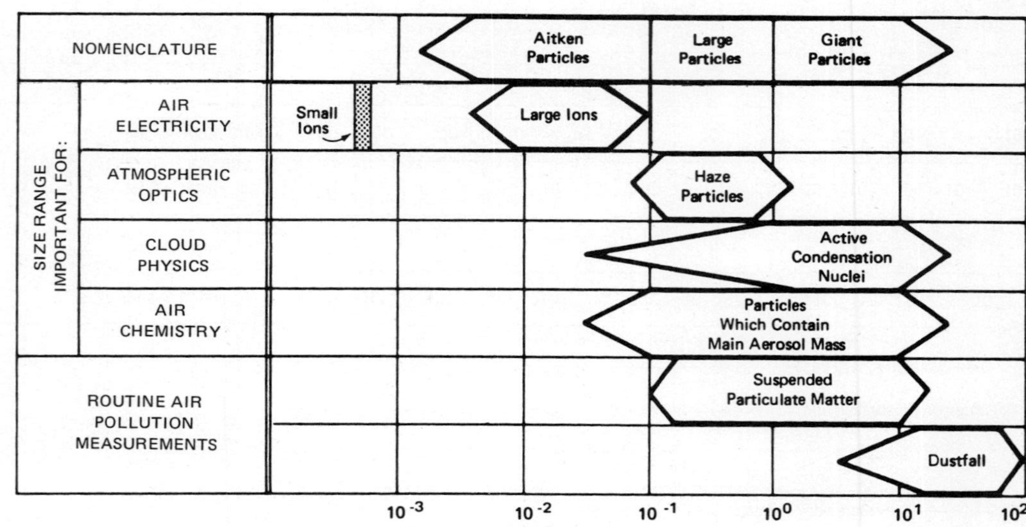

Fig. 10.2 Nomenclature of natural aerosols and the importance of particle sizes for various atmospheric phenomena. (*C. E. Junge, "Air Chemistry and Radioactivity," Academic Press, New York, 1963.*)

discussed here since the use of simple plug-in formulas for calculating transmission and scattering may result in large errors. Visibility is affected not only by the distribution of the particles by size but also by their shapes and surface finish. References to material on visibility are indicated at the end of this chapter.

Dust fall refers to particles of sufficient size that they fall quickly out of the air. It is measured by collection in dustfall jars and is expressed in units of tons/mile^2 per 30-day period or mg/cm^2 per 30-day period. The typical collection jar has, however, a 4-inch diameter size! Dust fall on the order of 30 tons/mile^2-month (1.08 mg/cm^2-month) might be typical for a large urban area (AP-51). Suspended particulate is smaller than dustfall particulate and remains in the air, falling only slowly because of gravitational force.

Pollen particles are not listed but range in size from about 15 to 90μ; most lie in the range of 20 to 40μ. Common ragweed pollen is from 16 to 20μ in diameter. Pollen grains are hygroscopic, that is, they absorb water, and thus vary in weight with changes in humidity. While not a serious air pollutant in terms of total quantity, pollen is an obvious and serious irritant to the many hay fever sufferers, including the author.

SIZE DISTRIBUTION

Solid particulates can range in shape from spherical, if formed by condensation and then solidification, to highly irregular. Initially spherical particles may increase in size by coagulation and become nonspherical. Dusts, for example, consist of particles of random irregular shape, whereas fly ash from coal-fired power plants may often be approximated as spherical. The shape and especially the size distribution of particulates are important parameters since the cleaning technique specified depends on the size of the particles to be removed. The size distribution for a typical sample is noted in Fig. 10.3(a). Figure 10.3(b) indicates a second way to present this information in which the log of the particle size is plotted as a function of the frequency. The ordinate has either units of number of particles or units of number of particles/unit log interval. If the log of the radius is used then the curve is often approximated by a gaussian distribution. This distribution is further plotted in Fig. 10.3(c) on a log-probability scale. The resulting straight line indicates that the size distribution is *log-normal* and it is a characteristic of many particle-size distributions that they are log-normal. In fact, it is characteristic of many gaseous pollutants that their concentrations, as a function of time, exhibit log-normal behavior. That is, the plot of the log of ppm versus percent

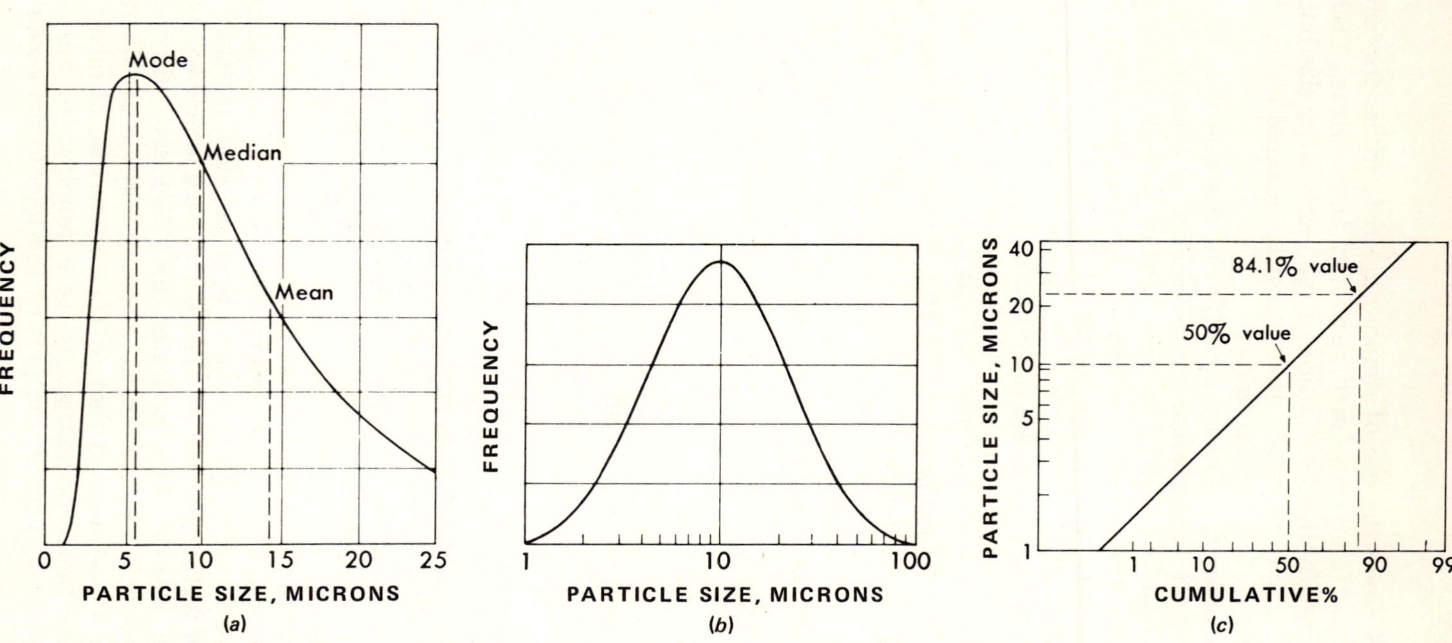

Fig. 10.3(*a*) Particle-size distribution. (*b*) Particle-size distribution, logarithmic size scale. (*c*) Logarithmic probability plot of particle size.

time that a certain concentration is exceeded shows a straight line distribution.

REMOVAL MECHANISMS

Particulate matter is removed from the air naturally by fallout, frequently referred to as sedimentation. For spherical particles in free fall the *terminal velocity* is given for *Stokes flow* by

$$V = \frac{2(\rho_{\text{part}} - \rho_\infty)R^2 g}{9\mu} \tag{10.1}$$

where R is the radius of the particle and μ the viscosity of the medium, here air. Stokes flow holds to Reynolds numbers of about 1. A more general expression for the terminal velocity for spherical particles is given by

$$V = \frac{2(\rho_{\text{part}} - \rho_\infty)Dg\mathcal{V}}{C_D A_p \text{Re}\,\mu} \tag{10.2}$$

where $\mathcal{V}$ is the particle volume
D the particle diameter
A_p the particle frontal area
Re the Reynolds number, $VD\rho_\infty/\mu$
C_D the drag coefficient

For Stokes flow $C_D = 24/\text{Re}$, $A_p = \pi R^2$, and $\mathcal{V} = \frac{4}{3}\pi R^3$ so

$$V = \frac{2(\rho_{\text{part}} - \rho_\infty)(2R)4R^3 g}{(3)(24)(R^2)(\mu)}$$

$$V = \frac{2(\rho_{\text{part}} - \rho_\infty)R^2 g}{9\mu}$$

as previously given. Problems 10.1 to 10.3 deal with settling velocities.

The drag coefficient is a function of the Re number and is available for spherical particles in any fluids text. For $\text{Re} < 1$ we may use $C_D = 24/\text{Re}$, for $3 < \text{Re} < 400$, $C_D = 24/\text{Re} + 4/(\text{Re})^{1/3}$ and for $1{,}000 < \text{Re} < 2 \times 10^5$, $C_D = 0.44$ and is independent of the Reynolds number. Interpolation can be used to fill in values between these expressions.

Example 1. Consider a particle of density = 1 g/cm^3 and 50μ diameter. Calculate the terminal velocity. We shall assume Stokes flow and then check the Re number. Neglecting the density of air relative to the particle density and using

$\mu = 1.8 \times 10^{-4}$ g/cm-sec we have

$$V = \frac{2}{9}\frac{\rho R^2 g}{\mu}$$

$$V = \frac{2}{9}\frac{(1)(25)^2(10^{-8})(980)}{(1.8 \times 10^{-4})}$$

$V = 7.5$ cm/sec

$V = 895$ ft/hr

Then checking for Stokes flow we can determine the Re as

$$\text{Re} = \frac{VD\rho_\infty}{\mu} = \frac{(7.5)(50 \times 10^{-4})(1.2 \times 10^{-3})}{(1.8 \times 10^{-4})} = 0.25$$

Thus our assumption is verified. For air at standard conditions and for Stokes flow we may simplify the velocity expression as

$$V = 0.003\, \rho D^2 \tag{10.3}$$

where V is in cm/sec, D is in microns, and ρ in g/cm^3. To convert the result to feet per minute multiply by 1.96. In the example problem this formula gives

$V = (0.003)(1)(2500) = 7.5$ cm/sec

Stokes flow holds to 5 percent in air for a particle of unit density with a diameter of 29 microns and becomes progressively less accurate as the diameter increases. When the particle diameter approaches the mean free path (mfp) of an air molecule then a correction must be applied because the particles "slip through" the molecules. This factor, called the Cunningham correction, is given by

$$C = 1 + \frac{\lambda}{R}\left[1.26 + 0.4 \exp\left(\frac{-1.1R}{\lambda}\right)\right] \tag{10.4}$$

where λ is the mfp and is thus a function of altitude. At 20°C (68°F) and 1 atmosphere, $\lambda = 0.65 \times 10^{-5}$ cm. Figure 10.1 includes terminal velocities, with this correction included, for a particle density of 2 g/cm^3. Since the terminal settling velocity goes as the density, a correction for other densities can be easily made.

Example 2. Calculate the Cunningham correction for a particle of radius equal to 0.1μ if $\lambda = 0.65 \times 10^{-5}$ cm. We have

$$C = 1 + \frac{0.065}{0.1}\left[1.26 + 0.4 \exp\left(\frac{-1.1 \times 0.1}{0.065}\right)\right]$$

$$C = 1 + 0.65\,(1.26 + 0.07) = 1.86$$

Thus for this size particle at sea level the effect of slip is roughly to double the settling velocity.

Most particles are nonspherical and this must be taken into account. A very rough correction is to use a velocity of 2/3 the velocity calculated for the spherical particle (Slade, 1968).

Two definitions used for nonspherical particles follow:

Stokes radius:

R_s = radius of a sphere with the same terminal velocity and same density as the material which formed the nonspherical particle.

Equivalent aerodynamic size, radius or diameter:

D_e = diameter of a sphere with the same terminal velocity but a density of 1 g/cm^3.

Nonspherical particles can thus be classified if their size and shape can be determined.

Brownian motion occurs when the particle size becomes so small that collisions with individual gas molecules impart a change in velocity to the particle. Particles smaller than 1μ will exhibit some Brownian motion superimposed on their settling velocity. Below about 0.25μ the Brownian motion will dominate, at 1 atmosphere and 20°C.

Particles in an aerosol cloud are affected by interaction with the other bodies and can move faster than independent individual particles. Fuchs (1964) discusses this phenomenon.

COAGULATION

When aerosol particles come into contact and coalesce or adhere the process is called coagulation. For example, simple Brownian motion may bring two particles sufficiently close to coalesce. Such coagulation is referred to as thermal since the random Brownian motion may be related to the ambient temperature. We can also force coagulation by hydrodynamic or electrical forces. Theory (see, for example, Fuchs) indicates that the basic equation for thermal coagulation, in which it is assumed that in each collision the particles adhere, is given by

$$\frac{dn}{dt} = -kn^2 \tag{10.5}$$

The form can be justified by considering that in an aerosol with n particles per cm^3 the probability of a collision between two of them is proportional to n^2. The rate of change of particle concentration is thus $-kn^2$. Solving this simple differential equation we have

$$\frac{1}{n} - \frac{1}{n_0} = kt$$

or

$$n = \frac{n_0}{1 + kn_0 t} \tag{10.6}$$

where n_0 is the initial particle concentration in particles per cm^3. The constant k can be found from theory but is also determined empirically for a particular distribution and type of particle. Values taken from Fuchs are noted in Table 10.3 and may be used as an indication of how rapidly coagulation occurs. These values are for spherical particles of similar size. Coagulation between a large and a small particle is governed by much larger k values than the above. Fine particles are collected by larger ones so that the distribution rapidly becomes one with few small particles. Information on coagulation of nonspherical particles is included in Fuchs (1964) and Green and Lane (1964). In a wind-borne cloud there is an additional effect due to coagulation produced by the turbulent mixing in the cloud. Thus the above results would be conservative since they do not include this effect.

Gravitational coagulation, important in some pollution situations, takes place when particles of unequal size and hence speed collide under the action of the gravitational field. This form of coagulation is

TABLE 10.3 Values for the coagulation constant, k cm^3/sec

Particle size					
R, cm	10^{-7}	10^{-6}	10^{-5}	10^{-4}	10^{-3}
k*, $\times 10^{10}$	4.5	12	5	3	3

*Includes correction for slip effects.
Source: Fuchs, 1964.

especially important in rainfall from clouds and is also a mechanism for cleaning aerosols out of the air by use of water spray in a wet scrubber.

Besides the removal mechanism of coagulation and fallout for smaller particles, or direct sedimentation for large particles, aerosols are also removed by washout from rain or snow. Slade (1968) includes an introduction to this method of particulate removal.

COH's

Transmittance of light through a deposit of particulate on filter paper can be used as a measure of the atmospheric air quality. The light transmittance through a clean and then a dirty filter is compared, and converted to "coh" units per 1,000 linear feet of air passing through the filter. A coh stands for the "coefficient of haze" and is defined as the quantity of light-scattering solids on the filter to produce an optical density equivalent of 0.01 when measured by light transmission. The optical density is the $\log_{10}$ of the opacity which, in turn, is the inverse of the fractional transmittance. An example should help to clarify this.

Assume that the transmittance of a particulate sample on a filter paper is measured to be 60 percent of that of a clean sample. The opacity is then 1/0.60 or 1.67. The optical density is $\log_{10}(1.67) = 0.22$. Since a coh is an optical density of 0.01 we have 22 coh units. However, this sample might have required running the filter for 60 minutes at a velocity at the filter of 2 feet per second. Thus $60 \times 60 \times 2 = 7{,}200$ linear feet of air have passed through the filter to produce this transmittance. The coh/1,000 linear feet is therefore $22/7.2 = 3.06$. Such a reading would represent a high level of particulates.

Consider a second example where a filter is operated for 6 hours in a rural area with a velocity of 1 fps at the filter. The transmittance relative to a clean filter is measured as 82 percent. Determine the coh/1,000 linear feet.

$$\text{opacity} = \frac{1}{0.80} = 1.25$$

$$\text{optical density} = \log_{10}(1.25) = 0.097$$

$$\text{coh units} = \frac{0.097}{0.01} = 9.7$$

$$\text{linear ft} = 6 \times 60 \times 60 \times 1 = 21{,}600 \text{ ft}$$

$$\frac{\text{coh}}{1{,}000 \text{ ft}} = \frac{9.7}{21.6} = 0.45 \text{ (slight pollution)}$$

PARTICULATE COLLECTION

The choice of collection equipment for a specific purpose depends upon a number of factors: the properties of the material, such as particle size, and physical and chemical properties; the concentration and volume of the particulate to be handled; the temperature and humidity of the gaseous medium and, most importantly, the collection efficiency required.

There are six types of collectors:

1. gravitational settling chambers
2. centrifugal separators
3. wet scrubbers
4. filters
5. electrostatic precipitators
6. ultrasonic agglomerators

The latter is used to agglomerate the particulate so that it may be collected in another device.

The mechanisms by which particles are collected depend upon the particle size and velocity. Gravitational settling is an obvious mechanism, as is centrifugal force which is really just enhanced gravitational settling. Other mechanisms are noted in Fig. 10.4. Inertial impact or inertial collection depends upon the particle *not* following a streamline as flow goes around a collection device such as a fiber or water drop. Instead the particle's inertia carries it into the collector. Direct interception refers to the case when the edge of the particle hits the collector even though the center would clear the obstruction. High velocities are necessary in order to get inertia effects combined with direct interception. Brownian collection or diffusion is important with small particles having a low velocity. The Brownian motion superimposed on the main stream flow may cause the particle to contact the collector as noted in Fig. 10.4. Other collection mechanisms include electrostatic effects as in the case of electrostatic precipitators and direct filtration if fibers are put sufficiently close together.

Gravitational Settling Chambers

Gravitational settling chambers are only used to collect large particles above approximately 50μ. The pressure drop is small, less than 1 inch of water, but the efficiency is low for particles below 50μ. If we assume Stokes flow, which is questionable for particles greater than this size, we can readily determine a formula for the minimum diameter of a particle collected at 100 percent theoretical efficiency in a chamber of length L. The geometry is shown in Fig. 10.5. We have a balance

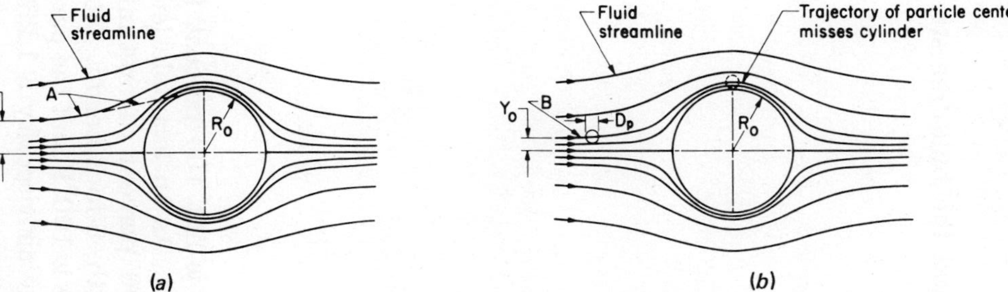

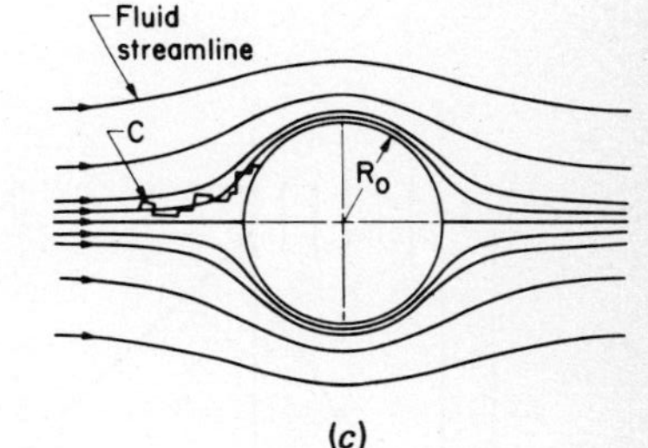

Fig. 10.4 Mechanisms of mechanical filtration. (*a*) Inertial impaction: A is trajectory of particle center which just touches cylindrical fiber. (*b*) Direct interception: B is trajectory of particle center and fluid streamline. Particle surface touches fiber at point of closest approach. (*c*) Diffusion: C is path of particle center due to fluid motion and random diffusion. (*From American Industrial Hygiene Assoc. "Air Pollution Manual," pt. 2, Control Equipment.*)

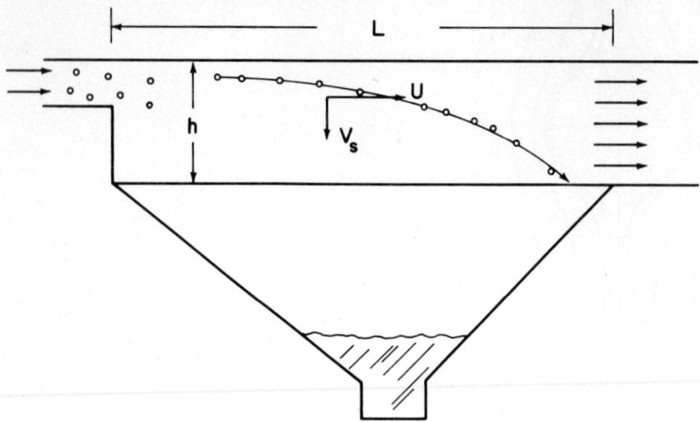

Fig. 10.5 A settling chamber.

between the settling velocity and the bulk-free stream velocity such that

$$\frac{V_s}{h} = \frac{U}{L} \tag{10.7}$$

and

$$V_s = \frac{2}{9}\left[\frac{(\rho_{\text{part}} - \rho_\infty)gR^2}{\mu}\right]$$

so

$$R = \frac{D}{2} = \left[\frac{9}{2}\frac{V_s\mu}{g(\rho_{\text{part}} - \rho_\infty)}\right]^{1/2}$$

and

$$D = \left[\frac{18\,U\,h\mu}{Lg(\rho_{\text{part}} - \rho_\infty)}\right]^{1/2} \tag{10.8}$$

In practice the efficiency is not 100 percent for such particles since some reentrainment occurs. As an example let us calculate the minimum size particle with 100 percent theoretical efficiency for a settling chamber of 30-foot length, 6-foot height, with a bulk velocity of 5 ft/sec if the air temperature is 100°F and the particulate density is 2 g/cm^3. At this temperature the air viscosity is 1.28×10^{-5} lb/ft-sec. Then substituting we have,

$$D = \left(\frac{18 \times 5 \times 6 \times 1.28 \times 10^{-5}}{30 \times 32 \times 124.8}\right)^{1/2}$$

$$D = 2.37 \times 10^{-4} \text{ ft}$$

$$D = 72\mu$$

Settling chambers are often used upstream of further collection devices so that larger particulate has already been eliminated before more sophisticated removal devices are used.

Centrifugal Separators

Figure 10.6 shows a drawing, roughly to scale, of a well-designed simple cyclone. Centrifugal action throws the heavy particles to the side of the cyclone where they can slide down into a collector. The separator operates with two vortexes as noted on the figure. The advantage of a cyclone, which is the most common dust collector used, lies in the high separation factor given as the ratio of the radial velocity in the cyclone

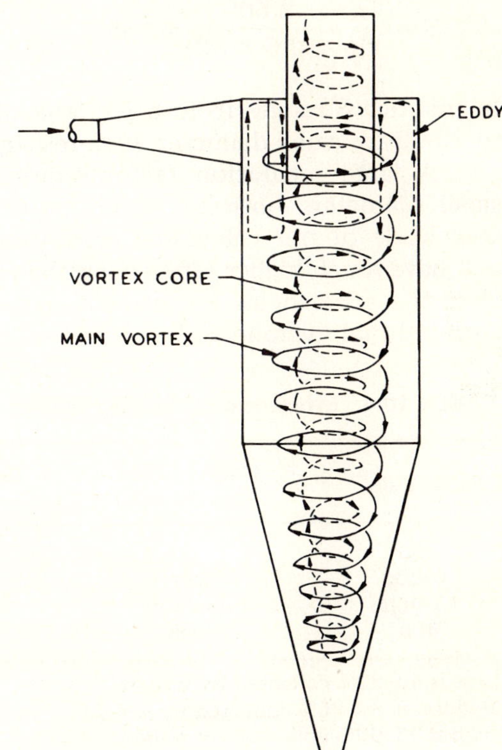

Fig. 10.6 A typical cyclone.

to the Stokes velocity in a simple settling chamber. We have

$$V_s = \frac{\rho D^2 g}{18\mu} \tag{10.9}$$

and for the radial velocity (see Prob. 10.10)

$$V_R = \frac{D^2 \rho V_T{}^2}{18R\mu} \tag{10.10}$$

where V_T is the tangential velocity. Thus

$$S = \frac{V_R}{V_s} = \frac{V_T{}^2}{Rg} \tag{10.11}$$

For typical parameters of R = 6 inches and V_T = 50 ft/sec we have a separation factor given by

$$S = \frac{V_R}{V_s} = \frac{2{,}500}{(0.5 \times 32)} = 155$$

In practice S ranges from 5 for large-diameter, low-resistance cyclones to 2,500 for small-diameter, high-resistance devices.

A large separation factor requires high tangential velocities and small diameters both of which result in a large pressure drop. The magnitude of our calculated separation factor indicates that cyclones will have much higher efficiencies than the settling chamber. Table 10.4 gives the efficiencies for percent particulate collected by weight in a conventional cyclone and in a high-efficiency cyclone. The latter is one

TABLE 10.4 Efficiencies of simple cyclones

Particle size	Conventional cyclone	High efficiency cyclone
5μ	—	low
5–20μ	low	medium
15–50μ	medium	high
40μ	high	high

Low is 50–80% collected by weight.
Medium is 80–95% collected by weight.
High is 95–99% collected by weight.
Source: AP-51, 1969.

with a diameter of 9 inches or less. At peak performance a high-efficiency single cyclone would only collect particles of about 2μ diameter at efficiencies of about 50 to 60 percent by weight.

The effect of changing parameters such as body length (Am. Ind. Hyg. Assoc., 1968, suggests typically a length giving 5 to 10 turns), entry velocity, and entry shape is not readily calculated and is usually measured. The important economic parameter in a cyclone is the pressure drop which is calculated from

$$\Delta P = \frac{KQ^2 P\rho_g}{T} \tag{10.12}$$

where ΔP is in inches of water
Q is cfm of gas
P is the absolute pressure in atmospheres
ρ_g is the gas density in lb/ft^3
T is the temperature in °R
and K is a function of cyclone diameter as given in Table 10.5.

The K versus diameter relation is a straight line on log-log paper. Cutting the diameter in half increases the pressure drop by about a factor of 10 if the other variables are held constant. The above equation is intended as a guide to indicate the effect of the variables. Data from manufacturers should be used for accurate calculations on specific designs. Typical pressure drops would be from 1 to 8 inches of water.

TABLE 10.5 Value of cyclone design variable K

Cyclone diameter inches	29	16	8.1	4.4
K	10^{-4}	10^{-3}	10^{-2}	10^{-1}

Scrubbers or Wet Collectors

Scrubbers or wet collectors come in a variety of shapes and include

1. spray towers, without beds
2. cyclone scrubbers
3. venturi scrubbers
4. packed bed or floating bed scrubbers

The latter are often used in combination particulate and gaseous removal applications. The principle of the scrubber is to remove the particulate, or gas, by absorbing the material into liquid droplets directly by contact. The contact mechanism may be inertial impingement or gravitational settling. After a simple cyclone has removed the larger particles, a scrubber might be used to remove material in the size range of 0.2 to 10μ. The advantages of the scrubber system are that it can remove simultaneously particulate and gases; it has high efficiencies for small particles; and there is no particle reentrainment. Scrubbers, however, do not function well where plume rise is important since a wet plume has little buoyancy. Handling of the dirty liquid and removal of the entrained material is difficult and may cause a water pollution problem. Settling ponds or centrifugal methods may be required to remove the particles from the liquid.

Spray Towers and Cyclone Scrubbers. In its simplest form the spray tower consists of a downward flow of water droplets sprayed into the tower and an upward flow of dirty gas. High pressure sprays produce small drops with more surface area per mass of water used and these are effective in collecting particles in the 1–2μ range.

In the cyclone scrubber the gas is tangentially swirled around just as in the dry scrubber. Water sprays are introduced in a variety of ways either across the cyclone from the outside wall to the centerline or down the cyclone from the top. The combined impingement and centrifugal forces clean the gas. The wet cyclone has the advantage of a higher efficiency than the dry cyclone. However, sludge removal is more difficult than dry particulate removal.

There are an enormous number of possible designs and variations in tower scrubbers. AP-51, 1969, indicates some of these and shows schematics. A particular design for a floating bed scrubber is shown in Fig. 10.7. In this case the liquid is both sprayed into the air and then allowed to drip or flow through a bed of inert spheres to promote contact between the liquid and the gas streams.

Venturi Scrubbers. Figure 10.8 notes a venturi scrubber in which water is injected upstream of the venturi throat. The curtain of water is broken up by the gas stream into drops which collect the dirt. Inertial impact is the primary collection mechanism; consequently, the faster the gas passes through the venturi, the higher the efficiency.

Figure 10.9 gives performance curves for venturi scrubbers. The disadvantage to this device is clearly the high-pressure drop. The advantage is the high efficiency for small particles. Once again engineering design tradeoffs are evident. Because of the high-pressure drop, power costs are high, but difficult cleaning situations may require this type of collector with its greater than 90 percent efficiency for submicron-size aerosols.

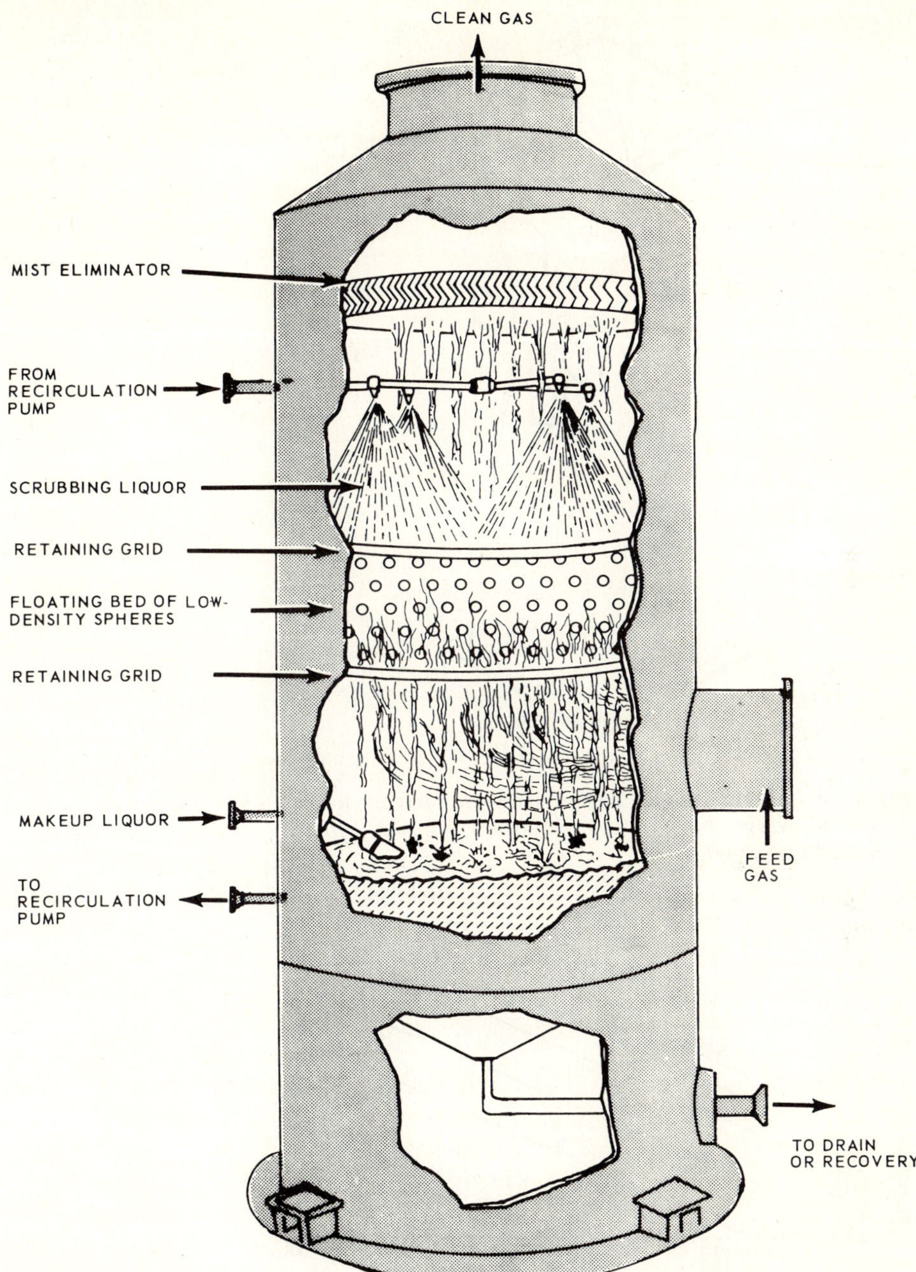

Fig. 10.7 Floating-ball (fluid-bed) packed scrubber. (*By permission of UOP Air Correction Division.*)

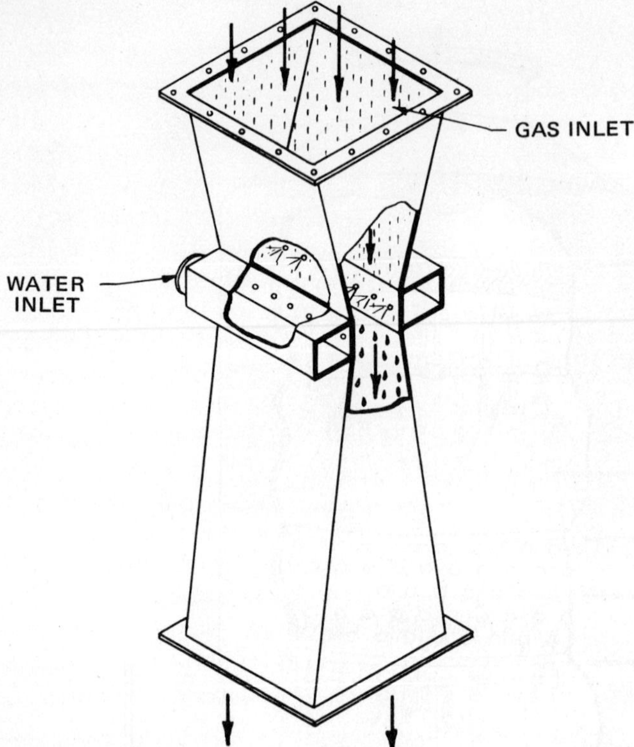

Fig. 10.8 Venturi throat design with jet feed of liquid. (*Am. Ind. Hyg. Assoc., "Air Pollution Manual," pt. 2, Control Equipment.*)

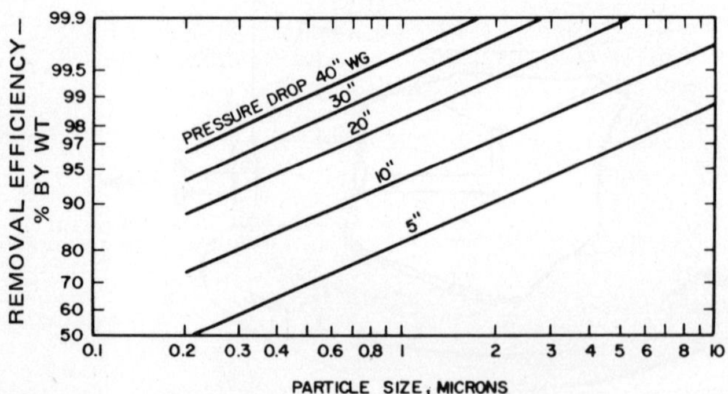

Fig. 10.9 Relationship between collection efficiency and particle size in venturi scrubbers. (*Am. Ind. Hyg. Assoc., "Air Pollution Manual," pt. 2, Control Equipment.*)

Filters

Filters are used in a wide variety of materials and geometries to remove particulates from gas streams. For example, *bag houses* consist of a large number of filter bags, arranged so that continuous removal of the collected material is possible. Since filters show an increase in pressure drop as material is deposited, some means for periodically removing the solids is required. In fact, the disadvantage to filtering is the high-pressure drop and the clogging of the filter. Advantages are high efficiency, even for very small particles, low capital costs, and ability to handle a wide range of operating conditions. The filter collection mechanisms are direct interception and inertial impact for the larger particles, and Brownian, or diffusion impact, for the smaller particulate.

A filter should be used so that it can be readily serviced and the collected particulate easily removed.

Figure 10.10(*a*) shows a bag house with a shaker mechanism for cleaning off the collected dust cake. Here the dirty air flows from the inside to the outside of the bags. In Fig. 10.10(*b*) a second system, again with air flow from inside to outside is shown. Here, however, a blow ring traverses the bag to remove continuously the collected matter. Filter materials may be made from practically any material depending upon the application. Some limits are noted in Table 10.6. Details on filter selection are given in AIHA, 1968 and AP-51, 1969.

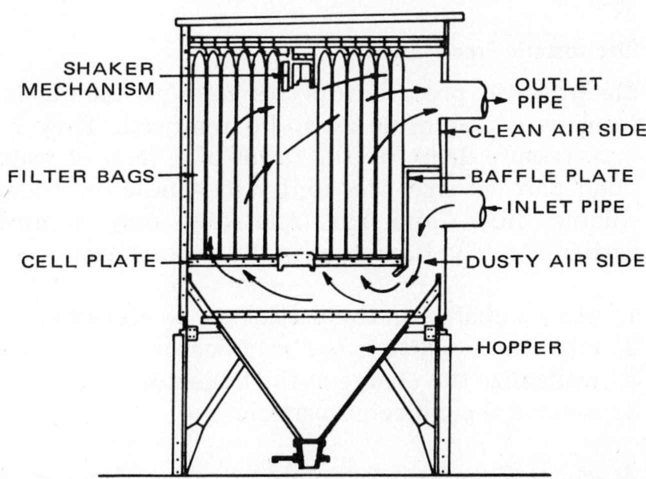

Fig. 10.10(*a*) Single-compartment baghouse filter. Mechanical cleaning is intermittently applied. (*Am. Ind. Hyg. Assoc., "Air Pollution Manual," pt. 2, Control Equipment.*)

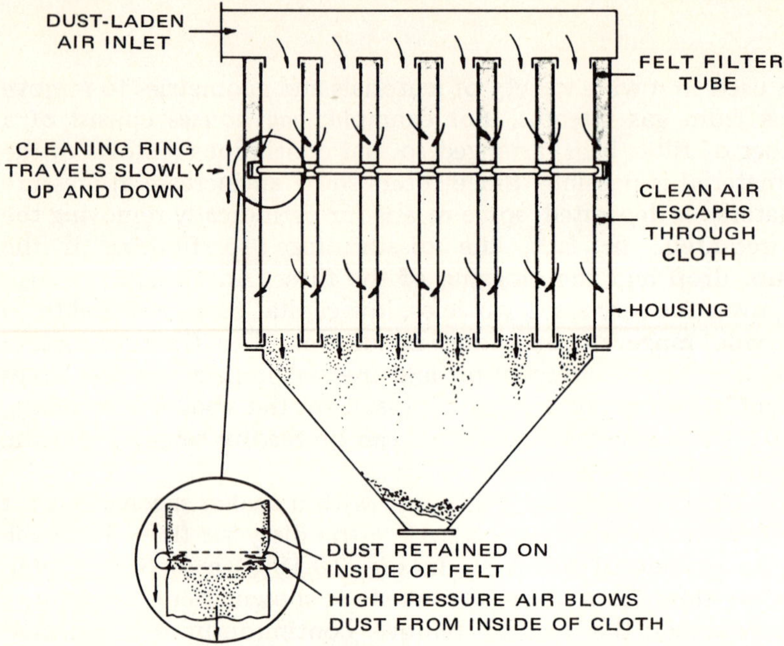

Fig. 10.10(*b*) Blow ring, which works on the reverse jet principle. (*Am. Ind. Hyg. Assoc., "Air Pollution Manual," pt. 2, Control Equipment.*)

Electrostatic Precipitators

Electrostatic precipitators are used to handle large volumes of gases from which aerosols must be removed. They have the advantage of low-pressure drop, on the order of 1 inch of water, high efficiency for small particle size, the ability to handle both gases and mists for high volume flow, and the relatively easy removal of the collected particulate. The four steps in the process are:

1. place a charge on the particle to be collected
2. migrate the particle to the collector
3. neutralize the charge at the collector
4. remove the collected particle

A schematic of a precipitator is shown in Fig. 10.11. The negatively charged center wire causes a discharge to the grounded, positive collector shown here as the outer cylinder. At the very high DC voltages used, 25 to 100 kv, a corona discharge occurs close to the negative

TABLE 10.6 Filter fabric characteristics

Fiber	Operating exposure °F.		Supports combustion	Air permeability* cfm/ft^2	Composition	Resistance†				Cost‡ rank
	Long	Short				Abrasion	Mineral acids	Organic acids	Alkali	
Cotton	180	225	yes	10–20	Cellulose	G	P	G	G	1
Wool	200	250	no	20–60	Protein	G	F	F	P	7
Nylon§	200	250	yes	15–30	Polyamide	E	P	F	G	2
Orlon§	240	275	yes	20–45	Polyacrylonitrile	G	G	G	F	3
Dacron§	275	325	yes	10–60	Polyester	E	G	G	G	4
Polypropylene	200	250	yes	7–30	Olefin	E	E	E	E	6
Nomex§	425	500	no	25–54	Polyamide	E	F	E	G	8
Fiberglass	550	600	yes	10–70	Glass	P-F	E	E	P	5
Teflon§	450	500	no	15–65	Polyfluoroethylene	F	E	E	E	9

*cfm/ft^2 at 0.5 in. W.G.
†P = Poor, F = Fair, G = Good, E = Excellent.
‡Cost rank, 1 = lowest cost, 9 = highest cost.
§Dupont registered trademark.
Source: AP-51.

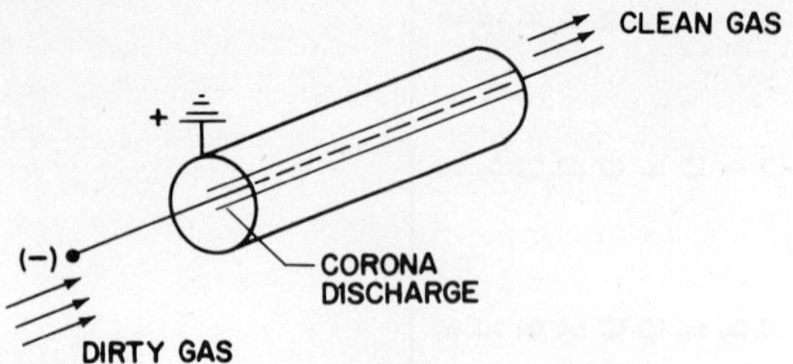

Fig. 10.11 An electrostatic precipitator.

electrode. The gas close to the negative electrode is thus ionized upon passing through the corona. As the negative ions and electrons migrate toward the collector electrode they, in turn, charge the passing particulate. The electric field then draws the particulate to the collector where it is deposited. The collected material is removed, usually by hitting the collector, a process which is called rapping. If a mist is being collected, then the material runs down the collectors and is removed at the bottom.

The theoretical efficiency, that is, weight fraction collected, of an electrostatic precipitator is given approximately by the relation

$$\eta = 1 - e^{-2VL/RU} \tag{10.13}$$

where V is the particulate velocity towards the electrode, typically on the order of 0.1 to 0.7 ft/sec
L is the length of collector
R is the radius of collector, if cylindrically shaped
U is the bulk gas velocity, 2 to 8 ft/sec.

Noting that the collector area, A, is equal to $2\pi RL$ and the volumetric flow rate, Q, is given as $\pi R^2 U$, we can rewrite this equation[2] as

$$\eta = 1 - e^{-AV/Q} \tag{10.14}$$

This expression, while approximate, indicates that higher efficiencies occur for low Q, high V, and large collector area. One can obtain high efficiency, if required, by increasing the surface area. However, in order to increase efficiency from 90 to 99 percent one must approximately double the area; to go from 90 to 99.9 percent requires tripling the area.

[2] There are other design equations for ESP's, some of which are given in AP-40, 1967. One such equation has the same form as Eq. (10.14), but with the exponent to the ½ power.

An increase in flow rate reduces the efficiency and even though only a small reduction in η might occur for a slight increase in Q, this means a large increase in emissions. A precipitator dropping from 99 percent efficiency to 97 percent triples the emissions and it is the emissions which are the important variable. In practice, the theoretical efficiency is never attained since some reentrainment of the collector material occurs. The collectors are usually plates rather than cylinders. Plate-type collectors are used for dry particulate collection as from power plants while the tube type is often used for wet collection.

The electrostatic precipitator system is not without problems. Ionization of the gas occurs only in a limited temperature range so that the device can suffer a loss in efficiency if sudden changes occur in the operating conditions. Build-up of collected material can cause "spark-over" between the electrodes which in turn causes a high current flow and excessive power use. Build-up of material on the negative electrode can suppress the corona discharge and reduce efficiency. The resistivity of the gas-particulate combination also affects the corona and the collection efficiency. Reentrainment of the collected material can interfere with the particle charging and result in direct release of particulate up the stack.

The resistivity of the deposited dust layer on the positive electrode is an important parameter in ESP design. As this increases, the insulating property of the dust increases and if it becomes sufficiently high ($\rho > 2 \times 10^{10}$ ohm-cm) the electrons have a difficult time penetrating the layer to reach the grounded electrode. It is ironic that adding sulfuric acid, SO_2, or SO_3 to the gas can then improve the efficiency of the precipitator. Thus, for electric power generation, the use of low sulfur coal can result in a reduction of particulate removal in the precipitator!

Figure 10.12 shows a full-scale precipitator with plate type collectors. Such a unit for 99 percent efficiency might measure 40 or more feet high and require a large floor space. Increasing the efficiency to 99.9 percent would require increasing the unit to 60 feet in height with a corresponding increase in capital and operating costs.

CHOICE OF EQUIPMENT

There are a large number of variables in selecting a particulate collector. It is necessary to consider

1. particulate-size distribution, including shape and density
2. required efficiency
3. allowable pressure drop versus the required flow rate

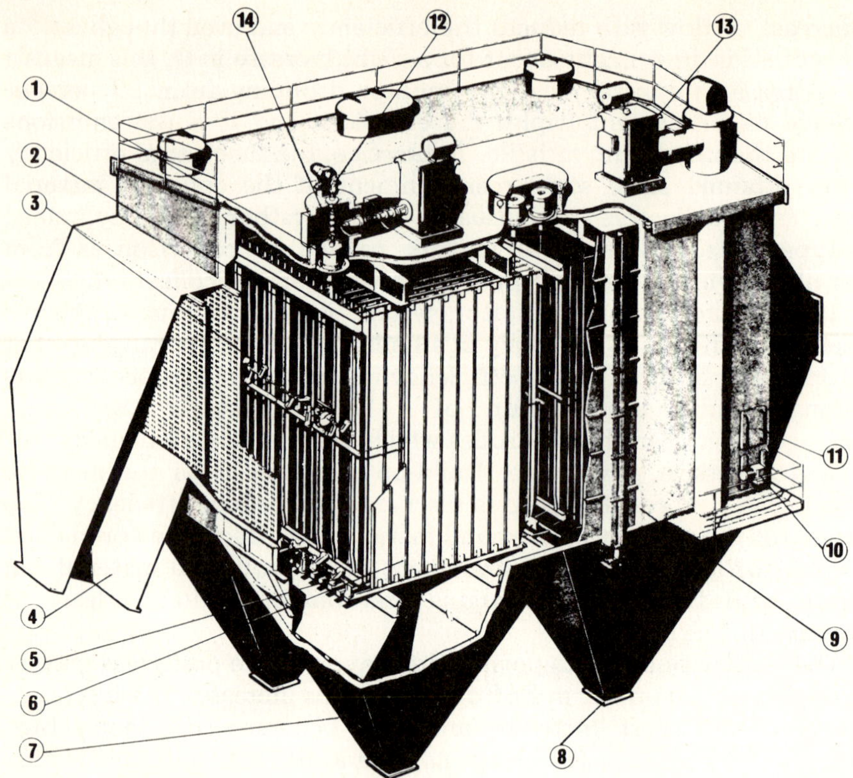

Fig. 10.12 An electrostatic precipitator.

1. Support insulators
2. Discharge system
3. Rapping mechanism for discharge system
4. Gas-distribution screens
5. Rapping mechanism for collecting system (one rapping mechanism for each section)
6. Collecting electrodes
7. Dust hopper (other types of discharge arrangements such as troughs and flat bottoms with scrapers are also available)
8. Dust-discharge opening
9. Heat insulation
10. Drive gear for the collecting system rapping mechanism (only one entry required for each section)
11. Insulated access door
12. Drive gear for the discharge-system rapping mechanism
13. Transformer rectifiers with constant current regulation
14. Insulator housing (electrically or hot air heated to prevent condensation on insulators)

4. particulate loading, often given in grains per ft^3 (1 gr/ft^3 = 2,300 milligrams/m^3 and 7,000 grains = 1 lb)
5. capital and operating costs of the equipment
6. properties of the carrier gas
7. contaminate disposal
8. ease of maintenance and reliability

A summary of the operating information for some mechanical collectors is given in Table 10.7. Some economic information is given in Table 10.8 for installed costs of control equipment. Not all of this equipment has been discussed in this chapter. Since the material in the table was developed in 1968 some correction for rising prices is required. However, the relative costs between the different methods of control are still accurate.

More detailed cost figures and formulas for calculating costs are given in AP-51, 1969, and Figs. 10.13 and 10.14 are taken from that reference. The results of Fig. 10.13 are for the following conditions:

nominal low efficiency 50 percent
nominal medium efficiency 70 percent
nominal high efficiency 95 percent
annual operating time 8,760 hours
(24 hours/day, 365 days/year)
collector pressure drop 3 inches of water
power cost 0.011 dollars/kw-hr
maintenance cost 0.015 dollars/actual cfm

The results of Fig. 10.14 are for nominal efficiencies of 90 percent (low), 95 percent (medium), and 99.5 percent (high). The assumptions are:

annual operating time 8,760 hours
electrical power requirements:
0.00019 kw/acfm for low efficiency
0.00026 kw/acfm for medium efficiency
0.00034 kw/acfm for high efficiency
power costs 0.011 dollars/kw-hr
maintenance costs 0.02 dollars/acfm

Since the cost of money changes with time these figures should be taken as approximate. Nevertheless these results indicate how costly particulate control can be. Rising interest rates have increased costs beyond those indicated here.

Example 3. Estimate the purchase costs, installed cost, and annual operating cost for the electrostatic precipitators of 99 percent efficiency for an 800-megawatt power plant to be built from 2-400 megawatt units. Such a plant might have a gas flow rate of 10^6 cfm per 400-megawatt unit.

From Fig. 10.14 we have for *each* 400-megawatt unit using the high efficiency curve

TABLE 10.7 Mechanical collection equipment

Collector type	Space require-ments	Volume range cfm	Efficiency by weight	Pressure loss* in inches H_2O	Temperature limitations	Power† hp per 1,000 cfm gas	Costs per cu ft gas	Application areas
Settling chambers	Large	Space available only limitation	Good above 50μ	0.2 to 0.5	700-1000°F limited only by materials of construction	0.04-0.12	.01-.05c	Precollector for fly ash, metal-lurgical dust, can be used for any large size dust particles above 50μ
Conventional cyclone	Large	Normal range up to 50,000 cfm	Approx. 50% on 20μ	1 to 3	700-1000°F limited only by materials of construction	0.24-0.73	.03-.10c	Woodworking, paper, buffing fibers, etc.; well suited for dry dust particles in 20μ and above range
High efficiency cyclone	Medium	Normal range up to 12,000 cfm	Approx. 80% on 10μ	3 to 5	700-1000°F limited only by materials of construction	0.73-1.2	.07-.15c	Woodworking, material con-veying, product recovery, etc.; well suited for dry dust parti-cles in 10μ and above range

Multitube cyclones	Small	Normal range up to 100,000 cfm	90% on 7½μ	4.5	700-1000°F	1.1	.09-.15c	Precollector for electrostatic precipitator on fly ash, product recovery, etc.; well suited for dry dust particles in 5μ and above range
Dynamic precipitator	Small	17,000 cfm	80% on 15μ	No loss (True fan)	700°F	Power consumption will depend on selection point, mechanical efficiency in usual selection range from 40-50%	.05-.15c	Woodworking, nonproduction buffing, metal-working, etc. Well suited for dry dust particles in 10μ and above range
Impingement separator	Small	Space available only limitation	90% on 10μ	1 to 5	700°F	0.24-1.2	.06-.10c	Certain types used for coarse particle collecting boiler fly ash, cement clinker coolers. Recent designs used for cleaning atmospheric air to diesel engines, and gas turbines

*Pressure drop is based on standard conditions.
†Power consumption figured from hp = cfm × t.p./6356 × M.E. (Mechanical efficiency assumed to be 65%.)
Source: Am. Ind. Hyg. Assoc., "Air Pollution Manual," pt. 2, Control Equipment.

TABLE 10.8 Installed costs of control equipment

Collector type	Approximate installed cost, in thousands of dollars — Gas flow rates (1,000 actual cubic feet per minute)						
	2	5	10	15	100	300	500
Gravity	0.5	1.2	2.6	15	28	—	—
Mechanical	—	—	4	13	23	80	—
Wet	—	7.5	10	30	55	150	—
High-voltage electrostatic precipitator	—	—	—	85	120	265	415
Low-voltage electrostatic precipitator	—	13	24	105	200	—	—
Fabric filter:							
High temperature (550°F)	—	—	30	88	155	430	720
Medium temperature (250°F)	—	—	15	45	82	225	375
Afterburner, direct flame catalytic	8.2	12	18	—	—	—	—
	16	20	29	—	—	—	—

Source: AP-51.

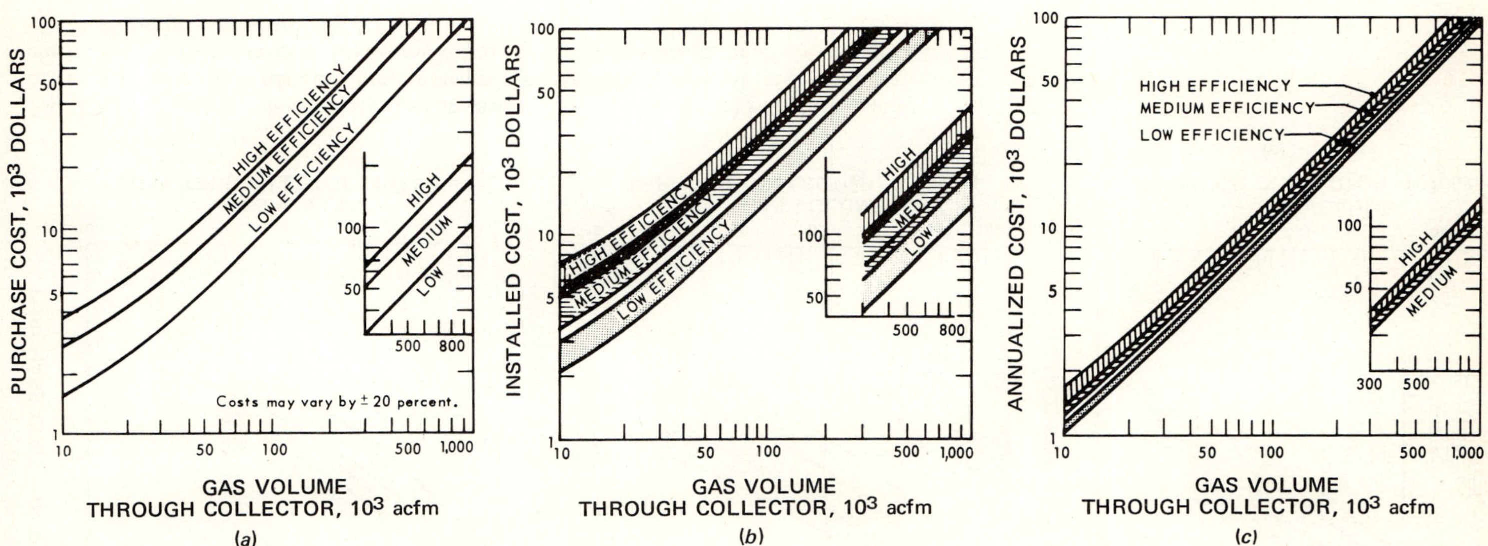

Fig. 10.13 Costs of centrifugal collectors. (*From AP-51.*) (*a*) Purchase cost of dry centrifugal collectors. (*b*) Installed cost of dry centrifugal collectors. (*c*) Annualized cost of operation of dry centrifugal collectors.

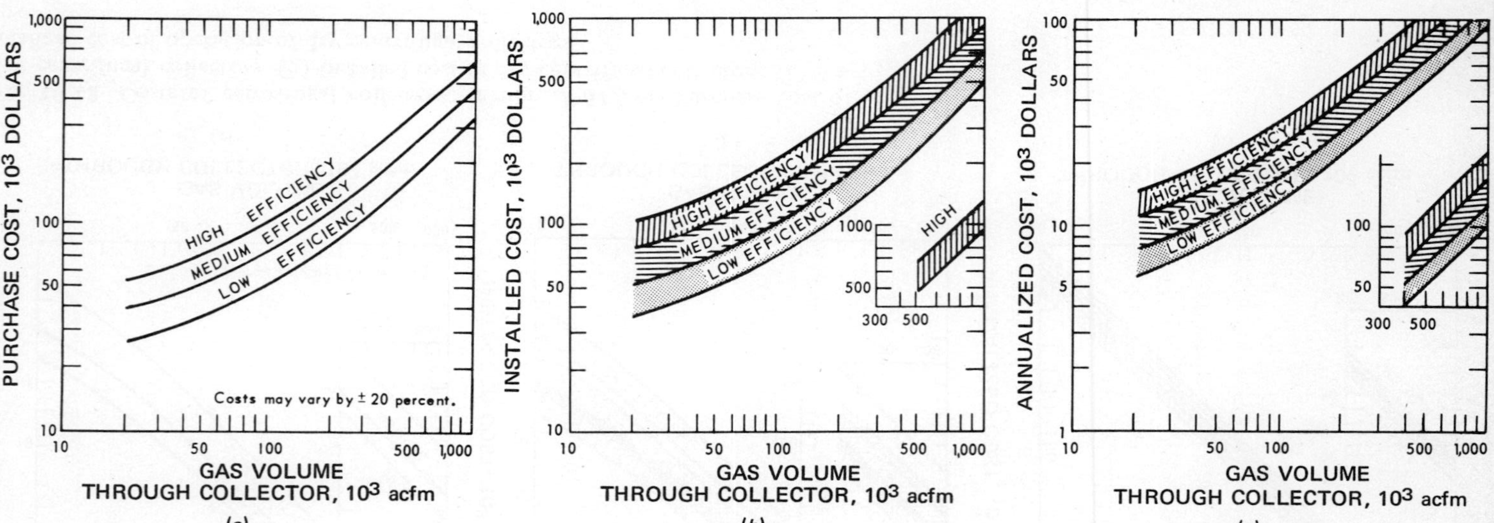

Fig. 10.14 Costs of electrostatic precipitators. (*From AP-51.*) (*a*) Purchase cost of high-voltage electrostatic precipitators. (*b*) Installed cost of high-voltage electrostatic precipitators. (*c*) Annualized cost of operation of high-voltage electrostatic precipitators.

purchase cost \$620,000
installed cost \$1,200,000
annualized cost \$200,000

Typical collection efficiency data for particulate control equipment is noted in Table 10.9 from Duprey, AP-42, 1968. These data have been based on a standard silica dust with a particle density of 2.7 g/cm^3 and with the following size distribution:

particle size range, microns	percent by weight
0–5	20
5–10	10
10–20	15
20–44	20
>44	35

TABLE 10.9 Collection efficiency of particulate control equipment

Collector type	Efficiency, %					
	Overall	0–5	5–10	10–20	20–44	>44
Baffled settling chamber	58.6	7.5	22	43	80	90
Simple cyclone	65.3	12	33	57	82	91
Long-cone cyclone	84.2	40	79	92	95	97
Multiple cyclone—12-in. diameter	74.2	25	54	74	95	98
Multiple cyclone—6-in. diameter	93.8	63	95	98	99.5	100
Irrigated long-cone cyclone	91.0	63	93	96	98.5	100
Electrostatic precipitator	97.0	72	94.5	97	99.5	100
Irrigated electrostatic precipitator	99.0	97	99	99.5	100	100
Spray tower	94.5	90	96	98	100	100
Self-induced spray scrubber	93.6	85	96	98	100	100
Disintegrator scrubber	98.5	93	98	99	100	100
Venturi scrubber, 30-in. pressure drop	99.5	99	99.5	100	100	100
Wet impingement scrubber	97.9	96	98.5	99	100	100
Baghouse	99.7	99.5	100	100	100	100

Source: AP-42.

Duprey comments that this standard dust is similar to that from coal fired furnaces. The ability of wet collectors to clean out small particles is evident from this table.

In concluding this chapter we should note that all of these clean up methods require the use of substantial amounts of power either to push the gas through the device or to provide the collection mechanism as in the precipitator. Yet the very production of power is a substantial source of air pollution both from particulates and gases. It reminds one of the Second Law of Thermodynamics: "you can't even break even!".

REFERENCES

AP-40: "Air Pollution Engineering Manual," National Center for Air Pollution Control, PHS Publication 999-AP-40, 1967.

AP-42: "Compilation of Air Pollutant Emission Factors," National Air Pollution Control Administration, PHS Publication 999-AP-42, 1968.

AP-49: "Air Quality Criteria for Particulate Matter," National Air Pollution Control Administration, Publication AP-49, 1969.

AP-51: "Control Techniques for Particulate Air Pollutants," National Air Pollution Control Administration, Publication AP-51, 1969.

AP-61: "Characteristics of Particulate Patterns," National Air Pollution Control Administration, Publication AP-61, 1970.

American Industrial Hygiene Association: "Air Pollution Manual," pt. 2: Control Equipment, 1968.

Fuchs, N. A.: "The Mechanics of Aerosols," Pergamon Press, Oxford, 1964.

Green, H. L., and W. R. Lane: "Particulate Clouds: Dusts, Smokes, and Mists," 2d ed., E. and F. N. Spon, London, 1964.

Slade, D. H. (ed.): "Meteorology and Atomic Energy," TID-24190, July, 1968.

Stern, A. C. (ed.): "Air Pollution," 2d ed., vol. 1, Academic Press, New York, 1968.

Visibility Bibliography

AP-49, chap. 3; AP-50, chap. 1 (see "References" at end of Chap. 11).

Conner, W. D., and J. R. Hodkinson: "Optical Properties and Visual Effects of Smoke Stack Plumes," PHS Publication 999-AP-30, 1967.

Charlson, R. J., N. C. Ahlquist, and H. Horvath: On the Generality or Correlation of Atmospheric Aerosol Mass Concentration and Light Scatter, *Atmos. Environ.*, vol. 2, pp. 455–464, 1968.

Green, H. L. and W. R. Lane: chap. 4 (see "References" above).

Horvath, H. and K. E. Noll: The Relationship Between Atmospheric Light Scattering Coefficient and Visibility, *Atmos. Environ.*, vol. 3, p. 543, 1969.

Middleton, W. E. K.: "Vision through the Atmosphere," University of Toronto Press, Toronto, 1963.

Noll, K. E., P. K. Mueller, and M. Imada: Visibility and Aerosol Concentration in Urban Air, *Atmos. Environ.* vol. 2, pp. 465–475, 1968.

Robinson, E.: chap. 11 in Stern (see "References" above).

PROBLEMS

10.1 Develop the expression for the free fall of a spherical particle for the case when the particle size is bigger than the mean free path. Compare your results to those of Eq. (10.1).

10.2 For the case of a 100-micron diameter particle of unit density calculate the terminal free-fall velocity assuming Stokes flow. Calculate the Reynolds number and check to see if Stokes flow holds. Is your calculated velocity too high or too low compared to that which would actually exist?

10.3 Plot up the Cunningham correction factor for sea-level conditions for $0.01 < R < 0.5$ microns. If we are interested in the fallout of particles greater than 0.1μ is this correction important?

10.4 Using the rate equation for coagulation as

$$\frac{dn}{dt} = -kn^2$$

determine the time for one-half of the particles to coagulate if the initial concentrations are 10^{11}, 10^9, 10^7, and 10^5 particles/cm^3. Take $k = 3 \times 10^{-10}$ cm^3/sec. What can you conclude about making measurements of concentration and size distributions in particulate clouds?

10.5 For the cases of 10^4 and 10^6 particles/cm^3, find the coagulation after 1 hour if $k = 3 \times 10^{-10}$ cm^3/sec.

10.6 Calculate the COH if there is a 50 percent transmittance of light for 20,000 linear ft of air through the filter. What is the COH if the transmittance is 90 percent for 1,000 linear ft of air? For comparison, a COH between 0 and 1 represents slight pollution, from 1 to 2 moderate, from 2 to 3 heavy, and above 3 represents very heavy pollution.

10.7 The terminal velocity for Stokes flow has been calculated in this chapter. Determine how this velocity is approached as a function of time for low Reynolds number flow, that is, Re less than 1. Does the particle travel far in this time if its initial velocity is zero?

10.8 Consider the gravitational fallout of aerosols in a settling duct with a laminar flow velocity profile. Calculate the particle trajectory and determine the length of tube required for complete fallout. Does this length depend on the velocity profile in the duct?

10.9 For a gravitational settling chamber determine the length required to collect 50-micron diameter particles if the flow rate is 30,000 cfm, the chamber depth is 15 ft, the chamber height is 5 ft, and the particle specific gravity is 2.5. Assume the air temperature is 80°F. Good design practice calls for a bulk velocity of less than 10 ft/sec. Does this design meet that requirement?

10.10 Develop the expression for the particulate radial velocity in a cyclone as given in Eq. 10.10.

10.11 The removal of dust collected in a cyclone must be done so that no atmospheric air enters through the collector into the low-pressure vortex to disrupt the swirl flow. Design a device to allow the dust to be removed without allowing any air leakage.

10.12 A "Little Wonder" cyclone has been specified for a small industrial operation to operate on 5,000 cfm of air at approximately 70°F and 1-atm pressure. The diameter is 16 in. and the capital cost is 20 cents per cfm. Electrical operating costs and initial costs are to be determined if the cyclone requires 0.50 hp/in. water pressure drop per 1,000 cfm; industrial electricity costs 1.5 cents per kw-hr, and the cyclone will operate 8 hours per day.

10.13 In the design of wet scrubbers the disposal of the waste is often a difficult problem. The following articles indicate methods to do this. Write a brief summary of such methods.

Conway, R. A. and V. H. Edwards: How to Design Sedimentation Systems from Laboratory Data. *Chem. Eng.*, vol. 68, pp. 167–170, Sept. 18, 1961.

Rickles, R. W.: Waste Recovery and Pollution Abatement *Chem. Eng.*, vol. 72, pp. 133–152, Sept. 27, 1965.

Busch, A. W.: Liquid-Waste Disposal System Design *Chem. Eng.*, vol. 68, pp. 83–86, Mar. 29, 1965.

10.14 For a large, electrostatic precipitator installation on a power plant the purchase cost is about 1 dollar per cfm for 95 to 97 percent efficiency. Estimate the cost of a unit for a 97 percent efficiency collector for a new 1,000-megawatt plant. How much would this cost increase if we desire to increase the collection efficiency to 99.9 percent?

10.15 In the theoretical efficiency equation for electrostatic precipitators $\eta = 1 - e^{-AV/Q}$, V, the drift velocity, can be treated as a parameter to be determined empirically. For fly ash from power plants typical values are 0.3 to 0.5 ft per sec. Plot up the efficiency of such an ESP plate-type unit which is 20 ft high with 10-ft wide plates for velocities through the unit between 2 and 8 ft per sec and a plate spacing of 2 in. Why aren't very low bulk flow velocities used for the air stream to maximize the efficiency?

10.16 The Tastee coffee company has hired you to aid in the design of their new plant which will include both ground coffee and instant coffee lines. Discuss the necessary air pollution control equipment which will be required for this plant assuming that it is to be placed near a residential area. (AP-40, pp. 746–750.)

10.17 The Hot Batch asphalt company desires to build a new plant near the airport to meet requirements for runway and other industrial construction. Discuss the air pollution control equipment which would be needed for such a plant. (AP-40, pp. 325–333.)

10.18 Estimate the purchase, installed, and operating costs of a "U-Clean-Em" high-efficiency cyclone to be used on a 50,000 cfm operation run on a 1-shift-per-day basis. The manufacturer has specified that the cyclone will operate with a 3-in. water pressure drop.

10.19 Plot up the free-fall velocity for particles of specific gravity of 1, 2, and 3 assuming Stokes law behavior. Compare to water droplet velocities given here as

diameter, μ	50	100	500	1,000
velocity, ft/hr	900	3,100	24,000	46,000

Also compare to the results given in Fig. 10.1. Where does Stokes law begin to fail?

10.20 An air sampling station is located at an azimuth of 203° from a cement plant at a distance of 1,500 meters. The cement plant releases fine particulates (less than 15 micron diameter) at the rate of 750 lb per hr from a 30-meter stack. What is the contribution from the cement plant to the total suspended particulate concentration at the sampling station when the wind is from 30° at 3 m/sec on a clear day in the late fall at 1600?

10.21 Estimate the total particulate one might breathe *in*, in a day, in a polluted atmosphere.

11

SULFUR OXIDES

I am sure you can appreciate our position at this time. We wanted to spend our pollution control money wisely, we wanted to be able to perform better than the State Rules and Regulations which were yet to be written, and we had to have a system that would reliably operate, since this plant addition would represent 30 percent of our system capacity, and we could not afford the unit off the line.

D. M. Miller, 1969

The sulfur oxides, another common atmospheric pollutant, are produced by the combustion of fuels, particularly those in fossil-fueled power plants. In this chapter we shall discuss the sources and control techniques for the sulfur oxides. The compounds of major interest are SO_2, sulfur dioxide; SO_3, sulfur trioxide; H_2SO_3, sulfurous acid; and H_2SO_4, sulfuric acid. The sulfur salts, for example, $CuSO_4$, $CaSO_4$, and $MgSO_4$ are also important compounds and can result from power-plant and industrial processes, notably smelters. Some sulfur salts are also end products of control techniques in use on power plants to reduce SO_x emissions.

Sulfur dioxide is a nonflammable, colorless gas. The concentration required for taste detection ranges from 0.3 to 1 ppm in air and the odor threshold is about 0.5 ppm. The gas has a "pungent, irritating" odor. Table 11.1 lists the properties of SO_2. The conversion between ppm by volume and $\mu g/m^3$ is

$$1 \text{ ppm} = 2620\ \mu g/m^3 \text{ at } 25^\circ C, 760 \text{ mm Hg}$$

TABLE 11.1 Physical properties of sulfur dioxide

Molecular weight	64.06
Density (gas), g/liter	2.927 at 0°C; 1 atm
Specific (liquid) gravity	1.434 at −10°C
Molecular volume (liquid), ml	44
Melting point, °C	−75.46
Boiling point, °C	−10.02
Critical temperature, °C	157.2
Critical pressure, atm	77.7
Heat of fusion, Kcal/mole	1.769
Heat of vaporization, Kcal/mole	5.96
Dielectric constant (practical units)	13.8 at 14.5°C
Viscosity, dyne sec/cm^2	0.0039 at 0°C

Source: AP-50.

$$1 \text{ ppm} = 2860\ \mu g/m^3 \text{ at } 0°C, 760 \text{ mm Hg}$$

$$1\ \mu g/m^3 = 3.5 \times 10^{-4} \text{ ppm (vol) at } 0°C, 760 \text{ mm Hg}$$

In power-plant combustion processes SO_2 forms in the ratio of 40–80 parts per part of SO_3 (AP-50, 1969). The major oxide emitted is therefore SO_2. The term SO_x is used to denote the mixture of sulfur oxides emitted into the atmosphere.

Once in the atmosphere, several reactions may occur to remove SO_2 from the air. For example, in the presence of metallic oxides

$$SO_2 \xrightarrow{1/2\ O_2} SO_3 \tag{11.1}$$

However, some metal oxides oxidize SO_2 directly to sulfate as for example

$$4\ MgO + 4\ SO_2 \rightarrow 3\ MgSO_4 + MgS \tag{11.2}$$

SO_2 also reacts with water to form H_2SO_3, sulfurous acid, as

$$SO_2 + H_2O \rightarrow H_2SO_3 \tag{11.3}$$

Sulfur trioxide reacts with water vapor to form sulfuric acid as

$$SO_3 + H_2O \rightarrow H_2SO_4 \tag{11.4}$$

The absence of high concentrations of SO_2 in the atmosphere at some distance from sources of this pollutant does not mean that little or no SO_2 has been emitted. It means that SO_2 has been converted to sulfates and sulfuric acid as noted above. So the concentrations of sulfate salts and H_2SO_4 also need to be measured. As mentioned in Chap. 5, the compound(s) which is responsible for oxidizing SO_2 to SO_3 is not yet known. It is known that the oxidation of sulfur dioxide is affected by the presence of metallic oxides and metallic ions, the relative humidity, and (for photochemical oxidation) the insolation. The presence of hydrocarbons and nitrogen dioxide also affects the photooxidation rate of SO_2. Photochemical processes can be affected by SO_2, and aerosol formation is especially a function of the amount of SO_2 present. One aerosol formed is sulfuric acid which absorbs water to form a sulfuric acid mist capable of greatly reducing visibility.

SULFUR DIOXIDE SOURCES

In 1968 an estimated 33×10^6 tons of sulfur oxides were emitted in the United States. Of this 24×10^6 tons were produced by the burning of fuels: 20×10^6 tons from coal combustion, and 4.3×10^6 tons from fuel-oil combustion. Power plants produced 16.8×10^6 tons, or 50 percent of the national total. Table 11.2 indicates the sources over the 3-year period 1966–1968. Estimates of future emissions vary. However, one estimate (Rohrman and Ludwig, 1967) suggests that the total sulfur oxide emission in the year 2000 might be 60×10^6 tons, of which about 40×10^6 tons would be from power plants. The concern with sulfur oxide emission is that at even moderate concentration both plants and animals (man) are adversely affected. As our power needs grow, the emission of sulfur pollutants also grows and areas of significant SO_2 levels in the country are expanding. On the other hand, plants do require sulfur as a necessary input, like nitrogen and phosphorous, and one recent paper suggests that large but diffuse emission of sulfur in *rural* areas may be beneficial (Ross, 1971). The effects of SO_2 on plants will be discussed in Chap. 13. It is sufficient to say here that while some sulfur is required by plants, a concentrated dose of SO_2, such as from a power plant or smelter plume, will cause severe damage. This is especially true of field crops such as alfalfa and cotton.

Coal

In Chap. 3 we calculated the SO_2 emitted from a large power plant using coal, residual fuel oil, and natural gas as the fuel. In that particular example a low sulfur coal, 0.5 percent S, was used for the 2,250-megawatt plant. The same amount of emission would occur for a

TABLE 11.2 Nationwide emissions of sulfur oxides by year *(10^6 tons)*

Source	1966	1967	1968	Change from 1966 to 1968
Transportation	0.6	0.7	0.8	+0.2
Motor vehicles	0.2	0.3	0.3	+0.1
Other	0.4	0.4	0.5	+0.1
Fuel combustion	22.5	23.1	24.4	+1.9
Coal	18.7	19.1	20.1	+1.4
Fuel oil	3.8	4.0	4.3	+0.5
Natural gas	N*	N	N	N
Wood	N	N	N	N
Industrial processes	7.1	7.2	7.3	+0.2
Solid waste disposal	0.1	0.1	0.1	N
Miscellaneous	0.6	0.6	0.6	N
Manmade	0.6	0.6	0.6	N
Forest fires	N	N	N	N
Total	30.9	31.7	33.2	+2.3

*N = Negligible.
Source: AP-73.

1,000-megawatt plant if the sulfur content of the coal was 1.12 percent, namely, 17,400 lb of SO_2 per hour or 76,000 tons per year. Typical coal usage in the United States by electrical utilities is 42×10^6 tons of < 1 percent S coal, 138×10^6 tons of 1.1 percent–3 percent S coal, and 48×10^6 tons of >3 percent S coal (Perry and Decarlo, 1967). On the average the coal used by the utility industry before enforcement of the 1970 law was about 2.5 percent sulfur.

Sulfur occurs in coal as iron pyrites, FeS_2, organic compounds, and sulfates. Only the first two forms are of major importance. After the coal is crushed, pyritic sulfur can be partially removed since the iron pyrites are separable by gravitational techniques. Figure 11.1 shows the coal sulfur content, by type, as a function of percent number of mines sampled. Note that only a small amount of sulfur can be removed short of gasification or liquefaction of the coal. However, the water wash gravity cleaning can reduce the sulfur content by about one third.

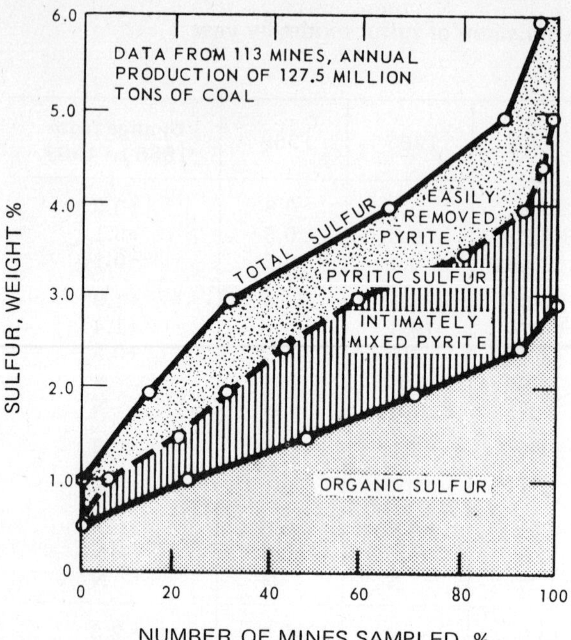

Fig. 11.1 Maximum sulfur content versus percent of mines sampled. (*AP-52.*)

Oil

Crude oil contains varying amounts of sulfur. Refining processes leave most of the sulfur in the heavier distillate fractions which have the highest boiling temperatures. Thus the sulfur content of residual fuel oil may be 4 to 6 times that of crude oil. Residual fuel oil, commonly burned in power plants, is grade 6; used in marine applications it is called Bunker C. The sulfur content would be typically between 0.75 and 2.5 percent. Table 11.3 indicates *crude* oil production in the United States as a function of sulfur content. The average sulfur content of residual fuel oil produced from domestic crude oil is 1.76 percent (AP-52). One can see the degree of concentration. Table 11.4 shows the sulfur content of imported *residual* fuel oils. Crude oil from Libya, Nigeria, and Indonesia are all low in sulfur and are not noted on this table. The point to be noted is that there is a limited amount of low sulfur oil available, without desulfurization, and the demand for such oil is increasing as the United States and other nations attempt to control emissions of sulfur compounds. In 1966 approximately 600 million barrels of residual fuel oil were burned in the United States and

TABLE 11.3 United States crude oil production by area and sulfur content category—1966

Area	Crude oil production, percent of U.S.	Annual crude oil production, 10^6 bbl					
		Sulfur, weight percent					
		0.00–0.25	0.26–0.50	0.51–1.00	1.01–2.00	2.00	Total
Alaska	0.47	14.3					14.3
Appalachia	0.68	20.4		0.1			20.5
California	11.34	25.7	40.7	110.1	122.5	45.4	344.4
Gulf Coast	29.52	568.4	328.4	0.4			897.2
Illinois, Indiana, Kentucky	3.02	67.3	24.2				91.5
Michigan	0.47	12.0	0.9	1.2	0.1		14.2
Mid-Continent	23.09	215.2	227.9	145.1	79.6	33.8	701.6
Rocky Mountain	9.08	140.4	23.1	47.7	14.0	50.7	275.9
Southeastern U.S.	2.14	13.8	20.0	1.5	7.1	22.5	64.9
West Texas and southeast New Mexico	20.19	150.2	107.4	90.2	171.3	94.4	613.5
Totals	100.0	1227.7	772.6	396.2	394.6	246.8	3037.9

Source: AP-52.

TABLE 11.4 Residual fuel oil imports into United States 1964–1966

Country of origin	Imports, 10^3 bbl			1966 average sulfur, percent
	1964	1965	1966	
Venezuela	142,256	180,538	194,676	2.2
Netherlands Antilles (Aruba, Curaçao)	95,182	103,645	100,101	2.46
British West Indies (Trinidad and Tobago)	36,527	37,600	44,614	1.93
Mexico	6,684	5,839	6,067	4.4
Italy	12	422	5,264	2.8
Puerto Rico	4,787	4,371	4,749	2.2
Argentina	1,290	2,945	4,346	1.0
Colombia	1,485	3,090	3,515	1.55
England	—	95	2,109	3.5
Canada	1,826	1,964	1,880	2.65
Netherlands	117	41	1,285	3.00
Panama	1,541	1,231	1,113	2.00
Kuwait	—	—	1,093	—
Others	4,184	3,406	5,983	—
Totals	295,891	345,187	376,795	—

Source: AP-52.

more than 80 percent of this contained at least 2 percent sulfur (AP-52). The largest uses were for space heating, industry, and power generation.

There are several chemical processes used to desulfurize crude oil. Some refineries have installed equipment which reduces residual fuel oil sulfur content to less than 1 percent. The simplest means, however, to produce such oil is to blend high sulfur residuals with low sulfur distillate oils; this scheme produces much of the low sulfur oil used on the East Coast. AP-52 discusses the state of the art of chemical desulfurization. Ironically, some studies indicate that desulfurization of fuel oil may bring the oil companies a profit since the sale of the by-products, H_2S or sulfur, and the increased production of number 2 distillate fuel oil offset the cost of running the desulfurization plants.

Natural Gas

Natural gas normally has a low sulfur content and consists of a mixture of the lighter hydrocarbons, particularly methane, CH_4. The power plant example of Chap. 3 indicated the desirability of using this fuel. Unfortunately, natural gas is the fuel in shortest supply in the United States. When appreciable sulfur does exist in the gas it is removed before sale.

Information in Table 11.2 indicates that while combustion of fuel, particularly in the generation of power, produces the largest part of the sulfur emissions, other significant sources do exist. The example in Chap. 8 of the copper smelter can be reviewed to indicate another source of SO_2 emissions. As a rule of thumb, with every pound of copper produced 2 pounds of SO_2 go up the stack.

NATURAL SOURCES

Some statistics comparing man-caused sulfur emissions to those from natural sources were given in Chap. 3. SO_2 world emissions have been calculated by Robinson and Robbins (1970) as 146×10^6 tons per year of which 70 percent is from coal combustion and 16 percent from petroleum combustion. Ninety-three percent of the SO_2 is emitted in the Northern Hemisphere. Total *sulfur* emissions are estimated at 220×10^6 tons per year from such gases as H_2S, SO_2, and marine sulfate compounds. The pollutant emissions of sulfur are only about 73×10^6 tons (half of the 146×10^6 tons) so natural sources of sulfur are still greater than emissions resulting from the combustion process. This is not, however, to minimize the adverse effects of concentrations of SO_2 particularly in urban areas. But on a global basis most sulfur emissions are still naturally caused.

According to Robinson and Robbins (1970), the lifetime of H_2S is about 2 hours in urban atmospheres and 2 days in remote areas. They estimate the lifetime of SO_2 to be about 4 days. The latter figure is difficult to determine since there are a number of removal mechanisms for SO_2. It can be washed out directly by rain or deposited directly on vegetation. Under the action of the nitrogen oxides and hydrocarbon compounds SO_2 can be converted into H_2SO_4, forming an aerosol haze. The presence of fine metallic particles will aid in oxidizing SO_2 to SO_3 which subsequently combines with a water molecule to form H_2SO_4. Lastly, there is evidence that in foggy or very moist atmospheres SO_2 can react with ammonia (present in the atmosphere in small amounts) to form ammonium sulfate, $(NH_4)_2SO_4$, and this is subsequently removed from the atmosphere by washout. All these various removal mechanisms cause a short lifetime for SO_2 in the atmosphere.

AMBIENT CONCENTRATIONS

Ambient sulfur dioxide concentrations for six United States cities are noted in Fig. 11.2 where the SO_2 concentrations in ppm are plotted versus the percent of time a given concentration is exceeded. The results follow a log-normal distribution, that is, a straight line on this plot. For example, from 1962 to 1967 Chicago had SO_2 readings of 0.09 ppm, averaged over 1 hour, for more than 50 percent of the time. Readings of 0.5 ppm were exceeded approximately 3 percent of the 1-hour periods and therefore 97 percent of the 1-hour periods had readings of *less* than 0.5 ppm. The actual *peak* 1-hour reading is not given on such a plot. For Chicago between 1962 and 1967 the peak 1-hour SO_2 concentration was 1.69 ppm. In contrast, San Francisco had a peak 1-hour concentration of 0.26 ppm during the period of 1962–1964. The concentrations, as noted on the figure, also indicate the lower SO_2 levels for San Francisco.

Data such as these can be presented in many ways. A second method is shown in Fig. 11.3, where Chicago data for the 1-year period December 1, 1963 to December 1, 1964 are given for different averaging times. First note that as the averaging time gets larger the spread in the data, given as the crosses and triangles, gets smaller. For a 1-hour averaging time the minimum concentration is about 2×10^{-2} ppm and the maximum concentration is about 1.5 ppm. The solid lines are the results of calculations, using statistical methods, to predict the concentrations as a function of the percent of time that these will be exceeded. The values along the top axis are calculated maximum concentrations during a 1-year period and are higher than those values actually observed. This plot is taken from AP-50 which describes it as follows:

> A plot of sulfur dioxide concentration data for Chicago from December 1, 1963 to December 1, 1964 is given in the figure. Averaging time is on a logarithmic scale, as is concentration in ppm, while frequency (percent) is given on a normal probability scale. The crosses and triangles in the figure denote observed values while the lines represent calculated concentrations. For example, for an averaging time of 1 hour, 30 percent of the 1-hour-duration samples obtained during the year exceeded a concentration of about 0.2 ppm, and 0.1 percent of these samples showed a concentration in excess of 1 ppm. (These two points are marked with asterisks in the figure.) The 13 lines in the figure represent the 11 percentiles and the maximum and the minimum lines. In addition, several expected maximum concentration values are listed at the top of the figure. Thus the middle line represents the expected 50 percentile (or median) line, i.e., the concentration values which one would expect to be exceeded half the time at different averaging times. For example, for a 1-hour-averaging time this concentration is about 0.12

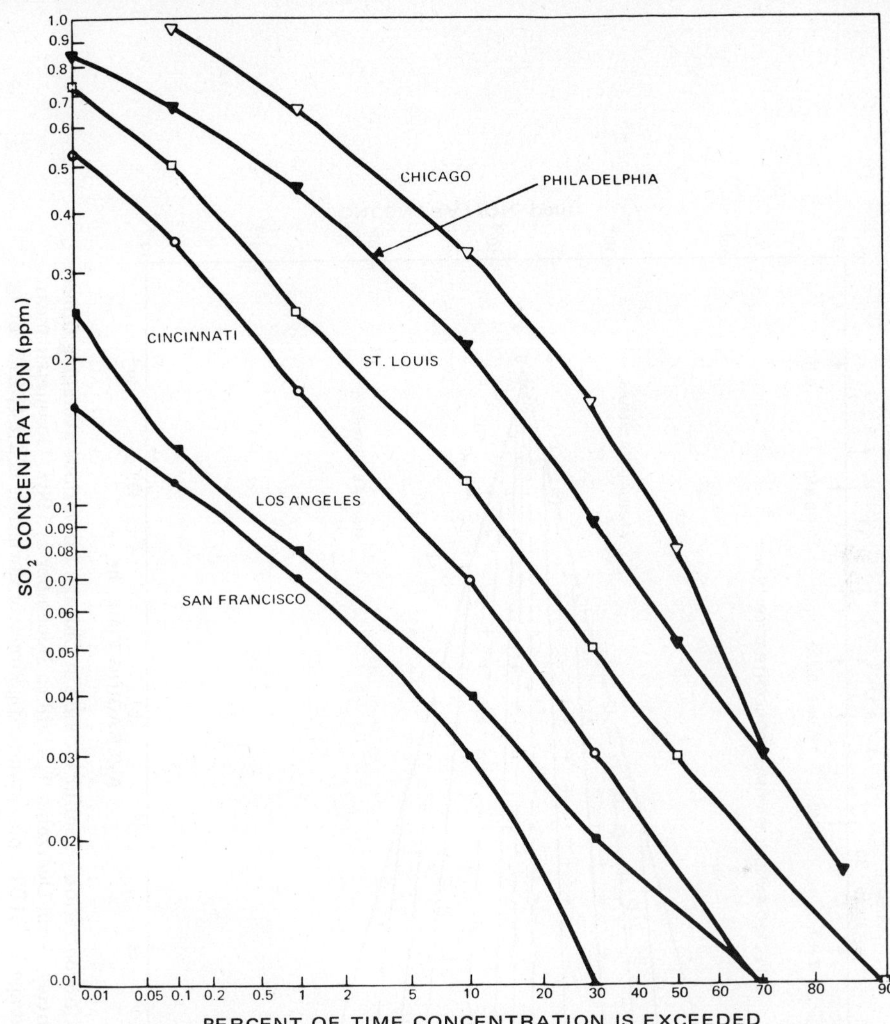

Fig. 11.2 Frequency distribution of sulfur dioxide levels in selected American cities, 1962 to 1967 (1-hour averaging time). The approximate log normality of the distribution of sulfur dioxide concentrations is shown by the rather straight lines of the distribution functions when these are plotted on a logarithmic scale (concentrations) against a normal distribution frequency. (*AP-50.*)

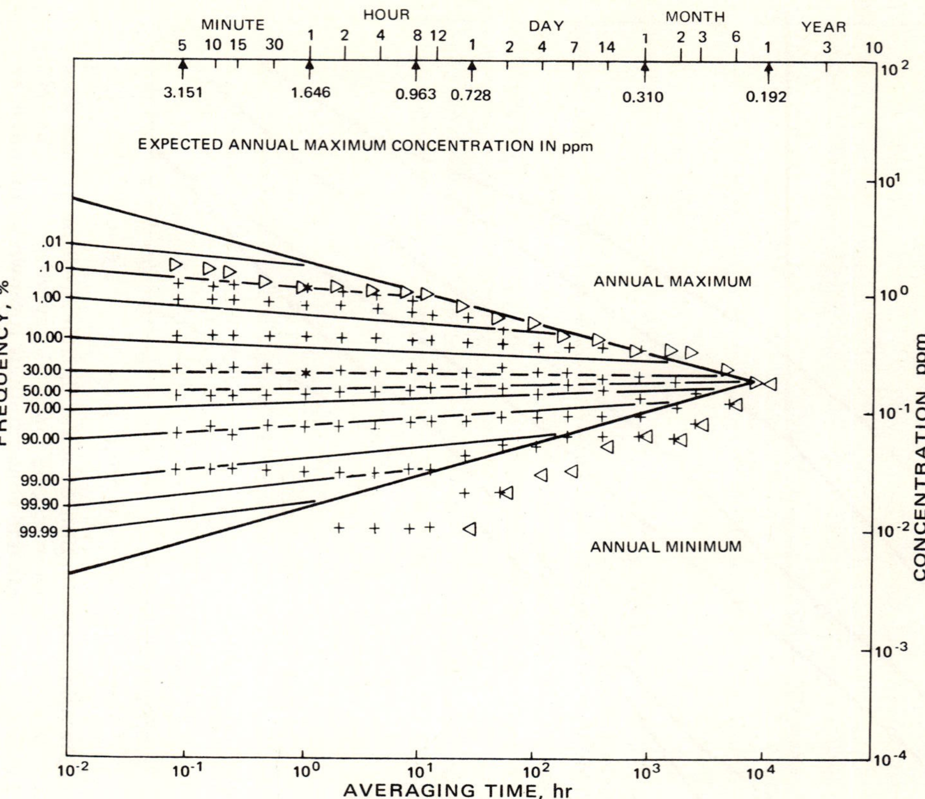

Fig. 11.3 Concentration of sulfur dioxide in Chicago for various averaging times and frequencies, from December 1, 1963 to December 1, 1964. Geometric mean for 1 hr average is 0.160 ppm; standard geometric deviation is 1.84; 87% of hr have data available. (*AP-50.*)

ppm. At several points, percentile lines will coincide with maximum or minimum concentration lines. For instance, when the averaging time is 0.88 hour and the total time period is 1 year, the 0.01 percentile will correspond to the maximum, since there are 10,000 of these time-units in a year; similarly, the 99.99 percentile corresponds to the minimum.

Put another way, the odds that one would find the maximum concentration if one averaged over any 1 hour of the year are 1/8760 = 0.011 percent; over 1 day, 24/8760 = 0.27 percent. Such a plot is valuable both because much information can be concisely presented, and the variation in concentration as a function of averaging time can be quickly determined. The latter is an important variable in setting ambient air quality standards. In the United States there is a requirement for SO_2 level for an annual average, a 24-hour average (not to be exceeded more than once a year), and a 3-hour average (not to be exceeded more than once a year). The 3-hour average is for the secondary standards.

Let us now compare these measurements to the new federal ambient air quality standards. These are noted in Table 11.5.

The measured results for 1963–1964 for Chicago (Fig. 11.3) indicate that the 1-year average is about 0.2 ppm which is a factor of 10 greater than the standard. The yearly averages for 1962–1967 were respectively 0.10 ppm, 0.14 ppm, 0.18 ppm, 0.13 ppm, 0.09 ppm, and 0.12 ppm. The data of Fig. 11.3 include the highest readings, but even the other years shown exhibit concentrations far higher than the federal

TABLE 11.5 SO_2 ambient air quality standards

	Primary standard	Secondary standard	Measurement method
Annual average	80 $\mu g/m^3$ (0.03 ppm)	60 $\mu g/m^3$† (0.02 ppm)	pararosaniline (West-Gaeke)
24-hour average*	365 $\mu g/m^3$ (0.14 ppm)	260 $\mu g/m^3$ (0.10 ppm)	
3-hour average*	—	1300 $\mu g/m^3$ (0.50 ppm)	

*Not to be exceeded more than once a year.
National primary standards: the levels of air quality necessary, with an adequate margin of safety, to protect the public health.
National secondary standards: the level of air quality necessary to protect the public welfare from any known or anticipated adverse effects of a pollutant.
†In May 1973 EPA proposed to revoke the secondary annual average and subsequently did so.

TABLE 11.6 SO_2 concentrations in selected cities

	Maximum 1-day average for the year in ppm						Percent time given value is exceeded	
	1962	63	64	65	66	67	1%	30%
Chicago	0.36	0.71	0.79	0.55	0.48	0.65	0.50	0.16
Cincinnati	0.11	0.11	0.14	0.18	0.10	0.13	0.10	0.04
St. Louis	—	—	0.26	0.19	0.18	0.21	0.14	0.05
Los Angeles	0.06	0.07	0.10	—	—	—	0.06	0.02

Source: AP-50, 1969.

standards. The 24-hour standard of 0.10 ppm was exceeded about 80 percent of the time. Clearly Chicago has a difficult task ahead to reach the goal set by federal standards. The data in Fig. 11.2 are for 1-hour averaging times and thus are not directly applicable to the federal standards. One-day-average values for several cities are noted in Table 11.6.

The data show that among these cities only Los Angeles meets the SO_2 daily standard. Cincinnati appears to meet the primary standard but will require some strategy to meet the secondary standard. St. Louis has some way to go and Chicago has the longest pull ahead. These SO_2 concentrations were determined by the conductivity method discussed in the next section.

TEST METHODS

There are a variety of sampling techniques for measuring ambient concentrations of SO_2. In the West-Gaeke method, also called the *colorimetric* or *pararosaniline* method, sulfur dioxide is absorbed in dilute aqueous sodium tetrachloromercurate to form the nonvolatile dichlorosulfitomercurate ion. This reacts with formaldehyde and bleached pararosaniline to form red-purple pararosaniline methylsulfonic acid. The color intensity of the dye is measured at 560 mμ (millimicrons) and is proportional to the concentration of SO_2. The reaction is specific to SO_2 and sulfite salts and can be used for SO_2 concentrations from 0.002 ppm to 5 ppm (5.7 $\mu g/m^3$ to 14,300 $\mu g/m^3$). Ozone and nitrogen dioxide reduce the apparent concentration and can significantly affect SO_2 determinations. Even so, EPA uses this technique as its reference method.

The *conductometric* method, specified for the California standards, uses the oxidation of SO_2 to H_2SO_4 by aqueous hydrogen peroxide and the subsequent measurement of the increased electrical conductivity of the solution as a measure of the SO_2 level. The method is not specific to SO_2 and other pollutants can affect the conductivity of the solution. This technique has the advantage of being applicable to continuous automatic sampling. It is used for concentrations from 0.01 ppm to 20 ppm. The Thomas autometer is a commercial instrument which uses this technique and is referred to often in the literature.

The method commonly used in Europe (sometimes referred to as the OECD standard) is the hydrogen peroxide-acid titration method and is applicable to SO_2 concentrations in the 0.01 ppm to 10 ppm range. Air is bubbled through 0.03 N hydrogen peroxide solution with a pH of 5. The SO_2 is oxidized to H_2SO_4 which is then titrated with a standard alkali. The pH of 5 is used to accommodate methyl-red bromcresol-green indicator, green above 5 and red below 5. The method is affected by the presence in air of other acidic gases, which cause high results, or alkaline gases, which produce low results.

Katz (1969) indicates a variety of methods for determining SO_2 concentrations. The three discussed above are the major ones used. The values obtained by different methods are usually not in complete agreement because of different interferences which occur with each method.

CONTROL TECHNIQUES

We shall concentrate on SO_x control techniques for power plants since these are the largest source of sulfur dioxide emissions. Control techniques for sources such as smelters, paper plants, and sulfuric acid plants are briefly outlined in AP-50.

The physical processes in controlling gaseous emissions depend upon adsorption, absorption, and catalytic conversion. *Adsorption* is the capture and retention of molecules from the gas phase by the *surface* of a solid adsorbent. Surfaces include the exterior of the surface, cracks, and crevices. Since physical adsorption is a surface phenomenon one must provide as much surface area, per unit volume, as possible. Activated charcoal is an example of an adsorbent. The material removed is called the adsorbate. Odors are commonly removed from gases by the adsorption process.

Absorption, or scrubbing, is a diffusion process in which gas molecules are transferred into a liquid phase. A large surface area of the absorber and a good contact between the absorber and the absorbent maximize the process. The packed tower is a device which provides a large surface area. Water flows down through a tower packed with inert material rings (used to promote a thin liquid film), and the dirty gas is bubbled in counter flow up through the tower. Absorption in the scrubbing liquid removes the contaminate and the clean gas is exhausted out the top of the tower.

Catalytic conversion refers to the process of chemically changing a pollutant to another product through the action of a catalyst. Catalytic combustion is simply burning the waste in the presence of a catalyst which does not become a part of the end product. The role of the catalyst is to lower the normal combustion temperature required for the process.

These physical processes have been incorporated into various techniques for removing SO_2 from flue gas. Let us look at the typical statistics for a modern 1,000-megawatt power plant and see what kind of concentrations of SO_2 must be removed. AP-52 suggests that a power plant of this capacity, burning coal with a sulfur content of 2.5 to 3 percent will emit 1.7 to 2 million (standard) cubic feet per minute of flue gas with an SO_2 concentration of between 0.2 and 0.3 percent by volume, that is, 2,000 to 3,000 ppm.

Let us verify these figures. A 1,000-megawatt plant of 35 percent thermal efficiency will require a thermal input of

$$Q = \left(\frac{1{,}000}{0.35}\right)(3.4 \times 10^6) = 9{,}740 \times 10^6 \text{ Btu/hr}$$

Assuming that the coal produces 12,000 Btu/lb we require

$$\text{Coal} = \left(\frac{9{,}740}{12}\right) \times 10^3 = 810{,}000 \text{ lb/hr}$$

If the coal is 3 percent sulfur then we have

$$S = 0.03 \times 810{,}000 = 24{,}300 \text{ lb/hr}$$

How much air goes through the plant? Typically coal has 10 percent ash and is possibly 75 percent carbon so we can estimate the exhaust gas output from the following analysis

$$C + O_2 + 3.76N_2 \rightarrow CO_2 + 3.76N_2$$

The carbon will be (810,000)(0.75) = 606,000 lb/hr or 606,000/12 = 50,500 lb-moles/hr. The exhaust will therefore have 50,500 moles of CO_2 and 3.76 (50,500) moles of N_2 per hour. The exhaust gas thus consists of

$$50{,}500(44) + 3.76(50{,}500)(28) = 7.53 \times 10^6 \text{ lb/hr}$$

The actual exhaust also contains H_2O, SO_2, and many other gaseous products. If we neglect these and assume that the gas approximates air at 275°F then the density of the flue gas at exhaust conditions is

$$\rho = \frac{P}{RT} = \frac{(14.7)(144)}{(53.3)(735)} = 0.054 \text{ lb/ft}^3$$

The flue gas volume flow is then

$$\text{Vol} = \frac{(7.53 \times 10^6)}{0.054(60)} = 2.3 \times 10^6 \text{ acfm}$$

(However, for standard conditions the density is about 3/2 the density at 275°F so the volume flow is about 1.7×10^6 scfm.) The SO_2 concentration is calculated as

$$24{,}300 \times \frac{2SO_2}{S} = 48{,}600 \text{ lb/hr}$$

$$\rho = \frac{P}{RT} = \frac{(14.7)(144)}{(1545/64)(735)} = 0.12 \text{ lb/ft}^3$$

The volume of SO_2 to volume of exhaust is

$$\frac{SO_2}{\text{flue gas}} = \frac{(48{,}600/0.12(60))}{2.3 \times 10^6} = 0.29 \text{ percent}$$

Thus the given data are indeed accurate. We would like to remove as much SO_2 as possible, but at least 85–90 percent from the flue gas stream in which the SO_2 is only 0.3 percent by volume to start with! In contrast sulfuric acid plants use as a feed, gas which is 7–14 percent sulfur dioxide. It is the combination of a large amount of exhaust and a very small amount of SO_2 which causes difficulty in cleaning power-plant flue gas.

One method to reduce the sulfur compounds from stack gas is to switch to a low sulfur fuel. Natural gas is attractive in this respect but it is also in short supply. We have already discussed desulfurization of coal and oil and this is done at present, for example, when coal is crushed and part of the pyrites is removed. We can get an estimate of the economics involved if we consider buying coal for a 1,000-megawatt plant at 35 cents/

10^6 Btu for 1 percent sulfur coal and 27 cents/10^6 Btu for 3.3 percent sulfur coal. Our previous example indicated that we needed $9{,}740 \times 10^6$ Btu/hr for this plant. Fuel costs are therefore \$3,400 per hr for low sulfur coal and \$2,620 per hr for "standard" coal. At first this difference may seem negligible, but look at the cost per year. We have

$$\begin{aligned} 3{,}400 \times 8{,}760 &= \$29{,}600{,}000 \\ 2{,}620 \times 8{,}760 &= \underline{\$22{,}800{,}000} \\ \text{cost difference:} \quad &\ \ \$\ 6{,}800{,}000 \text{ per year} \end{aligned}$$

This difference of \$6,800,000 per year becomes even more significant if we consider a 30-year lifetime for the plant. Then the difference is \$204,000,000—a staggering amount and one which represents close to the capital cost of a new 1,000-megawatt plant. These figures also show why spending \$20,000,000 on clean-up systems (so that the high sulfur fuel can be used) is good economics, assuming that such a system can be purchased and is reliable. This is not yet the case although the techniques to be discussed are promising. The reader should refer to the quote at the beginning of this chapter as a reminder of the difficulties power companies face. It should also be pointed out that the use of nuclear power results in no effluent containing either sulfur compounds, carbon dioxide, or oxides of nitrogen. The new generation of nuclear plants has about the same thermal efficiency as fossil-fueled plants so the thermal discharge is the same. Nuclear plants do produce radioactive waste which must be concentrated and stored.

Figure 11.4 shows typical layout and temperature levels in a coal-fired power plant. The economizer is used to preheat the boiler feed water and the air preheater is used to heat the incoming combustion air. As a result the stack gas temperature is typically lowered to 275–300°F. Sulfur removal devices for existing plants would have to fit into this design, and this presents difficulties for several of the control techniques.

We have seen that total SO_2 emissions from all sources in the United States were about 33×10^6 tons per year for 1968. Slack (1967) indicates that total consumption of H_2SO_4 in the United States is about 22×10^6 tons. SO_2 can produce sulfur or sulfuric acid at the rates of 2 lb SO_2 to 1 lb S to 3 lb H_2SO_4. If complete sulfur oxide removal could be achieved and if the product were H_2SO_4 there would be a great surplus of the acid. Some sulfur oxide removal techniques can produce sulfur and the present United States market for sulfur is about 12×10^6 tons per year. If we consider the more reasonable goal of sulfur oxide removal only from power plants, then the potential

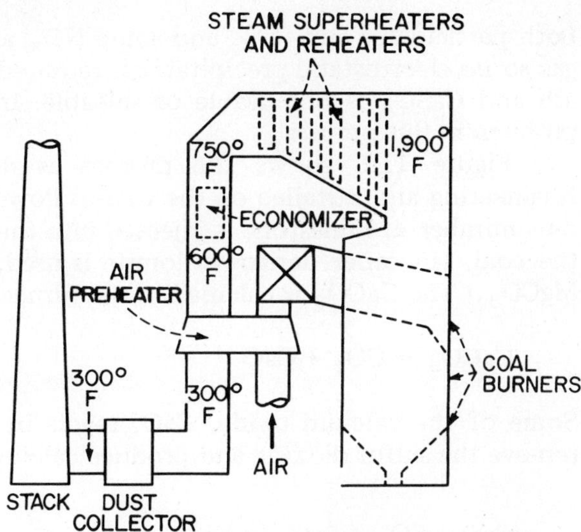

Fig. 11.4 Schematic of a coal-fired power plant.

sulfuric acid production is about 25×10^6 tons per year or just about the United States demand.

There are a bewildering variety of proposed SO_2 removal processes. Slack (1973) notes that 10 of these are in the testing stage on 100-megawatt or larger power plants; another 20 are in the prototype stage; and probably 75 pilot plant processes are being developed. We can roughly categorize the clean-up techniques as wet versus dry processes, and recovery versus throw-away processes. Ideally, SO_2 removal would produce a saleable product such as H_2SO_4 or sulfur. In the United States, however, the emphasis has been on throw-away methods in which the end product may itself become a problem for disposal. Some of the more promising techniques for SO_2 removal will now be discussed.

Limestone Injection Process

There are two limestone injection processes used to remove oxides of sulfur from flue gas. One is the dry process in which limestone, $CaCO_3$, is injected into the high temperature combustion zone where it produces calcium sulfate, $CaSO_4$. Unfortunately, this process appears to remove less than half of the sulfur oxides. The wet process is initiated like the dry process, but following the furnace there is a scrubber where the combustion flue gas is scrubbed with the lime, now a lime slurry, to complete the removal of the SO_2. In the wet process, both particulates and SO_x, and some NO_x are removed from the stack gas so no electrostatic precipitator is required. The waste product of fly ash and $CaSO_4$ is not usable or saleable. In fact, sludge disposal is a problem in this process.

Figure 11.5 shows the process as developed by Combustion Engineering and installed on the Kansas Power and Light 125-megawatt unit number 4. Limestone is injected into the furnace as a mixture with the coal. (In other designs dolomite is used, a mixture of $CaCO_3$ and $MgCO_3$.) The $CaCO_3$ is calcined in the furnace as

$$CaCO_3 \rightarrow CO_2 + CaO \tag{11.5}$$

Some of the calcium oxide, CaO, reacts in the furnace with SO_2 to remove the sulfur dioxide and produce calcium sulfate as

$$CaO + SO_2 + \frac{1}{2} O_2 \rightarrow CaSO_4 \tag{11.6}$$

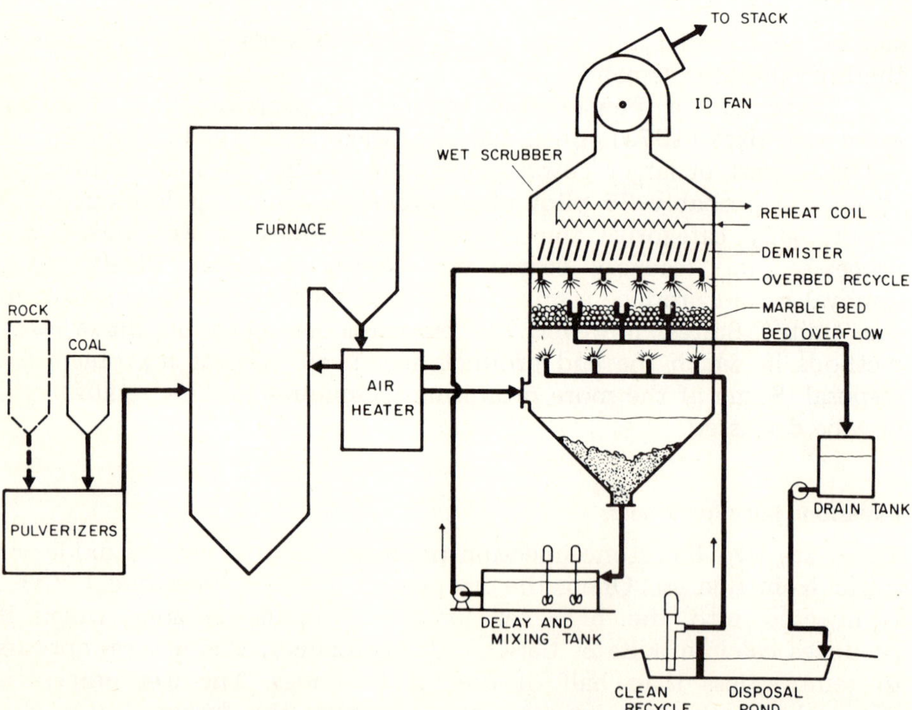

Fig. 11.5 Schematic diagram of air pollution control system for sulfur oxides and particulate removal, using limestone injection and a scrubber. (*Miller, 1969.*)

In the Kansas plant the system is designed to remove 99 percent of the particulates and 83 percent of the sulfur oxides (Miller, 1969). The coal is 3.4 percent sulfur and 12.5 percent ash. About one-third of the sulfur is removed by pyrite rejection in the pulverizer and 20–30 percent of the sulfur dioxide is removed directly in the furnace. In order to further the removal process the water scrubber is used. The stack gas with particulates, unreacted limestone, and reacted limestone travels up through the marble bed of the scrubber. A lime slurry flows down through the bed and the reaction between the slurry of reactive "milk-of-lime" and SO_2 is completed. In the final design of the Kansas plant the delay and mixing tank in Fig. 11.5 was added. The mixture falling into this is recycled to above the marble bed to use up unreacted limestone. With this recycle and some other modifications of the original design, the stack gas runs about 250 ppm SO_2 compared to perhaps 2,500 ppm with no controls.

Gifford (1972) discusses operating experience with a limestone scrubber system on the Commonwealth Edison Company's Will County unit 1. This system is designed to remove 98 percent of the fly ash and 76 percent of the SO_2 from a 4 percent sulfur coal. In this unit, flue gas passes through a venturi scrubber where fly ash is removed and then goes to the SO_2 absorber. Limestone slurry is the absorbing medium. Four percent of the gross electric capacity of the plant is required to operate the fans, limestone crushers, and pumps for the SO_2 clean-up system. As in all such units, scaling and plugging of various components has plagued operations. When one considers that calcium hydroxide, $Ca(OH)_2$, is the basic component of mortar and some of this compound is formed during the process, it is not surprising that plugging occurs. (Hydrated calcium sulfate, $CaSO_4 \cdot 2H_2O$, which can result from the scrubbing, is a component of plaster of Paris.)

The advantage of this sulfur oxide removal technique is that it is basically an "add-on" unit which can be used with existing plants. The process removes both particulate and SO_x, so no electrostatic precipitator is needed. On the minus side the process produces no saleable product and the stack gas must be reheated after the scrubber to give it sufficient buoyancy. At present there is no agreement as to whether this technique will result in the use of lower and therefore cheaper stacks. The argument is that with 90 percent SO_x removal the stack height can be reduced. The counter argument is that the whole idea of removing pollutant is to reduce the ground level concentration (as well as total emissions) so tall stacks should still be required. Another disadvantage of this technique is that the solids must be disposed of as a wet material. The ash alone from such a plant makes up a sizable heap over the years.

Catalytic Oxidation Process

The limestone process is based on absorption. In contrast the catalytic oxidation process converts sulfur oxides to sulfuric acid by passing the flue gases over a vanadium pentoxide catalyst which promotes the oxidation of SO_2 to SO_3. The SO_3 then reacts with water vapor to produce a dilute concentration of H_2SO_4. Figure 11.6 shows this process. The particulate must first be removed in a high temperature precipitator. After cleaning, the hot gas passes through the catalyst bed and enters the air preheater where it is cooled to just above the dew point. The gas enters a heat exchanger where condensation of the acid occurs. A demister removes acid carried over before the clean gas goes up the stack at the relatively low temperature of about 220°F. This system has the advantages of being simple and requiring no recycling of the agent, here V_2O_5. The acid is sold, although not necessarily at a profit. Disadvantages are that high-temperature gas cleaning is required to remove the particulates, corrosion problems occur in the system because of the dilute acid produced, and the process is feasible only on new plants.

Sodium-based Scrubber Process

The Wellman-Power gas process is a regenerative alkaline-absorption technique. The reaction between sulfur dioxide and sodium sulfite is

$$SO_2 + Na_2SO_3 + H_2O \rightarrow 2NaHSO_3 \tag{11.7}$$

In the regeneration section, heat is applied and the reaction is reversed to yield a concentrated stream of SO_2. The sodium sulfite is recovered and can be used again.

This process has been used on sulfuric acid plants and Claus plants to clean up stack gas. The Claus reaction is important since it is one way of producing sulfur from stack gas. Sulfur is produced by the reactions (the first one converts 1/3 of the H_2S):

$$2H_2S + 3O_2 \rightarrow 2SO_2 + 2H_2O \tag{11.8}$$

$$2H_2S + SO_2 \rightarrow 3S + 2H_2O \tag{11.9}$$

The Claus reactions have formed the basis for some clean-up techniques which attempt to convert SO_2 to H_2S via chemical means. Hydrogen sulfide can then be converted to sulfur.

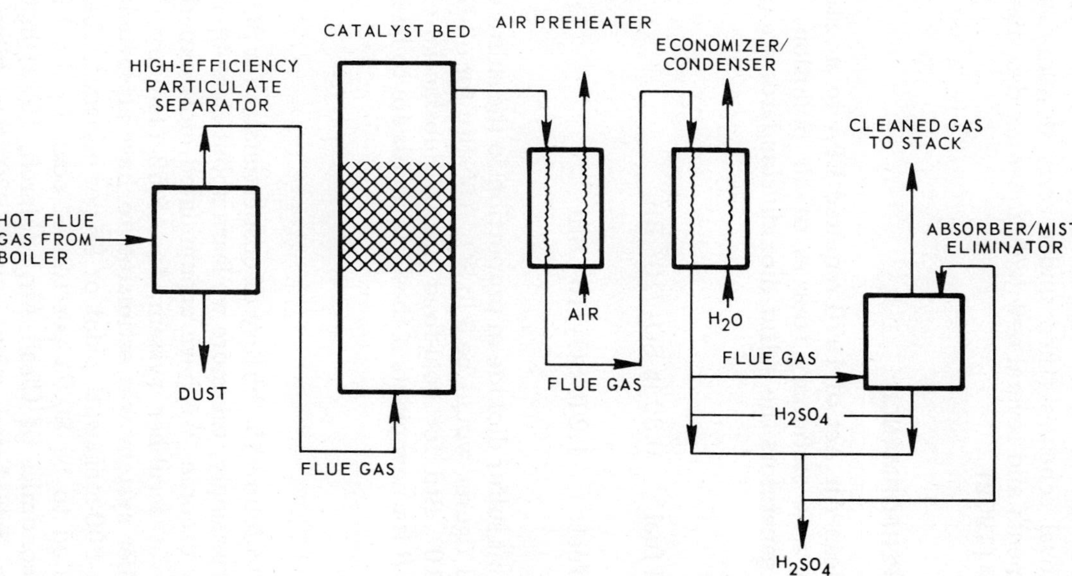

Fig. 11.6 The catalytic oxidation process for SO_2 removal. (*AP-52.*)

The sodium scrubbing process has been installed on a 60-megawatt oil-fired power plant in Japan. Some oxidation occurs during the process and produces sodium sulfate which must then be removed.

AP-50 (1969) and Maurin and Jonakin (1970) describe several other SO_2 removal schemes (then under early development). More mature developments and several new techniques are described in Slack (1973) and Davis (1972).

STANDARDS OF PERFORMANCE

The amended Clean Air Act of 1970 required EPA to set standards of performance for new stationary sources of air pollution. For fossil fuel-fired steam generators the sulfur dioxide standards are (maximum 2-hour average):

liquid fossil fuel	0.80 lb $SO_2/10^6$ Btu
solid fossil fuel	1.2 lb $SO_2/10^6$ Btu

Present plants emit sulfur dioxide in proportion to the sulfur content of the fuel. Typical figures would be 3 lb $SO_2/10^6$ Btu for oil-fired plants and 5 lb $SO_2/10^6$ Btu for coal-fired plants. Substantial clean up is therefore required for new plants if these standards are to be met.

COSTS

It is of interest to follow the history of cost estimates for SO_2 clean-up systems. Not surprisingly, the more we learn about clean-up devices the higher the costs become. We have accumulated most experience with the limestone wet scrubber system. In AP-52 (January, 1969) the limestone scrubber system was estimated to have an installed cost of $\$4 \times 10^6$ for an 800-megawatt plant or \$5 per kilowatt. The operating cost was estimated to be \$0.94 per ton of coal. In the EPA annual report, "The Economics of Clean Air," (March, 1972) installed cost estimates for a retrofitted plant were \$30 per kilowatt for a 200-megawatt plant and \$19 per kilowatt for a 1,000-megawatt plant. Operating costs were about \$1.70 per ton of coal. However, the report also states that "some recent experience with actual plant operators indicates 150–200 megawatt units are experiencing costs twice (these) figures." In a press release in January, 1973, the Tennessee Valley

Authority estimated that they would spend $\$42 \times 10^6$ to backfit a scrubber for a 550-megawatt unit. This cost is $76 per kilowatt.

Gifford (1972) discusses the experience of the Commonwealth Edison Company in back fitting a scrubber onto a 163-megawatt unit. The installed costs were $85 per kilowatt without sludge disposal cost. Operating costs were $\$0.32/10^6$ Btu ($6.31 per ton of coal) again without a sludge disposal system.

Capital costs for a fossil fuel power plant are about $250 per installed kilowatt (1973) and coal runs about $8.00 per ton, including the mechanical cleaning.[1] Thus the sulfur oxide removal equipment adds an appreciable additional cost to the present capital and operating costs for fossil-fueled plants.

Cost figures for other clean-up techniques are not available and those given in AP-52 are obviously out of date. It seems reasonable to suggest that $75 per kilowatt, installed costs for retrofitted units, is a likely figure. For new plants where the clean-up devices can be designed into the plant, costs should be reduced from this figure. However, the total dollars to be spent on clean-up equipment will still be staggering.

TVA estimates that from 1973 to 1978 they will spend $\$270 \times 10^6$ for upgrading their air pollution control facilities at coal-fired power plants. This averages $120 per customer! The Council on Environmental Quality (third annual report, August, 1972) states that the total investment necessary for meeting air control requirements for power plants from 1972 to 1976 will be $7.5 billion. They indicate that the cost to install control equipment on existing plants might be twice this figure.

AMBIENT DOWNWIND CONCENTRATIONS

Now that we have studied the sources of SO_x, the ambient concentrations, and some control techniques, we can consider what concentrations have, in fact, been determined downwind of large point sources. These concentrations should represent the levels which might be endured by individuals or vegetation living "under the plume." AP-50 gives some results for such measurements; others are in Stephens and McCaldin (1971) and Martin and Barber (1967).

The latter authors present SO_2 concentrations, determined by the conductometric method, downwind of a 1,000-megawatt power plant, High Marnham, in Britain. This plant has two 450-foot chimneys, a

[1] Some sense of inflation can be gained by noting that the original figure used here was $150 per kilowatt (1971).

stack exit velocity of 52 ft/sec, and a total sulfur dioxide emission of 7.2 lb/sec at full load.

The highest concentration during the 12-month period was a 3-minute maximum of 90 pphm or 2,670 $\mu g/m^3$ at a distance of 0.8 km from the stack. During the time that the plume was passing over the recorder, typical 3-minute concentrations 5 km downwind were between 5 and 25 pphm, or 143–715 $\mu g/m^3$. Daily mean values downwind of the stack were, as expected, much lower than the 3-minute values. The maximum daily mean concentration was in the range of 5.5–5.9 pphm, or between 157–168 $\mu g/m^3$, and occurred 0.8 km downwind. The next highest daily mean value at this distance was 97 $\mu g/m^3$. The new federal standards allow a 24-hour value of 365 $\mu g/m^3$ not to be exceeded more than once a year. If High Marnham were the *only* source of SO_x in this region, say to a radius of 40 km, then the plant would not appear to emit sufficient SO_x to cause the *ambient* air quality to be over the (United States) standards. However, in the same general area Martin and Barber note that the overall background pollution level is about 70 $\mu g/m^3$ which is right at the United States standard of 80 $\mu g/m^3$ annual average. In the recorder sites around the High Marnham area the 6-month average concentrations were all above 80 $\mu g/m^3$ and in the winter of 1963–64 the 6-month average was 148 $\mu g/m^3$.

One can see the problem of adding another pollution source to an area which already has a high background level of pollution.

AP-50 notes that in the United States, measurements half a mile from a coal-fired power plant, using a Thomas autometer, yielded an average concentration of 0.17 ppm (486 $\mu g/m^3$) for the period January to April. Tests using the West-Gaeke technique, however, showed only about half this much SO_2.

Stephens and McCaldin (1971), using SO_2 levels measured from an airplane, have tracked a plume from a power station 82 km downwind under very stable conditions. They also have determined SO_2 levels, in the plume, at 6 ppm at a distance 24 km downwind of a smelter at Douglas, Arizona.

All in all, these measurements, and the ambient concentrations given previously in this chapter, should indicate the difficulty faced in meeting federal ambient air quality standards for SO_2, especially in regions with oil or coal-fired power plants.

SUMMARY

There are many facets to the problem of emissions of sulfur oxides to the atmosphere. In Scandinavian countries there is concern for the acidifying trend in the soil and water caused by the acid rains. On the other hand, European and United States utility spokesmen urge the use of high stacks to control ground-level concentration of sulfur oxides stating that "this is a valid approach" (Crawford, 1972). The enforcement of the 1970 Clean Air Act has forced utilities to scramble for low sulfur fuel when such fuel is even available. And the implementation of the secondary standard for SO_2 would aggravate this situation.

A variety of clean-up processes to remove sulfur oxides from stack gas are under development. One or more of these may prove to be very effective. But the costs of available clean-up systems are high and will substantially increase the cost of electricity. The engineering problem now is to take the basic technology, which at least for some of the systems has been demonstrated, and convert this to a technology which offers long-term reliable operation.

REFERENCES

AP-50, "Air Quality Criteria for Sulfur Oxides," National Air Pollution Control Administration, Publication AP-50, 1969.

AP-52, "Control Techniques for Sulfur Oxide Air Pollutants," National Air Pollution Control Administration, Publication AP-52, 1968.

AP-73, "Nationwide Inventory of Air Pollutant Emissions, 1968," National Air Pollution Control Administration, Publication AP-73, 1970.

Crawford, W. D.: The Impact of National Emissions Standards on the Electric Power Industry, Paper No. 4D, Am. Inst. Chem. Eng. annual meeting, November, 1972.

Davis, J. C.: SO_2 Removal Still Prototype, *Chem. Eng.*, vol. 79, pp. 52-56, June 12, 1972.

EPA: The Economics of Clean Air, Annual Report of the Administrator of the EPA to the Congress, March, 1972.

Gifford, D. C.: Will County Unit 1, Limestone Wet Scrubber, Paper No. 2A, Am. Inst. Chem. Eng. annual meeting, November, 1972.

Katz, M.: "Measurement of Air Pollutants—Guide to the Selection of Methods," World Health Organization, Geneva, 1969.

Martin, A., and F. R. Barber: Sulfur Dioxide Concentration Measured at Various Distances from a Modern Power Plant, *Atmos. Environ.*, vol. 1, pp. 655-677, 1967.

Maurin, P. G., and J. Jonakin: Removing Sulfur Oxides from Stacks, *Chem. Eng.*, vol. 77, no. 9, pp. 173-180, 1970.

Miller, D. M.: Experience with Wet Scrubber for SO_2 Removal at the Lawrence Station of the Kansas Power and Light Company, *Kansas State Univ. Bull.*, Special Report 85, September, 1969.

Perry, H., and J. A. Decarlo: The Search for Low Sulfur Coal, *Mech. Eng.*, vol. 89, pp. 22-27, April, 1967.

Robinson, E., and R. C. Robbins: Gaseous Sulfur Pollutants from Urban and Natural Sources, *J. Air Pollution Control Assoc.*, vol. 20, pp. 233-235, 1970.

Rohrman, F. A., B. J. Steigerwald, and J. H. Ludwig: SO_2 Pollution, the Next 30 Years, *Power*, vol. 111, pp. 82–83, May, 1967.

Ross, F. F.: What Sulfur Dioxide Problem?, *Combustion*, vol. 43, pp. 6–11, August, 1971.

Slack, A. V.: Air Pollution: The Control of SO_2 from Power Stacks, pt. 3, Processes for Recovering SO_2, *Chem. Eng.*, vol. 74, no. 25, pp. 188–196, 1967.

Slack, A. V.: Removing SO_2 from Stack Gases, *Environ. Sci. Technol.*, vol. 7, 110–119, 1973.

Stephens, N. T., and R. O. McCaldin: Attenuation of Power Station Plumes as Determined by Instrumented Aircraft, *Environ. Sci. Technol.*, vol. 5, pp. 615–620, 1971.

PROBLEMS

11.1 Using the information in the table, estimate the SO_2 produced in a city of 500,000 which incinerates half of its solid waste and buries the rest. How does this figure compare to the emissions from a coal-fired power plant used to provide electricity for the city? (Assume 5 lb of solid waste per person per day.)

Emission factors for sulfur compounds from solid waste disposal

Source	Emission factor, lb of SO_2 per ton of refuse charged
Open-burning dumps and municipal incinerators	1.2–2.0
On-site commercial and industrial multiple-chamber incinerators	1
On-site commercial and industrial single-chamber incinerators	2
On-site residential single-chamber incinerators	0.4
On-site residential flue-fed incinerators	0.2

11.2 Prepare a local inventory of SO_x emissions for your area using the emission factors in Duprey, AP-42, 1968.

11.3 The problem of obtaining accurate measurements of SO_2 concentrations has been discussed in the following references. Write a brief summary of their conclusions:

Terabe, M. et al.: *J. Air Pollution Control Assoc.*, vol. 17, pp. 673–675, 1967.

Booras, S. G., and C. E. Zimmer: *J. Air Pollution Control Assoc.*, vol. 18, pp. 612–615, 1968.

Trieff, N. M., et al.: *J. Air Pollution Control Assoc.*, vol. 18, pp. 329-331, 1968.
Kuczynski, E. R.: *Environ. Sci. Technol.*, vol. 1, p. 69, 1967.

11.4 Estimate SO_2 emissions in the United States from fossil-fueled power plants in the year 2000 for the following possible conditions, assuming (1) no controls, and (2) SO_2 50 percent removal.

Possible electric generation in year 2000	% of plants fossil fueled	Average % sulfur content	SO_2 emissions
1.5×10^6 Mw	40-60%	1.5-3%	?
2.0×10^6 Mw			
2.5×10^6 Mw			
3.0×10^6 Mw			

11.5 Read Ross's paper in the August, 1971 issue of *Combustion* and write a brief review. Do you agree with his thesis that rural SO_2 emissions are good? You might consider the problems of acid rainfall and synergistic effects of SO_x and NO_x.

11.6 Compare the measured plume-dispersion coefficients, which are based on measured SO_2 concentration, of Stephens and McCaldin (1971) with those given in Chap. 8. What do you conclude about our accuracy in making dispersion calculations?

11.7 This chapter did not discuss sulfur oxide control techniques for such industrial processes as smelters, petroleum refineries, and paper or pulp mills. Choose one of these industries and write a review of the possible control techniques. Specify whether the technique is in use or still at the design stage.

11.8 For the High Marnham plant, calculate the SO_2 ground-level concentration for several different atmospheric conditions and distances from the stacks. Estimate a worst-case concentration. Are your results within a factor of 2 or 3 of the data? (Assume a stack diameter of 20 ft.)

11.9 Consider the economics of fuel substitution. Work out the annual cost of this form of pollution control in dollars/year and dollars/ton of SO_2 eliminated for the following conditions: 100,000 lb/hr steam from a boiler of 86 percent efficiency operating 80 percent of the time. (1,000 lb of steam requires approx. 10^6 Btu.)

coal	3.3 percent sulfur, available at 31 cents/10^6 Btu, 12,000 Btu/lb
coal	1 percent sulfur at 38 cents/10^6 Btu, 12,000 Btu/lb
no. 6 oil	2.2 percent sulfur at 49 cents/10^6 Btu, 19,000 Btu/lb
no. 6 oil	1 percent sulfur at 53 cents/10^6 Btu, 19,000 Btu/lb
no. 2 oil	0.2 percent sulfur at 72 cents/10^6 Btu, 19,000 Btu/lb
interruptible gas (90 percent gas, 10 percent no. 2 oil)	
gas	0 percent sulfur at 30 cents/10^6 Btu/21,000 Btu/lb

11.10 Determine the present prices for raw sulfur per ton and sulfuric acid per ton in a given section of the United States. If power plants with 10,000-megawatt capacity are located in this section and these use coal with 1.75 percent sulfur content, how much return could be achieved if sulfur or sulfuric acid could be recovered, assuming that 80 percent of the sulfur in the flue gas is removed?

11.11 When burning Utah King mine coal (ultimate analysis: carbon 72.3 percent wt; hydrogen 5.8 percent wt; nitrogen 1.3 percent wt; sulfur 0.5 percent wt; oxygen 14.9 percent wt; ash 5.2 percent wt), what will be the ppm (dry) SO_2 in the stack at 3 percent O_2 in the flue gas? What will be the ppm NO_x as NO if we assume all the chemically bound nitrogen is converted to NO? Calculate these emissions in lb/million Btu. The heating value of the coal is 12,000 Btu/lb.

11.12 A coal-fired power plant proposes to burn 3 percent sulfur coal with a heating value of 11,000 Btu/lb. If the plant is to meet the standard of performance what percent clean up is required for SO_2? What percent sulfur could the coal contain if the standards are to be met with no clean-up devices?

11.13 1 ppm SO_2 = 2620 $\mu g/m^3$ at 25°C, 760 mm Hg
1 ppm SO_2 = 2860 $\mu g/m^3$ at 0°C, 760 mm Hg
Derive these relations and therefore a general conversion from ppm to $\mu g/m^3$.

12

NITROGEN OXIDES

. . . NO, itself, must be controlled or even reduced from present ambient levels, because its oxidation product, NO_2, is currently at or near levels that are associated with adverse health effects in over 50 percent of the United States cities having a population greater than 50,000.

Air Quality Criteria for Nitrogen Oxides[1]

Like particulates and the oxides of sulfur, oxides of nitrogen are an important pollutant produced mainly in the combustion of fuel. In this chapter we shall point out that regulating the combustion process rather than adding clean-up devices (as for particulates and the sulfur oxides) is the principal form of nitrogen-oxide control. The effects of modest concentrations of NO_2 have been recently established. Nitrogen dioxide at levels of 118 to 156 $\mu g/m^3$ (0.063–0.083 ppm) over a 6-month period (AP-84) has been implicated in producing adverse effects on human health. It is also known that combined NO_2—SO_2 concentrations adversely affect plants, whereas concentrations of these gases taken singly would have to be much higher to affect the plants (Heck, 1968).

There are several oxides of nitrogen. The two important ones in air pollution are nitric oxide, NO, and nitrogen dioxide, NO_2. Nitrous oxide, N_2O, is also commonly present in the atmosphere but is not a

[1] AP-84.

pollutant. The term NO_x represents a composite atmospheric concentration of NO and NO_2. We shall see that the way to measure NO concentration is to oxidize the gas to NO_2 and then measure this. So the use of NO_x to account for both species is convenient.

To convert from the volume basis of ppm to the mass basis of $\mu g/m^3$ we have

$$1 \text{ ppm NO} = 1230\ \mu g/m^3 \text{ (at 25°C, 760 mm Hg)}$$

$$1 \text{ ppm } NO_2 = 1880\ \mu g/m^3 \text{ (at 25°C, 760 mm Hg)}$$

Nitric oxide is a colorless, odorless gas. It is formed under high-temperature combustion processes according to the Zeldovich mechanism

$$\begin{aligned} N_2 + O &\rightleftharpoons NO + N \\ N + O_2 &\rightleftharpoons NO + O \end{aligned} \tag{12.1}$$

which results in an overall reaction of

$$N_2 + O_2 \rightleftharpoons 2NO \tag{12.2}$$

The thermodynamics of this reaction will be discussed shortly.

Nitrogen dioxide is an absorber of visible light but absorbs more strongly in the short wavelength range, that is, blue-violet. Thus it appears as a reddish-orange gas. It is corrosive to materials and toxic to man. It is formed as

$$2NO + O_2 \rightleftharpoons 2NO_2 \tag{12.3}$$

Nitrous oxide is a colorless, odorless gas and is the "laughing gas" anesthetic. It is inert at normal atmospheric temperatures.

SOURCES

In Chap. 3 we established that the largest NO_x contributors are motor vehicles and fuel combustion in stationary sources. Table 12.1 indicates the emission of nitrogen oxides, here NO and NO_2, from 1966 to 1968. The increase in NO_x is largely explained because emissions from natural gas engines, used on the compressors for natural gas pipelines, were first included in the 1967 figures. The increase in NO_x from motor vehicles is attributed to the effect of using CO and HC exhaust controls. Since

TABLE 12.1 Summary of annual nationwide emissions of nitrogen oxides, 1966 through 1968 *(10^6 tons)*

Source category	1966	1967	1968	Change from 1966 to 1968
Transportation	7.6	7.6	8.1	+0.5
Motor vehicles	6.6	6.7	7.2	+0.6
Other	1.0	0.9	0.9	−0.1
Fuel combustion	6.7	9.5	10.0	+3.3
Coal	4.0	3.8	4.0	N*
Fuel oil	0.9	1.0	1.0	+0.1
Natural gas	1.6	4.2	4.5	+2.9†
Wood	0.2	0.2	0.2	N
LPG and kerosene	—	0.3	0.3	+0.3†
Industrial processes	0.2	0.2	0.2	N
Solid waste disposal	0.5	0.6	0.6	+0.1
Miscellaneous	1.7	1.7	1.7	N
Man-made	0.5	0.5	0.5	N
Forest fires	1.2	1.2	1.2	N
Total	16.7	19.6	20.6	+3.9

*N = Negligible.
†Apparent change. (Emission estimates added in 1968 from sources not included in 1966.)
Source: AP-84.

NO_x emissions are mainly from autos and fuel combustion they are concentrated in the urban areas.

We have seen that on a global basis the man-caused emissions of NO_x are far below natural sources. However, man-caused emission resulting in increased concentrations in urban areas is a cause for alarm. Figure 12.1 shows projected NO_x emissions assuming minimal controls (Bartok, et al., 1969). The data on this figure for 1968 are below the EPA estimates of 20.6×10^6 tons but the trends are thought to be correct. The figures indicate why mobile sources are to be controlled under the federal legislation of 1970 and why there is concern about the increase in NO_x from power plants. Table 12.2 indicates global emissions of nitrogen compounds as estimated by Robinson and Robbins (1970). They estimate NO_x man-caused emissions, shown here as NO_2, to be 52.9×10^6 tons per year. Emissions of NO due to biological

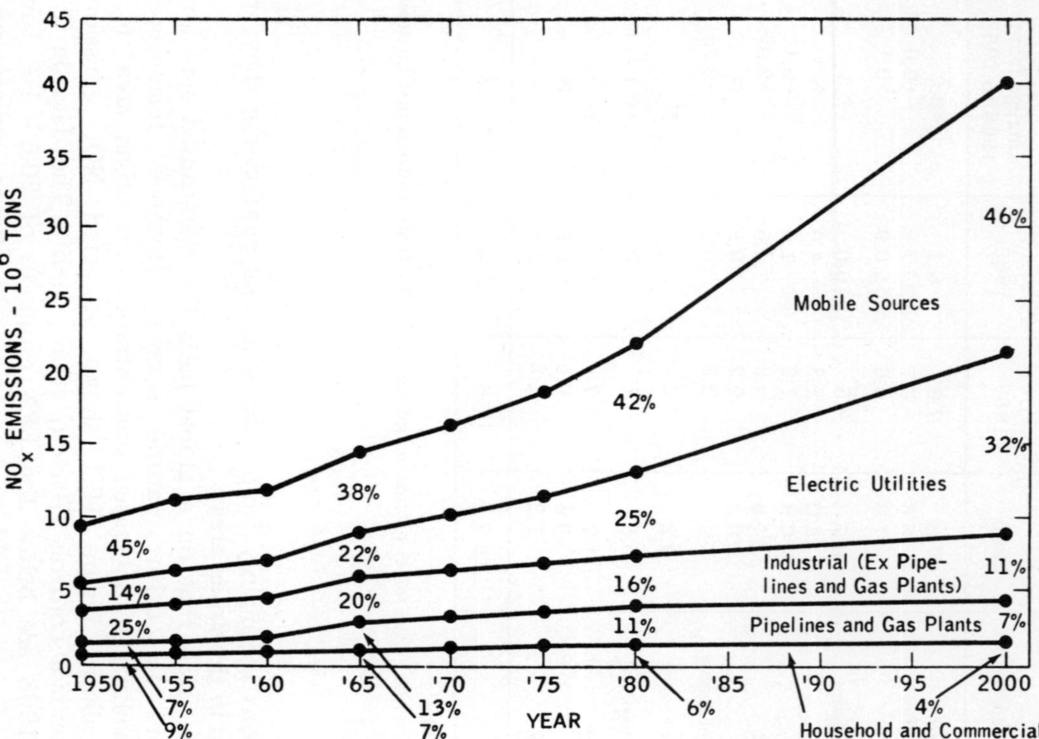

Fig. 12.1 Potential NO_x emissions in the United States assuming median economic trends. Numbers shown between curves represent the percentage contribution of the given sector to total estimated emission of NO_x for the year. (*Bartok et al., 1969.*)

TABLE 12.2 Estimated annual global emissions of nitrogen compounds

Compound	Source	Source (tn/yr)	Estimated emissions (tn/yr)	Emissions as nitrogen (tn/yr)
NO_2	Coal combustion	$3{,}074 \times 10^6$	26.9×10^6	8.2×10^6
	Petroleum refining	$11{,}317 \times 10^6$ (bbl)	0.7×10^6	0.2×10^6
	Gasoline combustion	379×10^6	7.5×10^6	2.3×10^6
	Other oil combustion	894×10^6	14.1×10^6	4.3×10^6
	Natural gas combustion	20.56×10^{12} (ft^3)	2.1×10^6	0.6×10^6
	Other combustion	$1{,}290 \times 10^6$	1.6×10^6	0.5×10^6
NO_2	Total		52.9×10^6	16.1×10^6
NH_3	Combustion		4.2×10^6	3.5×10^6
NO	Biological action		501×10^6	234×10^6
NH_3	Biological action		$1{,}160 \times 10^6$	957×10^6
N_2O	Biological action		592×10^6	378×10^6

Source: Robinson and Robbins, 1970.

activity are estimated to be 501×10^6 tons per year. In the atmosphere the NO is oxidized to NO_2. Robinson and Robbins suggest that the residence time of NO_2 in the atmosphere is about three days, whereas the NO residence time is approximately four days (AP-84). They also discuss the three cycles for nitrogen in the atmosphere: the N_2O cycle, the NH_3 cycle, and the NO_x cycle. The latter involves scavenging mechanisms both of gaseous deposition and oxidation to the nitrate aerosol, NO_3. This second removal mechanism can result in the production of HNO_3, nitric acid.

It is clear that on a global basis the man-caused production of NO_x is much less than that from natural sources. However, we have already acknowledged in Chap. 5 that nitrogen-oxide emissions from automobiles are a key element in photochemical smog. We cannot, therefore, ignore this pollutant.

EMISSIONS INVENTORY

Emission factors for NO_x are summarized in Table 12.3. It is of interest to convert these factors to an equivalent Btu basis to compare the emissions from a particular source which could be fueled with coal, oil, or gas. We know that coal produces about 12,000 Btu/lb; gas, about 1,000 Btu/ft^3; and fuel oil, about 18,000 Btu/lb (150,000 Btu/gal). Using emission factors appropriate to an electric utility we have

$$\text{coal:}\quad \left(\frac{20 \text{ lb NO}_x}{\text{ton}}\right)\left(\frac{1 \text{ ton}}{2{,}000 \text{ lb}}\right)\left(\frac{1 \text{ lb}}{12{,}000 \text{ Btu}}\right)=\frac{834 \text{ lb NO}_x}{10^9 \text{ Btu}}$$

$$\text{oil:}\quad \left(\frac{104 \text{ lb NO}_x}{1{,}000 \text{ gal}}\right)\left(\frac{1 \text{ ft}^3}{62.4 \text{ lb}}\right)\left(\frac{1 \text{ gal}}{0.1336 \text{ ft}^3}\right)\left(\frac{1 \text{ lb}}{18{,}000 \text{ Btu}}\right)$$

$$=\frac{694 \text{ lb NO}_x}{10^9 \text{ Btu}}$$

$$\text{natural gas:}\quad \left(\frac{390 \text{ lb NO}_x}{10^6 \text{ ft}^3}\right)\left(\frac{1 \text{ ft}^3}{1{,}000 \text{ Btu}}\right)=\frac{390 \text{ lb NO}_x}{10^9 \text{ Btu}}$$

These results suggest the general rule that nitrogen oxides are produced by combustion of fuels on an equivalent Btu basis in the order coal $>$ oil $>$ gas. This is also the order in which peak combustion temperatures and chemically bound nitrogen occur in the fuel. The latter is an important source of NO_x in the combustion process and is referred to as "fuel-nitrogen."

The emission factors given in Table 12.3 require careful application. We shall see that NO_x produced in a combustion process is very dependent upon the thermodynamics of the process; that is, the temperatures, air/fuel ratio, and the type of combustion equipment. Thus similar boiler units in power plants may have markedly different NO_x emissions because of minor variations in the combustion conditions. Likewise the NO_x emission from waste disposal processes is based on limited data and requires further measurements to define the uncertainty.

AMBIENT CONCENTRATIONS

Ambient concentrations of NO and NO_2 for Chicago, Los Angeles, and Washington, D.C. are shown in Tables 12.4 and 12.5. As we might expect, Los Angeles has the highest values. Robinson and Robbins (1970) have estimated the background concentrations of the nitrogen oxides as 2 ppb for NO and 4 ppb for NO_2 for land areas between the

TABLE 12.3 Emission factors for nitrogen oxides

Source	Average emission factor
Coal	
Household and commercial	8 lb/ton of coal burned
Industry	20 lb/ton of coal burned
Utility	20 lb/ton of coal burned
Fuel Oil	
Household and commercial	12–72 lb/1000 gal of oil burned
Industry	72 lb/1000 gal of oil burned
Utility	104 lb/1000 gal of oil burned
Natural Gas	
Household and commercial	116 lb/million cf of gas burned
Industry	214 lb/million cf of gas burned
Utility	390 lb/million cf of gas burned
Gas Engines	
Oil and gas production	770 lb/million cf gas burned
Gas plant	4300 lb/million cf gas burned
Pipeline	7300 lb/million cf gas burned
Refinery	4400 lb/million cf gas burned
Gas Turbines	
Gas plant	200 lb/million cf gas burned
Pipeline	200 lb/million cf gas burned
Refinery	200 lb/million cf gas burned
Waste Disposal	
Open burning	11 lb/ton of waste burned
Conical incinerator	0.65 lb/ton of waste burned
Municipal incinerator	2 lb/ton of waste burned
On-site incinerator	2.5 lb/ton of waste burned
Fuel Combustion (mobile sources)	
Gasoline-power motor vehicle	113 lb/1000 gal gasoline burned
Diesel-powered vehicle	222 lb/1000 gal oil burned
Aircraft: conventional jet	23 lb/flight per engine
Fan-type engine	9.2 lb/flight per engine
Chemical Industries	
Nitric acid manufacture	57 lb/ton HNO_3 product
Adipic acid	12 lb/ton product
Terephthalic acid	13 lb/ton product
Nitrations	
Large operations	0.2–14 lb/ton HNO_3 used
Small batches	2–260 lb/ton HNO_3 used

Source: AP-67.

TABLE 12.4 Nitric oxide concentration* at CAMP sites, by averaging time and frequency, 1962 through 1968
(ppm)

City and averaging time	Maximum for year							Data avail., %	% of time concentration is exceeded							
	62	63	64	65	66	67	68		0.01	0.1	1	10	30	50	70	90
Chicago																
5 min	0.66	0.63	0.97	0.67	0.74	0.69	0.66	79	0.67	0.52	0.33	0.18	0.11	0.08	0.05	0.02
1 hr	0.61	0.60	0.91	0.62	0.64	0.63	0.61	79	0.64	0.50	0.32	0.18	0.11	0.08	0.05	0.02
8 hr	0.41	0.39	0.49	0.34	0.42	0.41	0.40	83		0.40	0.27	0.17	0.11	0.08	0.06	0.03
1 day	0.36	0.35	0.35	0.28	0.34	0.26	0.21	86			0.23	0.15	0.11	0.09	0.07	0.04
1 mo	0.17	0.15	0.16	0.13	0.14	0.15	0.11	94				0.13	0.11	0.08	0.07	0.06
1 yr	0.10	0.10	0.10	0.10	0.10	0.08	0.07	100						0.10		
Los Angeles																
5 min	2.11	0.97	1.32					72	1.23	0.85	0.54	0.23	0.08	0.03	0.01	0.00
1 hr	1.42	0.79	1.24					72	1.12	0.80	0.53	0.23	0.08	0.03	0.01	0.00
8 hr	0.60	0.49	0.60					78		0.55	0.41	0.21	0.09	0.04	0.02	0.01
1 day	0.46	0.38	0.44					81			0.36	0.19	0.10	0.05	0.03	0.01
1 mo	0.25	0.18	0.19					92				0.16	0.09	0.06	0.03	0.02
1 yr	0.08	0.08	0.07					100						0.07		
Washington																
5 min	0.68	0.73	1.03	0.74	1.15	1.26	0.89	79			0.42	0.10	0.03	0.01	0.01	0.00
1 hr	0.65	0.68	0.88	0.62	1.02	1.14	0.69	80	0.88	0.62	0.33	0.09	0.03	0.01	0.01	0.00
8 hr	0.39	0.51	0.54	0.39	0.55	0.55	0.43	84		0.48	0.27	0.09	0.04	0.02	0.01	0.00
1 day	0.22	0.26	0.29	0.21	0.41	0.36	0.31	86			0.20	0.08	0.04	0.02	0.01	0.01
1 mo	0.07	0.08	0.07	0.06	0.09	0.08	0.10	93				0.07	0.04	0.03	0.02	0.01
1 yr	0.03	0.04	0.03	0.03	0.04	0.04	0.04	100						0.03		

*Determined by continuous Griess-Saltzman method.
Source: AP-84.

TABLE 12.5 Nitrogen dioxide concentration* at CAMP sites, by averaging time and frequency, 1962 through 1968
(ppm)

City and averaging time	Maximum for year							Data avail., %	% of time concentration is exceeded							
	62	63	64	65	66	67	68		0.01	0.1	1	10	30	50	70	90
Chicago																
5 min	0.26	0.23	0.79	0.21	0.35	0.29	0.29	79	0.28	0.18	0.13	0.07	0.05	0.04	0.03	0.02
1 hr	0.22	0.21	0.47	0.18	0.31	0.25	0.18	80	0.25	0.18	0.12	0.07	0.05	0.04	0.03	0.02
8 hr	0.16	0.17	0.22	0.15	0.20	0.15	0.12	84		0.16	0.12	0.07	0.05	0.04	0.04	0.03
1 day	0.12	0.13	0.15	0.09	0.15	0.11	0.10	87			0.10	0.07	0.05	0.04	0.04	0.03
1 mo	0.06	0.06	0.07	0.05	0.08	0.07	0.06	93				0.06	0.05	0.04	0.04	0.03
1 yr	0.04	0.04	0.05	0.04	0.06	0.05	0.05	100						0.04		
Los Angeles																
5 min		0.57	1.27					63	0.66	0.44	0.25	0.12	0.06	0.04	0.03	0.01
1 hr		0.52	0.68					63	0.55	0.40	0.24	0.11	0.06	0.04	0.03	0.01
8 hr		0.37	0.27					67		0.27	0.23	0.11	0.06	0.05	0.03	0.02
1 day		0.20	0.26					71			0.18	0.10	0.07	0.05	0.04	0.02
1 mo		0.10	0.07					79				0.09	0.05	0.04	0.04	0.03
1 yr		0.06	0.05					100						0.05		
Washington																
5 min	0.37	0.24	0.24	0.25	0.19	0.24	0.26	79	0.22	0.16	0.10	0.06	0.04	0.03	0.03	0.02
1 hr	0.30	0.22	0.23	0.23	0.18	0.21	0.24	79	0.21	0.15	0.10	0.06	0.04	0.03	0.03	0.02
8 hr	0.10	0.11	0.14	0.10	0.09	0.14	0.13	84		0.11	0.09	0.06	0.04	0.03	0.03	0.02
1 day	0.07	0.09	0.10	0.07	0.07	0.09	0.08	86			0.08	0.06	0.04	0.04	0.03	0.02
1 mo	0.04	0.04	0.04	0.04	0.04	0.05	0.05	95				0.05	0.04	0.04	0.03	0.03
1 yr	0.03	0.03	0.04	0.03	0.03	0.04	0.05	100						0.03		

*Determined by continuous Griess-Saltzman method.
Source: AP-84.

latitudes of 65°N and 65°S. Land areas north and south of 65° and all ocean areas are estimated to have concentrations of 0.2 ppb NO and 0.5 ppb NO_2. The effects of urbanization on ambient NO_x concentrations are therefore large.

The importance of the urban concentrations lies both in the direct effect of NO_x on plants and animals and in the role of NO_x in the photochemical smog cycle, already discussed in Chap. 5. There we found that the interactions between NO_x and the HC's led to a build-up of oxidants. Hydrocarbon radicals were able to promote oxidation of NO to NO_2 without using up the ozone. The concentration of ozone thus built up in time. Figure 12.2 indicates the variation of several pollutants in Los Angeles as a function of time of day. We see a peak in NO due to direct emission from automobiles. As evidence of this, note the correlation between NO and CO. As the sun rises, the NO_2 concentration increases; the NO concentration decreases. After a lag in time, the ozone concentration rises to peak out at about midday. The data on this particular day do not show an afternoon rise but other data do. Increased auto traffic in late afternoon increases the NO

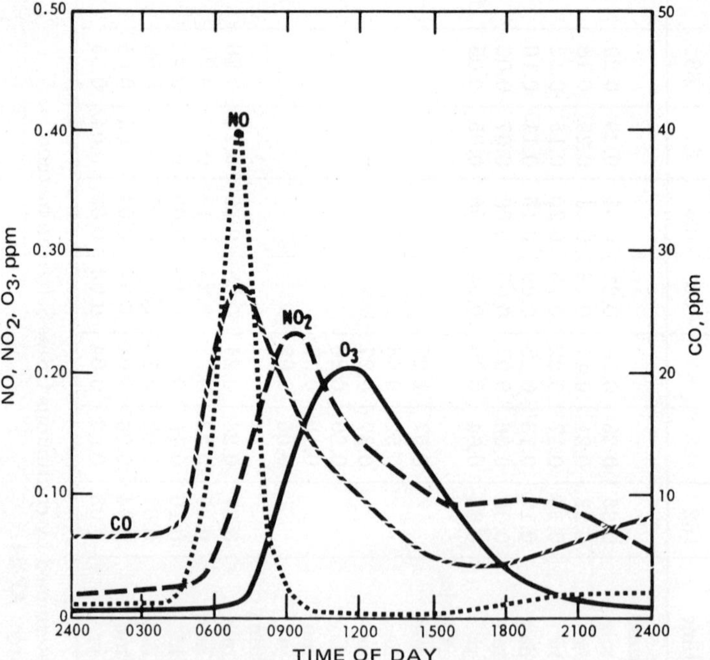

Fig. 12.2 Average daily 1-hour concentrations of selected pollutants in Los Angeles, California, July 19, 1965. (*AP-84.*)

concentration again. Ozone then reacts with NO to produce another rise in NO_2 while the ozone concentration decreases.

These data also indicate why long-term yearly averages may be similar for different locations while short-term averages are different. The NO_x concentrations are relatively low at night and the peaks during the day are roughly for 3-hour periods. A yearly average thus includes a great deal of time of low concentration and this explains why Los Angeles and Chicago have similar yearly average values.

Besides exhibiting a diurnal variation as noted on Fig. 12.2, NO also shows a seasonal variation. Higher mean values are noted in the late fall and winter months. This is true whether the primary winter source of NO is automobiles or power and space heating. NO_2 exhibits a less distinct pattern during the year. There is also a "weekend effect" when auto traffic is less concentrated during the 6 to 9 AM period.

Interaction between the nitrogen oxides, hydrocarbons, and oxidants is shown graphically in Fig. 12.3. The abscissa plots the 6–9 AM average NO_x concentration in ppm. The ordinate indicates the observed upper-limit, that is, maximum, 1-hour average oxidant concentration which results with the given NO_x concentration. The hydrocarbon envelopes, given as ppm of carbon, are superimposed on the figure. The HC values are calculated nonmethane hydrocarbon concentration. We see that for a given 6–9 AM NO_x concentration, the higher the nonmethane HC concentration the higher the oxidant level. The peak oxidant level is also more likely to be reached with higher HC values. It appears that if a 1-hour oxidant concentration of 200 $\mu g/m^3$ (0.1 ppm) is taken as a limit, then the NO_x levels must be kept below about 0.04 ppm (80 $\mu g/m^3$). In fact, the California standards are 0.1 ppm of oxidant but the newer federal standards are lower: 1-hour oxidant level of 0.08 ppm (160 $\mu g/m^3$), not to be exceeded more than once a year. To meet these standards NO_x concentrations must be well below 0.04 ppm.

The point is that one cannot set standards for each pollutant without considering the interaction occurring via the photochemical cycle. In regions of strong insolation and a dense auto population, the NO_x, HC, O_3 cycle forces us to look simultaneously at standards for the three pollutants.

What are the federal standards? The NO_2 primary standard is an *annual* average of 100 $\mu g/m^3$ (0.05 ppm). The federal secondary standard in this case is the same as the primary standard. (The oxidant standard is a *1-hour* limit of 160 $\mu g/m^3$; the hydrocarbon standard, corrected for methane, is a 6–9 AM 3-hour average of 160 $\mu g/m^3$ (0.24 ppm).) It is interesting that the only NO_2 federal standard is for a yearly average but the oxidant standard is for 1 hour. We have just seen

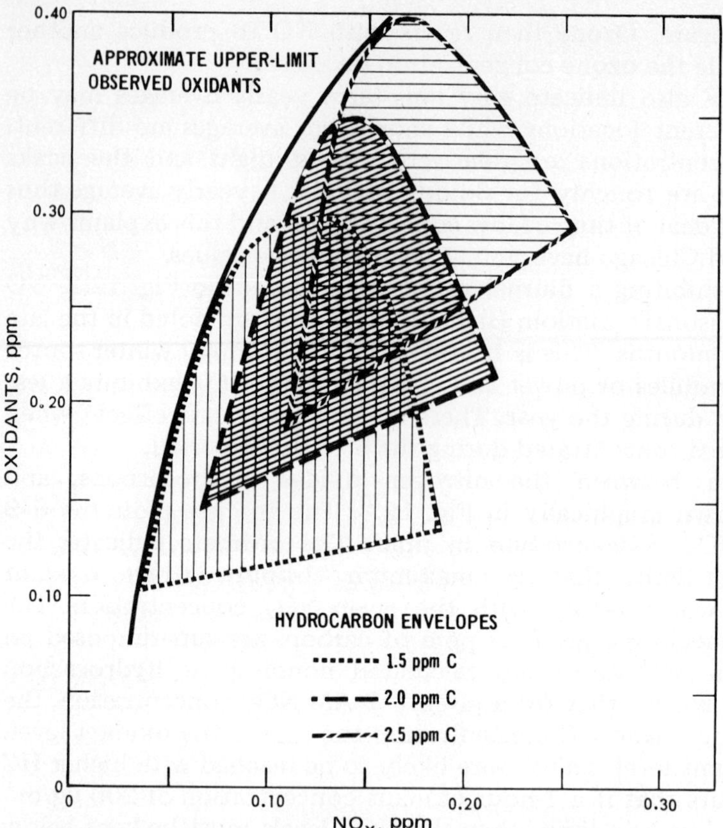

Fig. 12.3 Hydrocarbon-oxidant envelopes superimposed on maximum daily 1-hour-average oxidant concentrations as a function of 6 to 9 AM average of total nitrogen oxides in Pasadena, California, May through October 1967. (*AP-84.*)

that the *peak* oxidant level and the 6–9 AM average NO_x level can be correlated. This suggests that a 3-hour NO_x standard might also be appropriate if we are to meet the oxidant level. The peak oxidant level noted on Fig. 12.3 occurs about 1 percent of the days that the given NO_x level is reached.

AP-84 summarizes the results as: "Further observation of the data base shows that 80 $\mu g/m^3$ (0.04 ppm) NO_x occurs simultaneously with calculated nonmethane HC concentrations ranging from 200–930 $\mu g/m^3$ (0.3 to 1.4 ppm C). Similarly, observation of the 200 $\mu g/m^3$ (0.3 ppm C) nonmethane HC level shows NO_x in the range of 80–320 $\mu g/m^3$ (0.04 to 0.16 ppm)." Thus each HC level is associated with a range of NO_x values and each NO_x value is associated with a peak oxidant level which, based on the available data, occurs on about 1 percent of the days.

Do we meet the standards? The data of Table 12.5 indicate that Chicago, Los Angeles, and Washington, D.C. are right at the NO_2 standard of 0.05 ppm. These values were determined by the Griess-Saltzman method, whereas the new federal standard uses the colorimetric NaOH method. We will see that concentrations measured by the two techniques have not always been in agreement.

TEST METHODS

Two methods have been commonly used to determine concentrations of NO_2. The Griess-Saltzman method, used in the California standards, is based on the reaction of NO_2 with sulfanilic acid to form diazonium salt. This couples with N-(1-naphthyl)-ethylenediamine dihydrochloride to form a deeply colored red-violet azo dye. Air is sampled for 30 minutes or less through a fritted bubbler into a solution of the reagent. After the color develops, the absorbence is measured spectrophotometrically at 5500 Å. The NO_2 concentration is determined from a calibration curve preferably developed from accurately known concentrations of NO_2 gas. The method can be used for NO_2 concentrations in air of 40–1,500 $\mu g/m^3$ (0.02–0.75 ppm).

The second technique for determining NO_2 concentrations is the Jacobs-Hochheiser method which is used as the EPA standard. The air is passed through aqueous sodium hydroxide to form the nitrite ion from the NO_2. The resulting solution is treated with H_2O_2, to remove SO_2 interference, and acidified. The nitrous acid produced is measured by Griess-Saltzman diazotizing coupling procedure except that sulfanilamide is used in place of sulfanilic acid.

NO levels are determined by oxidizing this gas to NO_2 and then measuring the resulting concentration. The NO_2 can be determined initially, then NO is oxidized to NO_2 and a second measurement made. Alternatively, the one sample can be analyzed for NO_2 and a parallel sample analyzed for NO_x, that is, the combined NO and NO_2. The NO concentration should be the difference between the two measurements.

The Jacobs-Hochheiser technique for 24-hour samples has been used in the National Air Surveillance Network (NASN) which employs such a sampling period. Table 12.6 (AP-84) shows the ratio of NASN to CAMP (Continuous Air Monitoring Procedure) NO_2 measurements. The latter method uses the Griess-Saltzman technique. The measurements differ by a factor of 3!

AP-84 comments that there is no satisfactory explanation of the discrepancy but suggests that because of calibration to lower

TABLE 12.6 Ratio of average yearly NASN* to CAMP† NO_2 measurements

City	NO_2 ratio: NASN/CAMP		
	1967	1968	1969
Chicago	3.2	3.0	3.0
Cincinnati	3.3	3.2	—
Denver	2.0	2.7	2.3
Philadelphia	3.2	—	—
St. Louis	5.5	4.3	4.5
Washington	3.0	2.9	1.5
Median	3.2	3.0	2.7
Minimum	2.0	2.7	1.5
Maximum	5.5	4.3	4.5

*National Air Sampling Network.
†Continuous Air Monitoring Program.
Source: AP-84.

concentrations the NASN technique is the more reliable. The data of Table 12.5 which indicated that the selected cities were at or below the federal standards were based on the Griess-Saltzman technique. The moral of the story is simply that it is difficult to measure low concentrations accurately. The ability to reach ambient air-quality goals set in the federal standards should not, of course, depend on which measurement method is used. Unfortunately, because of technological limitations, this may well be the case. A discussion of EPA's position on problems in NO_x measurement is given in Hauser and Shy (1972).

THERMODYNAMICS AND KINETICS OF NITROGEN OXIDES

We need to understand something of the thermodynamics and reaction kinetics of the nitrogen oxides formation since, unlike SO_x, NO_x control starts at the combustion burner. SO_x control techniques are implemented before or after the combustion chamber, that is, desulfurization or flue-gas clean up.

Let us look at the formation of NO described as the "fixation" or fixing of nitrogen with oxygen. We have as the overall reaction

$$\frac{1}{2} N_2 + \frac{1}{2} O_2 \rightleftharpoons NO \tag{12.4}$$

The equilibrium constants for this reaction are given in Table 12.7.

In order to calculate the concentration *at equilibrium* we must pick a ratio of nitrogen to oxygen since the equilibrium concentration naturally depends on the amount of reaction material available. Let us calculate for an N/O atom ratio of 4/1 the equilibrium composition at 1 atmosphere. We will then calculate the composition for an N/O atom ratio of 40/1, which corresponds roughly to the composition of flue gas when only a little excess oxygen is available to react with the nitrogen. Our calculation will be only for the reaction of Eq. (12.4) and will totally ignore all the usual competing reactions involving the fuel hydrocarbons and oxygen. The trends we note, however, will be the same as those observed in the real case.

From the information in Chap. 4, we have for the equilibrium constant

$$K(T) = \frac{\chi_{NO}{}^{\nu_{NO}}}{\chi_{O_2}{}^{\nu_{O_2}} \chi_{N_2}{}^{\nu_{N_2}}} \left(\frac{P}{P_o}\right)^{\nu_{NO}-\nu_{O_2}-\nu_{N_2}} \tag{12.5}$$

Here $\nu_{NO} = 1$, $\nu_{O_2} = 1/2$, $\nu_{N_2} = 1/2$, and $P = P_o = 1$ atmosphere, so we have

$$K(T) = \frac{\chi_{NO}}{[\chi_{O_2} \chi_{N_2}]^{1/2}} \tag{12.6}$$

Furthermore, for our single reaction the mole fractions must add up to unity, so as a second condition we have

$$\chi_{NO} + \chi_{O_2} + \chi_{N_2} = 1 \tag{12.7}$$

TABLE 12.7 Equilibrium constants for reaction, Eq. (12.4)

T °K	T °R	T °F	$\log_{10} K$	K
1000	1800	1340	−4.068	8.59×10^{-5}
1400	2520	2060	−2.717	1.92×10^{-3}
2000	3600	3140	−1.703	1.98×10^{-2}
3000	5400	4940	−0.915	0.121
4000	7200	6740	−0.526	0.298

From our specified ratio of N/O we obtain a third condition as:

$$\frac{\chi_{NO} + 2\chi_{O_2}}{\chi_{NO} + 2\chi_{N_2}} = \frac{1}{4} \tag{12.8}$$

We will let $\chi_{NO} = a$; then we can set up a single equation in this unknown. From Eq. (12.7) we find

$$\chi_{O_2} + \chi_{N_2} = 1 - a$$

From Eq. (12.8) we have

$$\frac{a + 2\chi_{O_2}}{a + 2(1 - a - \chi_{O_2})} = \frac{1}{4}$$

or solving for

$$\chi_{O_2} = \frac{1}{5} - \frac{a}{2}$$

so

$$\chi_{N_2} = \frac{4}{5} - \frac{a}{2}$$

and

$$\chi_{NO} = a$$

Substituting these into Eq. (12.6) we find

$$K(T) = \frac{a}{\left(\frac{1}{5} - \frac{a}{2}\right)^{1/2} \left(\frac{4}{5} - \frac{a}{2}\right)^{1/2}} \tag{12.9}$$

The procedure is to pick an a and find $K(T)$. One iterates on the solution until the calculated $K(T)$ agrees with the known value. As an example let us try $a = 0.10$, then

$$K(T) = \frac{0.10}{(0.15)^{1/2}\,(0.75)^{1/2}} = 0.30$$

This value for the equilibrium constant turns out to be very close to $K(T)$ at 4000°K. Therefore at this temperature we have

$$\chi_{NO} = 0.10 \quad \text{(that is, 10 percent or 100,000 ppm)}$$

$$\chi_{O_2} = 0.15$$

$$\chi_{N_2} = 0.75$$

We can continue such calculations to fill in the results noted in Table 12.8.

These results are important in understanding the formation of NO during combustion. They show that substantial amounts of NO only occur (assuming equilibrium is reached) at high temperatures. Further, the amount of NO depends on the available oxygen (and nitrogen). Assuming, as is usually done, that the nitric oxide reaction is decoupled from the combustion reactions, the oxygen available to react with nitrogen is not in a 4/1 ratio but is a function of the excess air. Five percent excess air results in about 1 percent oxygen in the flue gas; 10 percent excess air results in about 2 percent oxygen in the flue gas. Since most of the nitrogen is unaffected by the combustion process the flue-gas composition for 10 percent excess air is approximately 40 parts of nitrogen to 1 part of oxygen. Figure 12.4 illustrates this point. Note that we are neglecting the CO_2 and H_2O in the exhaust gases.

Thus at equilibrium we would expect to produce nitric oxide at about 2,950 ppm for combustion at 2000°K (which represents the adiabatic flame temperature for about 30 percent excess air burning with methane). Since a boiler is designed to promote heat transfer the adiabatic flame temperature is not usually reached but is approached in the interior of the flame. It is already clear just from equilibrium considerations that we would like to minimize the peak combustion temperature in order to reduce the formation of NO. Also we should

TABLE 12.8 Equilibrium concentrations for NO

T °K	T °F	χ_{NO}*	χ_{NO}†	Ratio $\chi_{(4/1)}/\chi_{(40/1)}$
1000	1340	0.000034	0.0000132	2.56
1400	2060	0.000770	0.000298	2.58
2000	3140	0.007850	0.002950	2.66
3000	4940	0.045000	0.015200	2.95
4000	6740	0.100000	0.029000	3.45

*N/O = 4/1
†N/O = 40/1

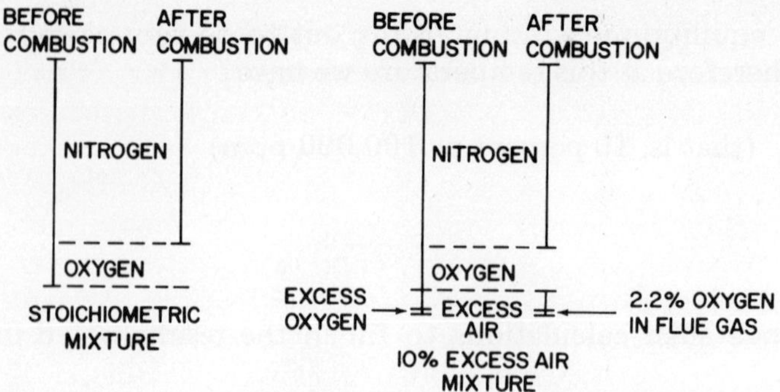

Fig. 12.4 Oxygen in flue gas.

operate with as little excess air, and therefore oxygen, as possible. Both of these considerations are basic in NO_x control techniques.

Do we get NO concentrations such as calculated here? The answer is no, since the real process does not have time to go to equilibrium. The *rate* of oxidation of nitrogen to NO is very temperature dependent, and the rate increases rapidly as the temperature increases. Table 12.9 (from AP-84) gives the time for 500 ppm NO to form as a function of temperature for a particular N/O ratio. Thus the residence time of the nitrogen in the high-temperature zone is another important control parameter. Actual residence times in power-plant boilers are about 2 seconds at or near the maximum temperatures (Heywood, 1971), and 4 to 6 seconds in all. At a typical maximum boiler temperature of 1900°K (2960°F) the time to reach equilibrium is on the order of 5 seconds. A residence time of 2 seconds at peak conditions means that our calculated equilibrium concentrations are higher than the concentrations actually reached.

We must now consider why combustion processes produce NO and not NO_2. Thermodynamic *equilibrium* at high temperature favors the

TABLE 12.9 Time for NO formation in a gas containing 75 percent nitrogen and 3 percent oxygen

Temperature, °F	Time to form 500-ppm NO, sec	NO concentrations at equilibrium, ppm
2400	1,370	550
2800	16.2	1,380
3200	1.10	2,600
3600	0.117	4,150

Source: AP-67.

formation of NO over NO_2. Table 12.10 from AP-67 indicates two sets of calculations illustrating this point. Both show the NO_2 concentration to be much less than that of NO at the higher temperatures. At the low temperatures, however, an equilibrium calculation shows the concentration of NO_2 to be greater than that of NO. This seems to suggest that as gases exhaust to the atmosphere the NO should react with excess oxygen to yield NO_2. However, once again, kinetics enters the picture. The time for the reaction to reach equilibrium is longer than the residence time available so that only 5 to 10 percent of the NO is oxidized to NO_2. But once in the atmosphere, the NO_2 concentration can rise because of the NO, O_3, NO_2 photolytic cycle.

Another possible reaction is the simple decomposition of NO back to N_2 and O_2. This reaction requires appreciable activation energy so it is a slow reaction, and its rate becomes vanishingly small below 2300-2400°F. This means that once the NO is formed it is quite stable, neither decomposing nor oxidizing to NO_2. We say that the NO concentration is "frozen" in that the changes require a long time compared to the actual time available. The frozen concentration thus is very much *higher than the equilibrium value for exhaust conditions but lower than the equilibrium value for peak-temperature conditions.* Since the concentration of NO produced is much greater than the NO_2 level, measurements of the NO_x level under *these* conditions will be close to the NO level.

In summary, the principal factors affecting NO_x emissions are thermodynamic and kinetic effects (although the chemically bound nitrogen in coal and oil is also a factor). The *peak flame temperature*, the amount of *excess air* available, and the *length of time* that the

TABLE 12.10 Calculated equilibrium concentration of nitrogen oxides*
(ppm)

Temperature, °F	Air		Flue gas†	
	NO	NO_2	NO	NO_2
80	3.4×10^{-10}	2.1×10^{-4}	1.1×10^{-10}	3.3×10^{-5}
980	2.3	0.71	0.77	0.11
2060	800	5.6	250	0.87
2912	6,100	12	2,000	1.8

*For the reactions: $N_2 + O_2 \rightleftharpoons 2\ NO$
$2NO + O_2 \rightleftharpoons 2\ NO_2$

†3.3% O_2, 76% N_2 in flue gas.

Source: AP-67.

combustion gases are at or near the peak-flame temperature are the controlling parameters for NO_x production and emission.

Our examples in this section have been based on simple models but illustrate the trends involved in NO_x formation. Bartok et al. (1969) have developed a more complete model and calculate the concentration of NO resulting from the combustion of methane, CH_4. Figure 12.5 illustrates the NO equilibrium concentration as a function of air/fuel ratio. The equilibrium concentration is highest at 10–15 percent excess air, and declines notably as the excess air is reduced to the stoichiometric value even though the adiabatic flame temperature increases. This suggests that limiting available oxygen will reduce NO formation, as we have already discussed. The NO_x concentration is also seen to decrease as the adiabatic-flame temperature decreases.

CONTROL TECHNIQUES

We have discussed the thermodynamics and kinetics of the nitrogen oxides. Now we are in a position to consider the control techniques to reduce the emissions of NO_x. Most of these control measures represent modification in the operating conditions of the combustion equipment. Modifications are made (1) to minimize the residence time at peak

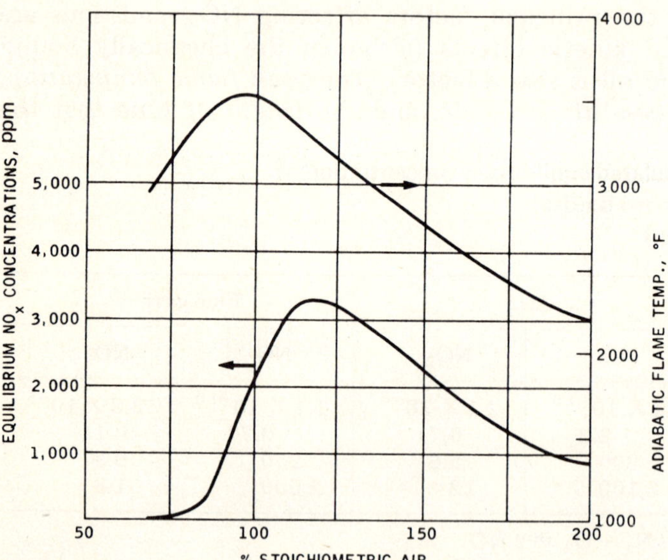

Fig. 12.5 NO_x equilibrium in methane-air flames. (*Bartok et al., 1969.*)

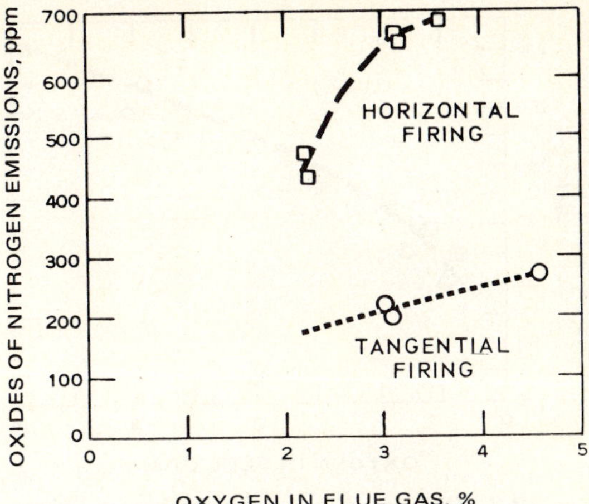

Fig. 12.6 Effect of excess air on oxides of nitrogen emissions from oil-fired boilers.

temperature, (2) to design so that peak temperatures are reduced, possibly by increasing the rate of heat transfer, and (3) to minimize the availability of oxygen for reaction with nitrogen.

1. Low Excess Air Combustion

Most combustion processes use excess air, that is, more than is required just to complete the combustion process. In this manner the effects of poor mixing between the air and fuel can be overcome and complete combustion can be assured. "Normal" boiler units operate at 10 to 20 percent excess air. This means that there is still an ample supply of oxygen available, after reactions with the fuel, to react with the nitrogen. Figures 12.6 and 12.7 (AP-67) indicate how NO_x emissions can be reduced on oil-fired burners by using lower excess air than normal. Note that 1 percent oxygen in the flue gas corresponds to about 5 percent excess air. This technique requires careful control so that emissions of CO and unburned HC are not increased. At present it is most proven for gas- and oil-fired boilers but data for coal-fired boilers also indicate a reduction in NO_x using this technique (Plumley, 1971).

2. Two-stage Combustion

This type of combustion attempts to reduce the peak-flame temperature. In burners of this design only about 95 percent of the

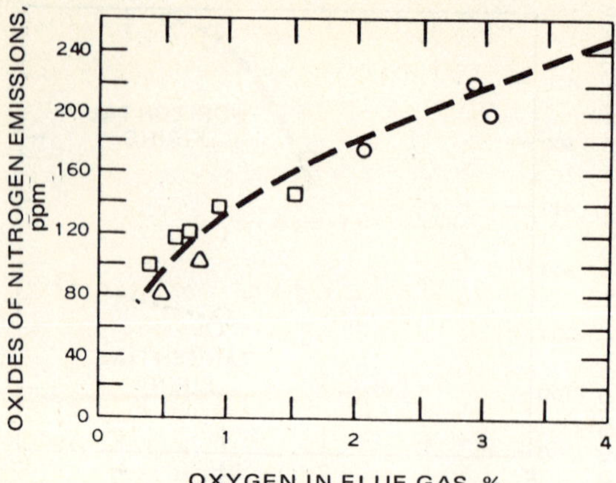

Fig. 12.7 Oxides of nitrogen emissions from oil-fired boilers at low excess air.

stoichiometric air is admitted at the burner. The remainder of the air is injected above the burner to complete the combustion in the second stage. Reductions of NO_x, on the order of 35 percent, have been observed under these conditions in gas- and oil-fired plants. The technique entails reducing available oxygen in the first stage of combustion by operating at less than stoichiometric conditions. Some heat is removed between stages so that the peak temperature in the second stage is lower than in single-stage combustion. Lastly, there may be a reduction in the residence time at the peak-temperature conditions reached in the second stage. Figure 12.8 shows an example of two-stage combustion.

3. Flue-gas Recirculation

Here a portion of the flue gas is recirculated back to the combustion chamber. Since this gas is deficient in oxygen it acts to reduce peak temperatures (as would excess air) without increasing NO_x formation. Many large steam boilers are now designed for some flue-gas recirculation as a control technique under changing load conditions. Figure 12.9 indicates calculated results from the Esso combustion model of Bartok et al. (1969) for the effects of two-stage combustion and flue-gas recirculation. In the two-stage model the first stage excess air was 0 percent. The effect of these control measures is very large, here a 90 percent reduction in NO_x. This model includes the kinetics of the fuel combustion as well as NO formation and decomposition processes. Thus, unlike our simple calculations in this chapter, it is not limited to equilibrium conditions. Table 12.11, from Tomany et al., 1970, indicates actual results for NO_x reduction using these techniques on gas-fired utility boilers. Substantial improvement is noted.

4. Modification of Burner Design

Variations in NO_x level occur with different burner configurations. Figure 12.6, while showing the reduction of NO_x with reduced excess air, also notes the effect of two different forms of boiler firing. In tangential firing the burners are located in the corners of the boiler, firing on a tangent to a circle at the center of the furnace. The flames radiate to the surrounding cooling surface and have little interaction with each other. As a result, peak temperatures are lower and NO_x emissions therefore are lower.

For power plants it appears that a combination of the above techniques will be most effective. Fortunately these control methods are relatively simple, particularly as compared to SO_x removal techniques, and do not require large capital costs.

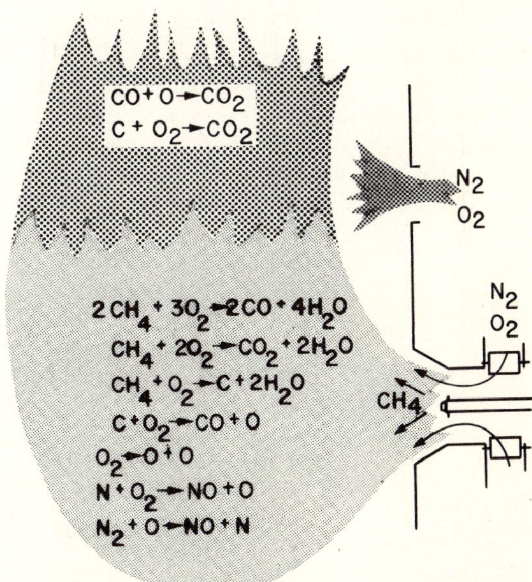

Fig. 12.8 A schematic of two-stage combustion. (*After Rawdon and Sadowski, 1972.*)

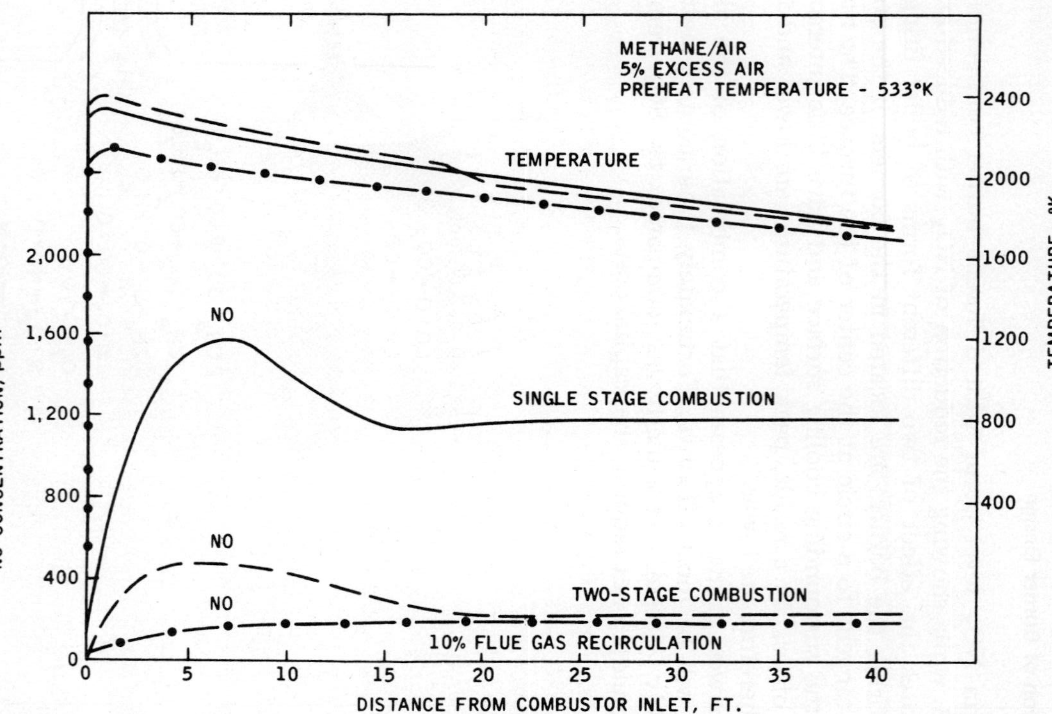

Fig. 12.9 Typical NO-temperature profiles predicted by the mathematical model: effects of two-stage combustion and flue-gas recirculation. (*Bartok et al., 1969.*)

TABLE 12.11 NO_x reduced emission values for various control techniques gas-fired utility boilers

Boiler rating	Control technique	NO_x concentration* Initial	Reduced
220 Mw	Two-stage combustion and off-stoichiometry operation	475	180
320 Mw	20 percent combustion products recirculation	350	150
320 Mw	Reduction of air preheat temperature	250	200

*All concentrations reported dry at 3 percent excess oxygen.

5. Flue-gas Cleaning

Because NO is a relatively stable gas it is difficult to remove from flue gases. The amount of NO_x in flue gas is on the order of 0.07 percent by volume which is even less, by a factor of 3, than typical SO_x concentrations. The wet scrubbing limestone process does remove some NO_x, about 20 percent, while cleaning out the SO_2. At present only limited attempts are being made to find a suitable flue-gas clean-up technique for NO_x, unlike the case for SO_x. The most promising reduction techniques appear to lie in combustion modifications.

STANDARDS OF PERFORMANCE

The amended Clean Air Act of 1970 required EPA to set standards of performance for new stationary sources of air pollution. These standards are to reflect "the degree of emissions limitation achievable through the application of the best system of emission reduction which . . . has been adequately demonstrated." For new fossil fuel-fired steam generators these standards have been set. The nitrogen oxide standards are:

gas-fired	$0.20\ \text{lb}/10^6$ Btu
oil-fired	$0.30\ \text{lb}/10^6$ Btu
coal-fired	$0.70\ \text{lb}/10^6$ Btu

The emissions are maximum 2-hour averages and are expressed as NO_2. EPA[2] has converted these to approximate ppm as:

[2] *Federal Register*, vol. 36, Aug. 14, 1971.

gas-fired 175 ppm
oil-fired 230 ppm

The heating value of coal varies widely between different samples but 0.70 lb/10^6 Btu is roughly 575 ppm.

Comparing these *new* plant standards with emissions from existing plants, we see that something like a 50 percent reduction in NO_x will be required.

TABLE 12.12 Estimated control costs and reduction of NO_x emissions, for a 1,000-megawatt boiler used 6,120 hours/year by selected methods and by ton

Control method	NO_x reduction, %	Fuel used	Control cost/yr, $	NO_x reduction, tons/yr	NO_x control costs, $/ton
Uncontrolled	0	Gas	0	53,000*	0
(base case)	0	Oil	0	30,000*	0
	0	Coal	0	30,000*	0
Low excess air	33	Gas	− 95,000†	17,500	− 5†
	33	Oil	−297,000†	9,900	−30†
	25	Coal	− 79,000†	7,500	−11†
Two-stage	50	Gas	0	26,500	0
combustion	40	Oil	0	12,000	0
	35	Coal	299	10,500	29
Low excess air	90	Gas	− 95,000†	47,700	− 2†
plus two-stage	73	Oil	−297,000†	21,900	−14†
combustion	60	Coal	220,000	18,000	12
Flue-gas	33	Gas	202,000	17,500	12
recirculation	33	Oil	202,000	9,900	20
	33	Coal	202,000	9,900	20
Low excess air	80	Gas	107,000	42,400	3
plus flue-gas	70	Oil	− 95,000†	21,000	− 5†
recirculation	55	Coal	123,000	16,500	8
Water injection	10	Gas	144,000	5,300	27
	10	Oil	179,000	3,000	60
	10	Coal	143,000	3,000	48

*Uncontrolled method, total NO_x emitted; no reduction.
†In instances wherein the installation of control equipment results in a savings, control costs are listed as negative numbers.
Source: AP-67.

COSTS

Bartok et al. (1969) devote a large part of their report to cost analysis for NO_x removal. Table 12.12 indicates the estimated control costs for a plant of 1,000 megawatts electric output. These costs include capital charges, operating costs, and operating credits from fuel savings. Note that for low excess air the control cost is negative. An actual fuel savings occurs, with a decrease in expenses as well as a decrease in NO_x. It thus appears that NO_x reduction in power plants can be achieved without large increases in capital or operating costs.

SUMMARY

NO_x control techniques for stationary sources are at present confined to changes in the combustion process via such methods as flue-gas recirculation, off-stoichiometric or two-stage firing, and low excess air firing. Flue-gas recirculation lowers the NO_x formation from thermal fixing of nitrogen but does not appear to be effective in reducing NO_x resulting from fuel nitrogen conversion. Two-stage combustion, however, appears to reduce both fuel nitrogen oxide and thermal nitrogen oxide. Control techniques for NO_x require a thorough understanding of concepts of thermal equilibrium and reaction kinetics. Results of calculations using both of these concepts were presented in this chapter.

We have not discussed control techniques for such industrial sources of NO_x as nitric acid plants, petroleum plants, or metallurgical processes. Nor have additional sources and their control methods from combustion processes been described, including domestic and commercial heaters, incinerators, and stationary engines (particularly those in use on gas pipelines). Discussion of these subjects may be found in AP-67 (1970).

REFERENCES

AP-67: "Control Techniques for Nitrogen Oxide Emissions from Stationary Sources," National Air Pollution Control Administration, Publication AP-67, 1970.

AP-84: "Air Quality Criteria for Nitrogen Oxides," Air Pollution Control Office, Environmental Protection Agency, Publication AP-84, 1971.

Bartok, W., A. R. Crawford, A. R. Cunningham, H. J. Hall, E. H. Manny, and A. Skopp: "Systems Study of Nitrogen Oxide Control Methods for Stationary Sources," PB192789, 1969.

Hauser, T. R. and C. M. Shy: Position Paper: NO_x Measurement, *Environ. Sci. Technol.*, vol. 6, pp. 890–894, 1972.

Heck, W. W., in O. C. Taylor: Effects of Oxidant Air Pollutants, *J. Occupational Med.*, vol. 10, pp. 485–499, 1968.

Heywood, J. B.: "Fluid Physics of Air Pollution," AIAA short course, Palo Alto, California, June 1971.

Plumley, A.: Fossil Fuel and the Environment—Present Systems and Their Emissions, in "Energy, the Environment, and Education Symposium, April, 1971," University of Arizona Press.

Rawdon, A. H., and R. S. Sadowski: An Experimental Correlation of Oxides of Nitrogen Emissions from Power Boilers Based on Field Data, Paper 72-WA/Pwr-5, Am. Soc. Mech. Eng., 1972.

Robinson, E., and R. C. Robbins: Gaseous Nitrogen Compound Pollutants from Urban and Natural Sources, *J. Air Pollution Control Assoc.*, vol. 20, pp. 303–306, 1970.

Tomany, J. P., R. R. Koppang, and H. L. Burge: A Survey of Nitrogen-Oxides Control Technology and the Development of a Low NO_x Emissions Combustor, Paper 70-WA/Pwr-2, Am. Soc. Mech. Eng., 1970.

PROBLEMS

12.1 A terephthalic acid plant produces 20 tons of acid per day. NO_x controls used are 75 percent effective in removing the nitrogen oxides from the stack gas. Estimate the release of NO_x per day from this plant.

12.2 Estimate the NO_x emission from burning waste for a city of 500,000. Assume 5 lb of waste per person per day and an additional 3 lb per person per day from industrial sources. The waste is burned in a municipal incinerator with no pollution control. Compare your answer for this NO_x source to the emissions from a gas-fired power plant for the same city.

12.3 A power plant of 1,000-megawatt electrical capacity is to run on residual fuel oil of 1.5 percent sulfur content. Estimate the NO_x and SO_x emissions per year from such a plant if the oil produces 18,000 Btu/lb and has a density of 1 g/cm^3. The plant is on line 90 percent of the time.

12.4 Prepare a local inventory of NO_x emissions from stationary sources for your area, using the emission factors given in this chapter.

12.5 Calculate the NO_x emissions from a city of 75,000 homes assuming each is heated with a gas home furnace rated at 100,000 Btu/hr. Compare this source of NO_x to the emissions from the local gas-fired power plant rated at 500 megawatts.

12.6 This chapter did not discuss NO_x control techniques for such industrial processes as nitric acid plants, petroleum refineries, and metal processing. Write a review of the sources and possible control techniques for NO_x from one of these processes.

12.7 Write a brief discussion of the effects of *kinetics* on the reaction Eq. (12.4) in determining NO levels in combustion flue gas.

12.8 The EPA standard of performance for new power plants for natural gas-fired heaters and boilers is 0.2 lb NO_x/million Btu. What is their standard in terms of ppm (dry basis) in the flue gas at 3 percent oxygen measured in the stack? The heating value of the gas is 1000 Btu/scf. (Treat NO_x as NO_2.)

12.9 NO concentrations are often referred to as "frozen" into the products of combustion. Explain physically why this occurs.

12.10 Instruments using chemiluminescence are finding increasing acceptance as a means of measuring NO_x levels. Write a brief description of how this technique works and what reactions are important to the NO_x determination.

13

SOME EFFECTS OF AIR POLLUTION

> **Real estate operators who are very active in selling resort property at first denied any smog damage in the mountains. One, however, who asked anonymity, said, "Don't use my name, the other realtors would crucify me. But, you're darned right that the smog has killed trees and it has hurt business.**
>
> *The Press, Riverside, California, November 15, 1971*

What does air pollution do? We have already seen some of the global effects of air pollution, but what are its results on a more localized basis? In this chapter we shall deal with the effects of air pollution on plants, on the corrosion of materials, and on deterioration of works of art. We shall also examine the estimates available as to the costs of air pollution to society.

PLANT DAMAGE

Air pollution has long been known to have an adverse effect on plants. Industrial pollution, particularly from smelters, has on occasion caused complete destruction of vegetation as at Ducktown, Tennessee (Seigworth, 1943). The international dispute between Canada and the United States over damage caused by the Trail, British Columbia copper smelter, is also well documented (NRC Canada, 1939). More recently, a result of Los Angeles smog has been widespread damage to certain crops and forests in southern California. Two publications have printed numerous pictures of damaged plants (Hindawi, 1970; Jacobson and Hill, 1970).

If we first examine the physiology of the leaf we can appreciate some of the reasons why damage occurs. Figure 13.1 shows a leaf and its internal tissue. The leaf veins function much as the blood vessels do in animals, acting as the transport system for water, minerals, and food. The leaf tissue is in layers with a skin or epidermis layer on top and bottom and the photosynthesis cells in between. The stomata are entrances in the leaf bottom (and in some leaves in the top) through which CO_2 enters to take part in the photosynthesis process. These openings are protected by two guard cells which open and close to

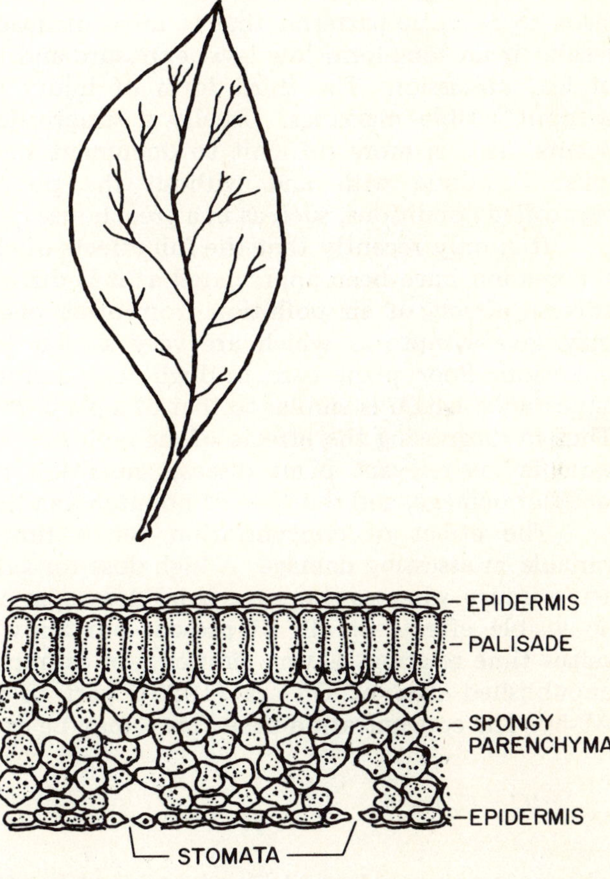

Fig. 13.1 A leaf and its internal tissue.

allow gases to enter or leave the leaf. Such gases can, of course, also include pollutants.

Damage to leaves takes several forms. *Necrosis* is a killing of tissue and can result from any of several air pollutants. *Chlorosis* is the loss or reduction of the green plant pigment, chlorophyll, with a resulting yellowing of the leaves. Leaf *abscission* is a dropping of leaves and leaf *epinasty* is a downward curvature of the leaf due to a higher rate of growth on the upper surface. Phytotoxicant is the name given to plant damaging substances.

Three kinds of injury are commonly considered important. Acute injury results from short-time exposure to relatively high concentrations, such as might occur under fumigation conditions. The effects are noted within a few hours to a few days and could, for example, be visible markings on the leaves due to a collapse and death of cells. This leads to necrotic patterns, that is, areas of dead tissue. Chronic injury results from long-term low level exposure and usually causes chlorosis or leaf abscission. The third form of injury is an effect on growth without visible markings. Usually a suppression of growth or yield occurs. This is more difficult to document since careful experiments must be done with and without the pollutants under carefully controlled conditions, such as in a greenhouse.

It is only recently that the full effects of chronic injury or growth suppression have been appreciated as it is difficult to separate out the adverse effects of air pollution from other phenomena. Plant diseases may give symptoms which are very similar to those caused by air pollution. Poor plant care or high temperatures may also cause an appearance which is similar to that of a plant damaged by air pollution. Thus in diagnosing the effects of air pollution one must consider such variables as relevant plant disease, nutrition history, insect damage, weather damage, and the type of pollutants in the area.

The effect of concentration versus time is also an important variable in assessing damage. A high dose for a short time may cause an acute injury whereas the same total dose over a longer time may cause no visible effects at all. At present our knowledge of concentration versus time relations is very limited. An example of such results is the unpublished data of O'Gara (1922), part of which are presented in AP-50. The concentration-time "law" may be expressed as

$$t(C - C_o) = K$$

where t = time in hours to produce a certain effect on a certain species
C = concentration of a specific gas in parts per million

C_o = the threshold concentration of the gas, in parts per million, to cause injury

K = an experimentally determined constant.

If we rewrite this equation as

$$C = \frac{K}{t} + C_o$$

then plotting C versus $1/t$ gives a straight line whose intercept C_o is the threshold for injury. With the exception of some results for SO_2 and O_3 there is not sufficient data to determine these parameters assuming that a linear model is even accurate. In addition, the results are complicated because for a given species different varieties can exhibit markedly different reactions to the same air pollutant.

Figure 13.2 shows some dose relations for a tobacco plant exposed to ozone. This particular variety, Bel W3, is very sensitive to oxidants. The complicated relationship is obvious. Note particularly the steep slope after 0.5- and 1-hour exposure times at moderate concentrations. Also note that the relations are highly nonlinear. These results are for an exposure to a single pollutant. In the field, however, exposures are usually not due to individual gases but to mixtures of pollutants whose relative concentrations change as functions of time. There are thus an almost infinite possible number of combinations of pollutants and plant species which could be tested. Certain tobacco plants are particularly susceptible to ozone damage and are often used as an indicator of the existence of this pollutant. Another plant monitor is gladiolus which is used as a measure for fluorides.

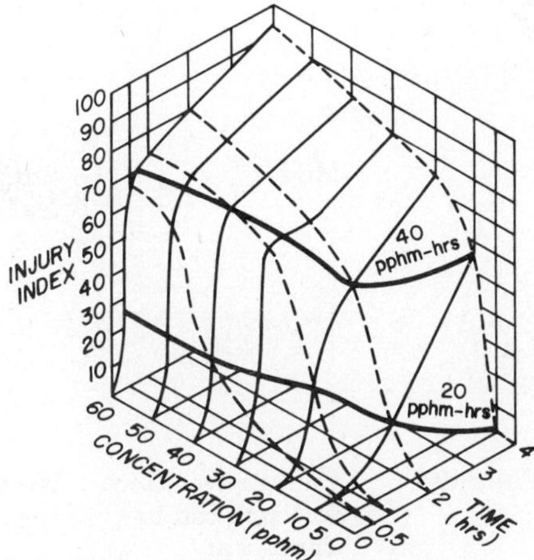

Fig. 13.2 Interrelations of time and concentration on the sensitivity of tobacco (Bel W_3) plants to ozone. (*Reprinted from W. W. Heck, et al.,* Science, **151**, *577–578, February 4, 1966. Copyright © 1966 by the American Association for the Advancement of Science.)*

TABLE 13.1 Effects of pollutants on plants

Pollutant	Dose	Effect	Specific cases*
Ozone	Mild	Flecks on upper surface, premature aging, suppressed growth	0.06 ppm for 3–4 hr damages alfalfa, white pine. Ponderosa pine badly damaged in S. California
	Severe	Collapse of leaf, necrosis and bleaching	
PAN	Mild	Bronzing of lower leaf surface (upper surface normal), suppressed growth. Young leaves more susceptible	0.03 ppm acute damage to sensitive plants: spinach, romaine lettuce, certain flowers
SO_2	Mild	Interveinal chlorotic bleaching of leaves	1 ppm for fumigation times causes damage to alfalfa, cotton, barley. 0.3 to 0.5 ppm for several days on sensitive plants causes chronic injury
	Severe	Necrosis in interveinal areas and skeletonized leaves	
Ethylene	Mild	Epinasty, leaf abscission	0.1 ppm for several hours affected tomatoes and pepper plants. Flowers affected at 0.1 ppm, i.e., orchids, carnations after a few hours
NO_2	Mild	Suppressed growth, leaf bleaching	0.5 ppm for 10–12 days suppressed growth on pinto beans and tomatoes. Navel orange yield greatly reduced at 0.25 ppm for 240 days, 0.5–1 ppm for 35 days.
Fluorides	Cumulative effect so 1 ppb can be significant	Necrosis at leaf tip	Gladioli, peaches, and grapes are particularly sensitive

TABLE 13.1 *(Continued)*

Pollutant	Dose	Effect	Specific cases*
Mixtures:			
Menser & Heggestad (1966)	$O_3—SO_2$ 0.028–0.28 ppm, 2–4 hours	Leaf damage on tobacco plants	No damage at these levels for single pollutants
Heck (1968)	$O_3—SO_2$ 0.03–0.10 ppm, 4 hours	Tobacco leaf damage in Bel W3 variety	Higher levels affected radish, tomato, broccoli, alfalfa
	$NO_2—SO_2$ 0.1–0.1 ppm, 4 hours	Moderate leaf injury to Bel W3 tobacco	NO_2 at 2 ppm alone OK for this time, SO_2 at 0.7 ppm OK for this time

*These examples are included to indicate some specific results. The results were obtained under a particular set of conditions; other experiments, even though similar, may not yield the same results.

Specific pollutants and their effects are noted in Table 13.1. A more detailed result for ozone and an indication of the level required for damage is given in Table 13.2. Remember that the typical ambient levels of ozone discussed in Chap. 5 were often above the levels indicated here, at least for short periods of time.

The results of Table 13.1 also include some mixtures of pollutants which exhibit synergistic effects (1 + 1 = more than 2). Such arithmetic may explain why different studies of individual pollutants have not correlated well.

The mechanisms by which these pollutants affect the plants are not yet well understood. AP-50 suggests for SO_2 that

> The mechanism by which acute injury occurs apparently involves the plant's ability to transform absorbed sulfur dioxide into sulfuric acid and then into sulfates which are deposited at the tip or edges of the leaf; with a high rate of absorption, sulfite is thought to accumulate, and sulfurous acid is then formed which subsequently attacks the cells. Chronic injury, on the other hand, results from the gradual accumulation of excessive amounts of sulfate in the leaf tissue. Sulfate formed in the leaf from sulfur dioxide absorbed from the atmosphere is additive to sulfate absorbed through the roots.

For SO_2 we see that levels greater than 0.5 ppm for short times can cause plant damage. The results in Chap. 11 indicated that a few large cities in the United States are over this level on a significant number of days per year. The high SO_2 days, however, tend to be in the winter and thus out of the growing season.

TABLE 13.2 Effects of ozone, from Heck (1968)
*Projected concentrations of ozone producing injury to vegetation**

Exposure, hours	Ozone concentrations (ppm)		
	Sensitive plants†	Intermediate plants‡	Resistant plants§
0.1	0.40	0.55	0.90
0.2	0.30	0.40	0.75
0.5	0.15	0.25	0.50
1.0	0.10	0.20	0.35
2.0	0.07	0.15	0.25
4.0	0.05	0.10	0.20
8.0	0.03	0.08	0.15

*Plants grown under sensitive conditions.
†Sensitive plants: spinach, radish, muskmelon, oat, pinto bean, white pine, potato, tomato, and tobacco.
‡Intermediate plants: begonia, onion, spinach, chrysanthemum, dogwood, sweet corn, wheat, and lima bean.
§Resistant plants: zinnia, beet, radish, poinsettia, black walnut, bean, strawberry, and carrot.

It is hard to assess the economic cost of air pollution to agriculture, largely because reduced yields and reduced plant growth are difficult to document. The California Department of Agriculture estimated damage in that state to crops in 1969 totaled 44.5 million dollars. This figure included invisible damage to citrus, though not to other crops. Most of this loss, 33.5 million, was to citrus, and of the total, 39.5 million was for damage in southern California.

The large national forests in southern California are being severely reduced by smog damage, particularly those of Ponderosa pine. The oxidant fraction of smog causes a yellow chlorotic mottling of the needles and subsequently an early dropping. Eventually the tree lacks sufficient needles to produce the required foodstuffs and the root system is affected. In turn this reduces growth and needle development. Damaged trees are also more susceptible to insect attack and it is the bark beetle that often kills the tree. An estimated 1.3 million trees are affected in the San Bernardino National Forest alone. At 20 dollars a tree (the bid rate in 1969–1970) the direct cost is 26 million dollars. Indirect cost in effects such as possible future soil erosion, real estate losses, and aesthetic losses cannot be calculated.

The total dollar cost to United States agriculture depends on whose estimate is used. Benedict et al. (1971) have estimated costs owing to oxidants, sulfur oxides, and fluorides as 132 million dollars per year. EPA has a figure of 500 million dollars per year from all pollutants. Both sources include losses to crops, forests, and ornamental plants. The figures reflect the costs of reduced growth and reduced yields as well as direct costs.

CORROSION

Corrosion usually means metallic corrosion. Our definition will be more general and include effects on metals, building materials, and textiles. The prime air pollutant responsible for metallic corrosion is SO_2 which, when converted to H_2SO_4, causes electrochemical corrosion of the metal. This is noted in experiments as a weight loss of metal from the test panel. Carbon soot appears to accelerate this process, probably because of the absorption or adsorption of the gaseous pollutants on the particulate matter. Other important variables are relative humidity and temperature. Figure 13.3 notes the weight loss of 100-gram mild steel test panels exposed in Chicago at seven different sites (Upham, 1967). The correlation between time of exposure and concentration is obviously good. Metals adversely affected include, but are not limited

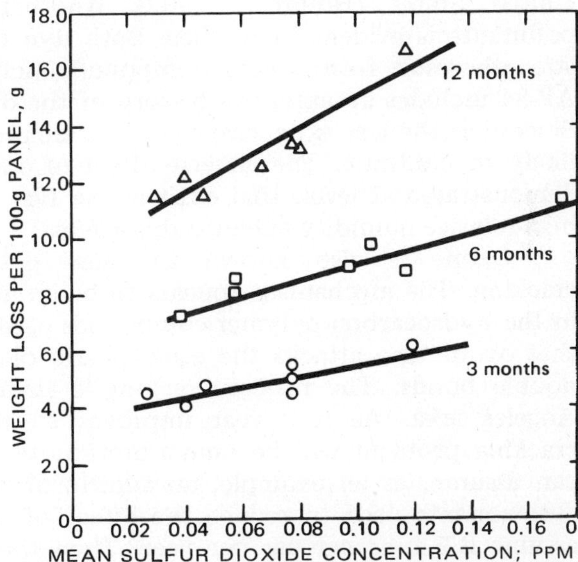

Fig. 13.3 Relationship between corrosion of mild steel and corresponding mean sulfur dioxide concentration for 3-, 6-, and 12-month exposures at seven Chicago sites (Sept. 1963–1964).

to, steel, aluminum, copper, zinc, and iron. Studies similar to the Chicago study have been done in Japan (1963) and Sweden (1971) with like results. Swedish results indicate a difference by a factor of 15 between the corrosion rate of carbon steel in Stockholm and in a subarctic rural atmosphere with low temperatures and very low SO_2 and sulfate levels. The cost of this effect of polluted air must include deterioration of such diverse objects as steel railroad rails, overhead wires, outdoor electrical contacts, metal on bridges, roofs, and so on. Such costs can only be estimated and no definite figures are available. Additional costs such as painting, replacement, and substitution of more expensive materials for corroded ones would also have to be included.

Building stones, too, are affected by corrosive air pollutants. In particular, limestone ($CaCO_3$) and dolomite ($CaCO_3$ and $MgCO_3$) are attacked by sulfuric acid. Staining of the stone by a combination of sulfur compounds and particulates requires additional cleaning expense. Since one of the proposed clean-up techniques for SO_2 in flue gas is to react it with $CaCO_3$, it is not surprising that the same reaction takes place, undesirably, in nature.

Paint is also affected by pollution: areas of high pollution require more frequent painting. The drying of paint is adversely affected by SO_2 levels that are just above those existing in major cities of the United States (Holbrow, 1962). Another corrosive effect of air pollutants is evident in textiles. Both dye fading and reduced tensile strength result from several compounds including SO_2, NO_x, and O_3. AP-84 includes an extensive history of the dye fading caused by NO_x. Since it is the less expensive dyes that are most affected the problem is likely to continue. The adverse effect of ozone on materials has been demonstrated at levels that exist in the Los Angeles area. Sunlight and high relative humidity enhance this effect.

Ozone is also known to cause deterioration of rubber by cracking. The mechanism appears to be an attack at the double bonds in the hydrocarbon polymer compounds used in the rubber. Remember that ozone also attacks the atmospheric olefins and diolefins at their double bonds. The rubber cracking is already a problem in the Los Angeles area. As tire wear improves so that tires last longer, the cracking problem will become a more noticeable economic effect. We can assume, as an example, an additional cost of 1 dollar a tire for antiozonants used to reduce the effect of ozone on the rubber, and assume 2.5 tires per car per year. Then the cost of inhibiting ozone attack in the greater Los Angeles area alone is some $2.5 \times 6 \times 10^6 = 15$ million dollars per year.

Soiling and cleaning costs are not a corrosion effect but we may note in this section that the effect of both gaseous and particulate air pollutants is to increase markedly such cleaning costs both for buildings and individuals. A visit to most large eastern cities in the United States quickly demonstrates the effect of particulate fallout on personal cleaning costs. Some of these costs will be described shortly.

ART TREASURES

Air pollutants have caused great damage to art objects throughout the industrialized world. The damage to Cleopatra's needle, moved to London from Alexandria, Egypt, is one such example of accelerated deterioration. More damage has occurred in the last 70 years in London than in the previous thousands of years. The Parthenon in Athens has similarly been affected in recent decades. Nitrate, sulfate, chloride, and other salts grow upon crystallization and if initiated in a small crack the force of this growth can enlarge the crack and cause the stone to crumble.

Paint dyes are also attacked by pollutants and color changes or fading occur. H_2S is particularly active in affecting pigments, especially some of the older yellows and reds. Frescoes from pre-Renaissance times have been badly damaged by air pollutants in some southern European cities.

Less recognized but also important is SO_2 damage to paper and leather. Paper products produced from about 1750 on are embrittled by sulfur compounds as they are converted to sulfuric acid. Thus old books which are not stored in sealed cases undergo a gradual deterioration. Leather bindings are similarly affected.

What is the cost of such deterioration? In many ways damage to the world's art treasures is a good example of the difficulties encountered in estimating the costs of pollution. The intangible assets of art are impossible to price, just as it is difficult to put cost tags on many other adverse effects of pollutants. One may argue that the costs are, after all, not as important as other effects we must deal with. The effort to clean up the air is accelerated, however, and often only initiated, when the costs of not cleaning up become evident.

The reduction of insolation due to particulate concentrations (Randerson, 1970 and AP-49) is another important effect of air pollution. In some areas this reduction may be sufficient to affect plant growth although data to indicate this is lacking. The acid build-up in lakes and rivers in Scandinavia is another example of a pollutant effect, one which results from long-distance transport of sulfur compounds. Reduction of forest growth is also suspected in Sweden because of a gradual acidification of the forest land (Sweden, 1971).

COSTS

As we have seen it is difficult to estimate the costs of air pollution as regards some of its damaging effects. Certain costs, such as those to agriculture, have been estimated, but many of the economic effects of pollution are indirect or hidden and thus difficult to determine. Examples would be the reduced growth of crops because of low level, but chronic, exposure from combinations of air pollutants; additional cleaning costs (how much of the dirt in the air is man-made air pollution and when is the building or shirt dirty enough to clean?); and damage to art work. In spite of these difficulties some estimates have been made. One study has compared the economic losses resulting from air pollution in a relatively heavily polluted area to those in a less polluted city. Steubenville, Ohio, with a suspended particulate level of 235 $\mu g/m^3$ was compared to Uniontown, Pennsylvania, where the particulate level was 115 $\mu g/m^3$. Table 13.3 shows the *difference* per person and the difference in total costs incurred by living in Steubenville. If we remove the category of hair and facial care which would seem to be a rather subjective activity, we find that the additional costs attributed to air pollution by living in the more polluted area were 74 dollars per year per person. Thus we should expect that total air pollution costs per person *in polluted areas* would be greater than 74 dollars per year.

The staff report of the Committee on Public Works (1963, printed in Wolozin, 1966) comments: "Accurate data are not available on the extent and cost of air pollution damage to property. Various cost estimates have been made. One frequently employed is 65 dollars per capita

TABLE 13.3 Differences in cleaning costs incurred at Steubenville and at Uniontown

Activity	Gross cost differences (Steubenville over Uniontown)	
	Annual	Per capita*
Outside maintenance of houses	$ 640,000	$17
Inside maintenance of houses and apartments	1,190,000	32
Laundry and drycleaning	900,000	25
Hair and facial care	370,000	10
Total	$3,100,000	$84

*Based on estimated 1959 population of 36,400.
Source: AP-49, after Michelson and Tourin (1966).

per year. This would represent an annual cost to the nation of over 11 billion dollars." Note that this figure is an indication only of costs to property and does not include health effects. Rossano (1969) suggests that the total estimated economic loss is 12 billion dollars per year without including health effects.

Lave and Seskin (1970) have attempted to estimate the economic effects of air pollution on human health. They estimate "the total annual cost that *would be saved* by a 50 percent reduction in air pollution levels in major urban areas, in terms of decreased morbidity and mortality, to be 2.08 billion dollars." This is not the health-related cost of air pollution, which would be greater in their estimate, but the possible savings based on the *earnings lost* due to sickness, disability, and early deaths attributed to air pollution. The major part of this figure comes from respiratory disease. The figure does not include health costs from smoking.

More recently, the Council on Environmental Quality, in its second annual report (CEQ, 1971), suggested a cost figure of 16 billion dollars, comprising health costs of 6 billion, materials and vegetation damages of 4.9 billion, and reduced property values of 5.2 billion.

Using all of these estimates, we suggest that the per capita cost of air pollution is on the order of that given in the 1963 estimate, plus some health-related cost. This would suggest a figure of 70–80 dollars per person per year, or a total United States national cost of about 14–16 billion dollars per year. It is of course true that some of these costs provide employment for window washers, painters, medical personnel, etc. Thus one man's pollution may well result in another man's income. The cost figure is important, however, in providing a possible guide to reasonable costs of cleaning up. Figure 13.4 indicates a balance between the cost of pollution and the cost of control. This

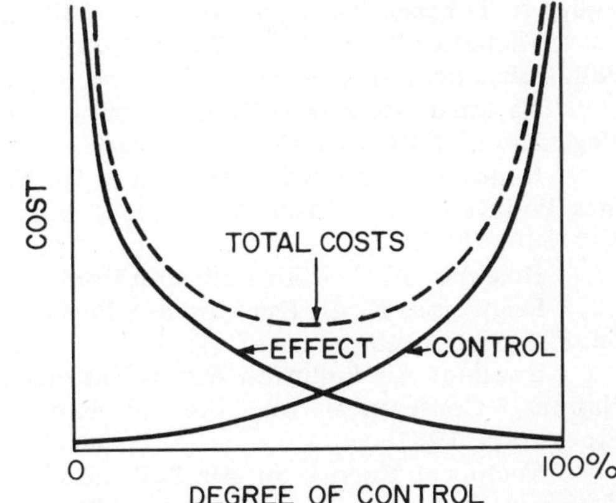

Fig. 13.4 Cost of air pollution effects and control vs. degree of control.

figure from Yocum and McCaldin (Stern, 1968), illustrates the point that the first costs of pollution can be greatly reduced with only a small expenditure, but the reduction of the last part of the costs requires very large expenditures. We would like to be at least on the right-hand side of the minimum total cost. The figure of 15 billion dollars per year gives an indication of what we might expect to pay in clean-up costs if economics were the sole criterion.

REFERENCES

AP-49: "Air Quality Criteria for Particulate Matter," National Air Pollution Control Administration, Publication AP-49, 1969.

AP-50: "Air Quality Criteria for Sulfur Oxides," National Air Pollution Control Administration, Publication AP-50, 1969.

AP-84: "Air Quality Criteria for Nitrogen Oxides," Air Pollution Control Office, Publication AP-84, 1971.

Benedict, H. M., C. J. Miller, and R. E. Olson: "Economic Impact of Air Pollution on Plants in the United States," Stanford Research Institute, November, 1971 (PB 209 265).

Environmental Quality, the Second Annual Report of the Council on Environmental Quality, August, 1971.

Heck, W. W., discussion to O. C. Taylor: Effects of Oxidant Air Pollutants, *J. Occupational Med.*, vol. 10, pp. 485–499, 1968.

Heck, W. W., J. A. Dunning, and I. J. Hindawi: Ozone: Nonlinear Relation of Dose and Injury in Plants, *Science*, vol. 151, pp. 577–578, 1966.

Hindawi, I. J.: "Air Pollution Injury to Vegetation," National Air Pollution Control Administration, Publication AP-71, 1970.

Holbrow, G. L.: Atmospheric Pollution: Its Measurement and Some Effects on Paint, *J. Oil Colour Chemists Assoc.*, vol. 45, pp. 701–718, 1962.

Jacobson, J. S. and A. C. Hill (eds.): "Recognition of Air Pollution Injury to Vegetation: A Pictorial Atlas," Air Pollution Control Association, 1970.

Lave, L. B. and E. P. Seskin: Air Pollution and Human Health, *Science*, vol. 169, pp. 723–733, 1970.

Menser, H. A. and H. E. Heggestad: Ozone and Sulfur Dioxide Synergism: Injury to Tobacco Plants, *Science*, vol. 153, pp. 424–425, 1966.

Michelson, I. and B. Tourin: Comparative Method for Studying Cost of Air Pollution, *Public Health Rept. U.S.*, vol. 81, no. 6, pp. 505–511, 1966.

National Research Council of Canada: "Effects of Sulfur Dioxide on Vegetation," NRC 815, Ottawa, Canada, 1939.

Randerson, D.: A Comparison of the Spectral Distribution of Solar Radiation in a Polluted and a Clear Air Mass, *J. Air Pollution Control Assoc.*, vol. 20, pp. 546–548, 1970.

Rossano, A. T.: "Air Pollution Control," E.R.A. Inc., Stamford, Conn., 1969.

Seigworth, K. J.: Ducktown-A Postwar Challenge, *American Forests*, vol. 49, pp. 521–523, 1943.

Sweden: Air Pollution Across National Boundaries . . ." Report to the United Nations Conference on the Human Environment, Utrikesdepartementet, Stockholm, 1971.

Technical Report on Air Pollution in Yokohama-Kawasaki Industrial Area, Kanagawa Prefectural Government, 1963.

Upham, J. B.: Atmospheric Corrosion Studies in Two Metropolitan Areas, *J. Air Pollution Control Assoc.*, vol. 17, pp. 398–402, 1967.

Wolozin, H. (ed.): "The Economics of Air Pollution," Norton, New York, 1966.

Yocum, J. E., and R. O. McCaldin: Effects of Air Pollution on Materials and the Economy, in Stern, "Air Pollution," vol. 1, Academic Press, New York, 1968.

14

EFFECTS OF AIR POLLUTION ON HUMAN HEALTH

We are in somewhat the same position in regard to polluted air as the fish are to polluted water. We live in it.

A. V. Kneese[1]

Because of the large number of variables involved in studies of air pollution and human health, it is difficult to prove that air pollution has a clearly demonstrable effect on human health at "normal" urban concentrations. We shall see that several studies show (but do not prove) such effects. Of course it is clear at the elevated levels occurring in air pollution disasters that air pollution effects can cause severe health changes and a quick death. It is also clear that air pollution has adverse effects on those who already have respiratory disease; it is more difficult to show that air pollution is the basic cause of the disease. We shall discuss in this chapter some epidemiological studies which indicate the adverse effects of ambient levels of air pollutants.

There are a large number of studies of effects of gaseous pollutants and particulates on animals (the so-called lower animals as contrasted to man). Most of these have been done at very high concentrations and thus have no direct meaning for humans living in lower urban

[1] H. Wolozin, (ed.), "The Economics of Air Pollution," Norton, New York, 1966.

concentrations of these pollutants. There is, furthermore, a wide variety of response by different animal species to the same concentration of pollutant. This complicates any effort to extrapolate these effects to humans. The guinea pig may be the laboratory animal most susceptible to sulfuric acid aerosols but it withstands concentrations which would be intolerable to man (AP-50).

Without having nearly as much data as we would like, we are forced to do the best we can in predicting the effects, if any, of a given dose of pollutant over a given time. We must also recognize that, despite the desire to clean up the air, we cannot yet implicate most ambient air pollution levels as a source of illness. As we learn more about man's response to toxic materials, however, it is usually the case that increasingly smaller doses can be shown to have an effect.

TABLE 14.1 Some important terms

Alveoli (plural)	Small, sac-like dilations at the innermost end of the airway, through whose walls gaseous exchange takes place
Bronchiole	One of the finer subdivisions of the bronchial tree
Bronchitis, chronic	A long standing inflammation of the bronchi, manifested by cough and the production of sputum
Bronchi (plural)	One of the larger air passages in the lung
Cilia	Small, hairlike process attached to the free surface of a cell, capable of rhythmic movement
Emphysema	A condition in which there is overdistension of air spaces and resultant destruction of alveoli and loss of functioning lung tissue
Epidemiology	A science dealing with the factors involved in the distribution and frequency of a disease process in a population
Morbidity	The occurrence of a disease state
Mortality	The ratio of the total number of deaths to the total population, or the ratio of the number of deaths from a given disease to the total number of people having that disease
Synergism	A situation in which the combined action of two or more agents acting together is greater than the sum of the action of these agents acting separately, also called potentiation
In vivo	Within the living body
In vitro	Within a test tube or artificial environment

Let us first present a brief description of the anatomy of the human respiratory system and the mechanisms by which it cleans itself. The famous, or infamous, episodes of air pollution will then be discussed. Lastly, we shall describe the effects of air pollution on human health as determined by epidemiological studies and other research. Table 14.1 presents some of the vocabulary to be used in this chapter.

THE RESPIRATORY SYSTEM

The human respiratory system functions to take in oxygen needed by the metabolic processes and to remove carbon dioxide produced in the body. The exchange of these two gases takes place in the lung, and blood acts as the carrier to transport oxygen to the tissue and to bring carbon dioxide to the lungs. We breath some 15–18 times per minute at rest and take in about 0.5 liters of air per breath. Most of this air is mixed in the lungs with air that is already there but a portion remains in the "dead space", which is the name given to the large air spaces in the respiratory system where no gaseous exchange takes place.

Most of the oxygen and carbon dioxide transported by the blood are not dissolved in the plasma but instead combine with the hemoglobin to move around the body. Oxyhemoglobin is the name given to hemoglobin to which oxygen is attached. Carboxyhemoglobin COHb, is the name given to hemoglobin in which carbon monoxide has become attached. Since CO has some 200 times the affinity of O_2 for hemoglobin it is easy to see why low levels of CO can still produce significant amounts of COHb.

The respiratory system is shown in Fig. 14.1. The nasopharyngeal section, the tracheobronchial system, and the pulmonary structure are the three main sections. The nasal passages lead through the nasopharyngeal structure (throat) to the trachea and bronchi. The bronchi break up into smaller bronchioles which then terminate in the alveolar sacs where exchange of oxygen and carbon dioxide occur between the air and the blood.

There are an incredible number of these tiny sacs and they provide a very large surface area for this exchange. As a rough idea of the size we can consider the surface area for gaseous exchange to approximate the surface area of a tennis court. The alveoli themselves are between 150μ and 400μ in diameter.

The airways narrow as the air goes from the trachea to the alveoli. However, the increase in cross-sectional area more than compensates, and the velocity of the air slows as it penetrates into the lungs. Figure 14.2 shows the bronchial and alveolar structure of the lungs.

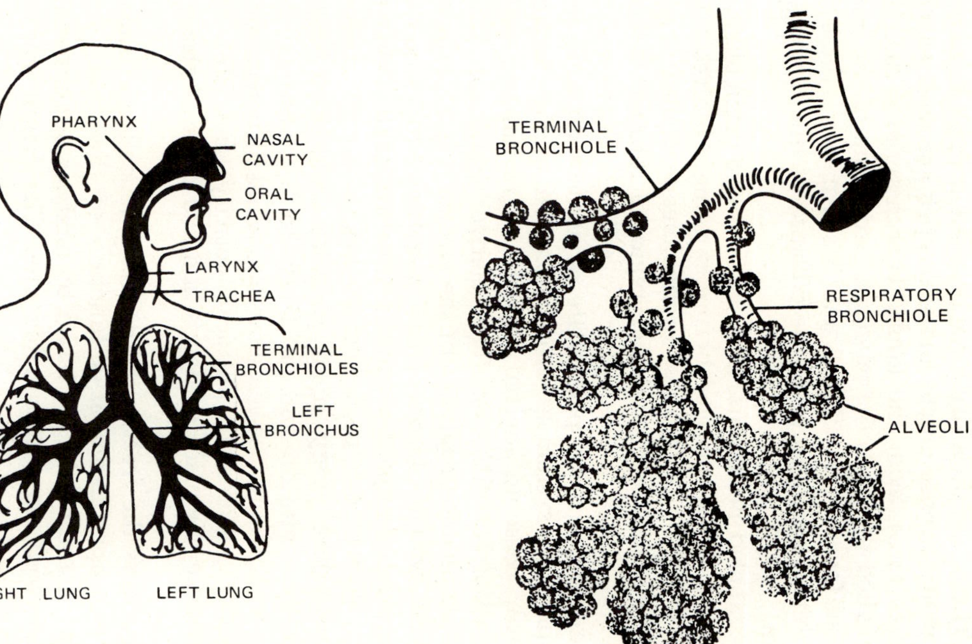

Fig. 14.1 The major anatomical features of the human respiratory system. (The diagram shows the major divisions of the human respiratory tract into nasopharyngeal, tracheobronchial, and pulmonary compartments.)

Fig. 14.2 The terminal bronchial and alveolar structure of the human lung. (The diagram shows the pulmonary structure of the respiratory tract.)

Another very important part of this structure is the cilia noted in Fig. 14.3. The nasopharyngeal and the tracheobronchial tubes have a layer of cells which have many small hair-like projections on them called the cilia. These beat in a rhythmic movement to produce a wave motion upward which carries mucus and entrapped particulates, germs, and dissolved gases away from the lungs. This material is then either swallowed or coughed up when the mucus is carried out of the trachea. The pulmonary surface is not ciliated and has a moist layer of cells covered by a material which prevents the collapse of the alveoli at the end of respiration.

We see that inhaled pollutants, both gaseous and particulate, have several opportunities to be removed before entering the alveoli. Gases can be absorbed along the pharynx or bronchioles and particulate can be deposited in the mucus layer and then removed through the action of the cilia.

The same mechanisms for particulate or aerosol droplet removal exist in the respiratory system as those discussed in the chapter on particulate control. These are inertial impaction, gravitational settling, and Brownian motion or diffusion. The relative significance of each mechanism depends on the properties of the aerosol and on its position in the respiratory system. Large, heavy particles will tend to be removed through inertial impact where the direction of flow changes at branching points. We would expect most of these to be removed in the nasopharyngeal structure. Very small particles, less than 0.5μ, are affected by Brownian motion and are likely to be deposited in the lower pulmonary tract where low velocities and small distances enhance effects of diffusion. Increased physical activity and thus rapid respiration will change the place of deposition and rates of absorption of aerosols.

Figure 14.4 shows the fraction of particles deposited in the three respiratory tract compartments as a function of the (mass median particle) diameter. These results are based on analysis but experiments indicate that the model is a reasonable one. The figure shows the

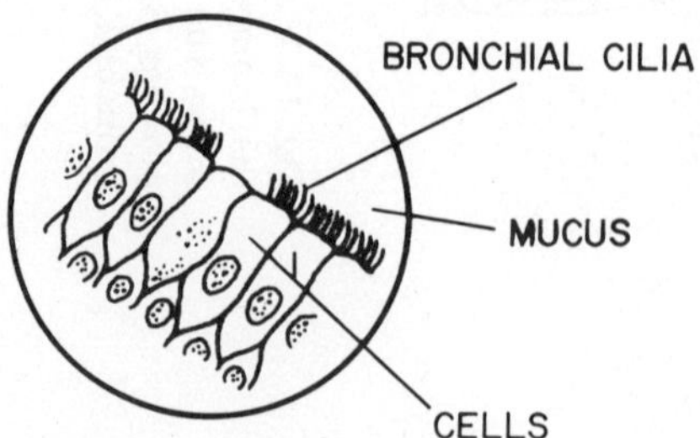

Fig. 14.3 The cilia.

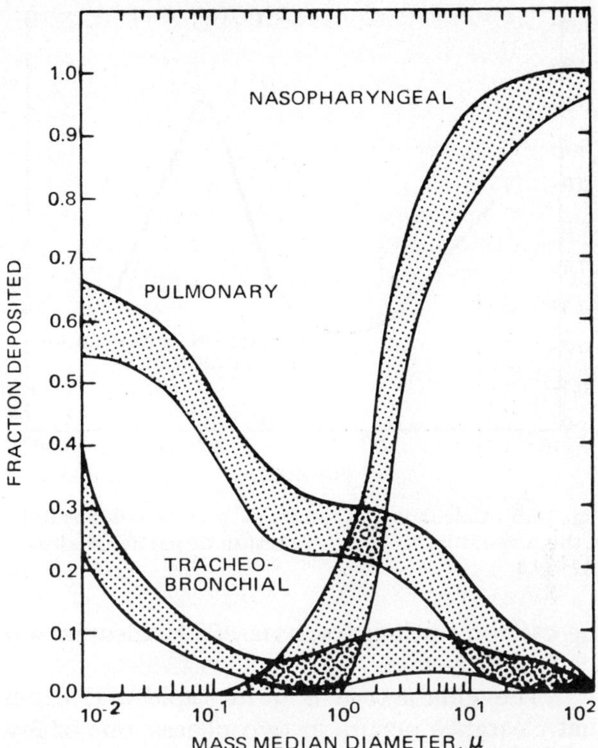

Fig. 14.4 Fraction of particles deposited in the three respiratory tract compartments as a function of particle diameter. (*AP-49.*)

deposition in each compartment of the particles inhaled. Note, however, that the alveolar region never sees most of the large particles since these are already removed in the nasopharyngeal system. Even a significant fraction of the smaller particles are removed before reaching the alveolar. On Fig. 14.4 the mass median diameter refers to the particle diameter above which half of the *mass* in the particulate population lies. It is, of course, much larger than the median diameter since a few large particles can contain a great part of the weight in a given sample. Figure 14.5 indicates the fraction of particles deposited in the alveolar region as a function of actual particle radius. We see that at about 1μ there is a peak in the removal efficiency.

The removal mechanism or clearance processes are of obvious importance in considering the effect of air pollution on the respiratory system. The material removed may be taken up by the blood stream, by

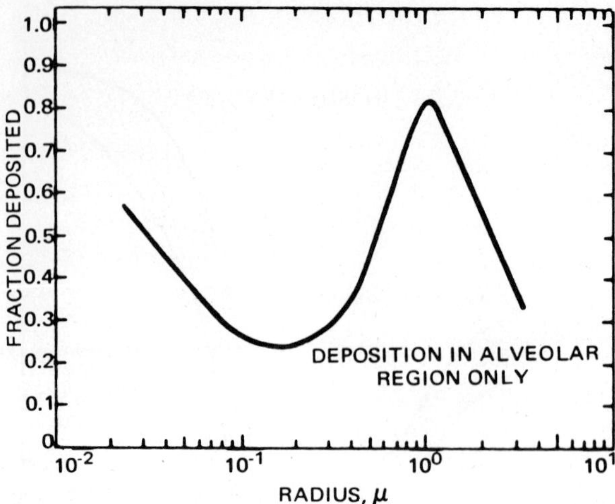

Fig. 14.5 Calculated fraction of particles deposited in the respiratory tract as a function of particle radius. (*AP-49.*)

the gastrointestinal tract via ciliary clearance of the bronchioles, or by the lymph system.

The cilia action is quite rapid and experimental results indicate that clearance occurs in two phases, one of few hours and one of many hours. It is presumed that these phases represent clearance from the near and far regions of the bronchial tree, not the alveolar. The point is, however, that the respiratory system can clean itself to a large extent. Even some of the deposits on the alveolar surface can eventually reach the ciliary mucus transport system (via the action of the macrophage cells) or the blood stream, by penetrating through the alveolar wall.

It is the continuous exposure to high concentrations that overloads the clearance mechanisms and results in high equilibrium concentrations of deposited material in the respiratory system. The high efficiency of the lower pulmonary system to absorb particles, on the order of 1μ in diameter, is important since these and smaller particles are difficult to remove from stack gas. While a 99 percent efficient collector seems very impressive we must remember that this efficiency is by weight, and the remaining 1 percent of the particulate can and does contain an enormous number of particles of 1μ in diameter and less.

We shall see that most of the effects of air pollutants on health are related to respiratory disease, that is, bronchitis, lung cancer, etc. Animal studies of individual pollutants have also shown that at high concentrations specific effects can occur; for example, the cilia may stop beating and thus cut off the main cleaning mechanism of the system. In short, there are definite relationships between common air pollutants and physiological effects on the respiratory system.

THE EPISODES

There were undoubtedly local episodes of air pollution in antiquity from sources such as volcanos. London also suffered episodes of air pollution in the 19th century. The first episode of "modern" times was the disaster in Belgium's Meuse Valley. This began on December 1, 1930, and resulted in the death of some sixty people. The valley contained a large number of industrial pollution sources including steel works, sulfuric acid plants, glass factories (which usually emit fluorides), and zinc works. Because of a thermal inversion a variety of pollutants became trapped in the low lying valley. Deaths occurred mainly on December 4 and 5. Older people with previous heart or respiratory disease suffered the highest death rate. Cattle as well as people were killed. The chemical concentrations were not measured but were later estimated. SO_2 concentrations were assumed to be between 9 and 38 ppm which, if true, would be very high indeed (Firket, 1931). Sulfuric acid mists must also have been involved as were possibly fluorides or hydrofluoric acid. It is interesting to note that Donora also had a zinc works in use when that disaster occurred, so zinc particulates may also have been important, that is, zinc ammonium sulfate.

Certainly one ironic finding to come from this episode was the remark by Firket (1936) that "the public services of London might be faced with the responsibility of 3,200 sudden deaths if such a phenomenon occurred there."

Donora

Donora, Pennsylvania, lies some 28 miles south of Pittsburgh in a horseshoe-shaped valley on the Monongahela River. On October 26, 1948, a fog closed over the town as a temperature inversion set in. The weather did not clear until October 31 by which time twenty people had died. Bowen (1970) has described a recent revisit to Donora which is of interest in studying the attitudes of people toward air pollution. In his view the people of Donora are still for business as usual. Business in 1948 included a steel mill, a zinc production plant, and a sulfuric acid plant. Out of a population of 14,100 people, 43 percent were made ill and 10 percent were "severely affected" (Schrenk, 1949). Symptoms included irritation of the eyes, nose, and throat; coughing; respiratory irritation; headache; and vomiting. The ages of those who died were 52 to 84 with a mean of 65. Most of the deaths occurred on the third day. An interesting subjective response was noted in the article by Bowen who quotes a town doctor: "Everything was black with gas and soot; you could even taste it. The odd thing was that two days later some of the people whom I had treated while they were gasping for breath denied they had been ill." Again no measurements were taken during the episode but later estimates of SO_2 concentrations were 0.5 to 2 ppm. Particulate matter was also present in high concentrations.

Amdur and Corn (1963) studied effects of pollutant concentrations postulated to have existed during this episode. They investigated the combined effect of sulfur dioxide at 2 ppm and zinc ammonium sulfate at 0.25 mg/m^3, with a 0.29μ mass median diameter, on the increase of pulmonary resistance of guinea pigs. The effect of the combination demonstrated synergism; that is, a greater effect occurred with the two pollutants then the sum of the individual effects. Sulfuric acid aerosols must also have been present during the Donora disaster.

London

The December 5-9, 1952, London fog is the best known air pollution episode in England. Table 14.2 shows that there have been several recent acute episodes in London. Note that the dates on this table, taken from AP-50, are not consecutive but rather the table is developed according to the number of "excess deaths." Excess deaths represent those which are above the normal number of deaths occurring at that time of the year, and which are based on several years' statistics.

The "killer-smog" began on Thursday, December 4, as a high-pressure air mass created a subsidence temperature inversion over southern England. A white fog formed in the London area. As particulate and sulfur dioxide levels built up, because of extensive use of coal as fuel for space heating and electric production, the fog became a black fog. At the same time the high-pressure area stalled and became stationary. The buildup of pollutants combined with the fog resulted in essentially zero visibility. By Saturday pollutants were sufficiently concentrated to cause deaths. Woolf (1969) has described the condition: "You couldn't see one's hand in front of one's face—a white shirt collar became almost black within 20 minutes—smog was intensely irritating to eyes, throat, and bronchi. A cough soon developed. The cough was to remain with me for months."

The old and respiratory cripples died first, but younger people exposed to the outside atmosphere by virtue of their work were also

TABLE 14.2 Survey of selected acute air pollution episodes in Greater London

	Dec. 1952	Jan. 1956	Dec. 1962	Dec. 1957	Dec. 1956	Jan. 1955	Jan. 1959
Duration of the cumulation period in days	5	5	5	5	10	11	5
Number of days with maximum pollution	2	2	1	1	5	1 X 3*	1
SO_2 level preceding episode	500	300	400	300	300	300	300
SO_2 maximum	4000	1500	3300	1600	1100	1200	800
SO_2 increase per day	1200	500	1000	325	400	450	250
Soot level preceding episode	400	500	200	400	400	500	400
Soot maximum	4000	3250	2000	2300	1200	1750	1200
Soot increase per day	1200	1300	600	500	400	600	400
Number of excess deaths	3900	1000	850	800	400	240	200
Number of days with excess mortality	18	10	13	10	6	6	6
Daily mortality expected under normal circumstances	300	330	310	300	270	320	325
Average daily mortality in the period (excess mortality as a percent of normal)	170	130	120	125	125	112	110

Remark: The SO_2 and soot concentrations mentioned are average values over 24 hours expressed in $\mu g/m^3$.
*Maximum pollution values of one day's duration occurred three times.
Source: AP-50.

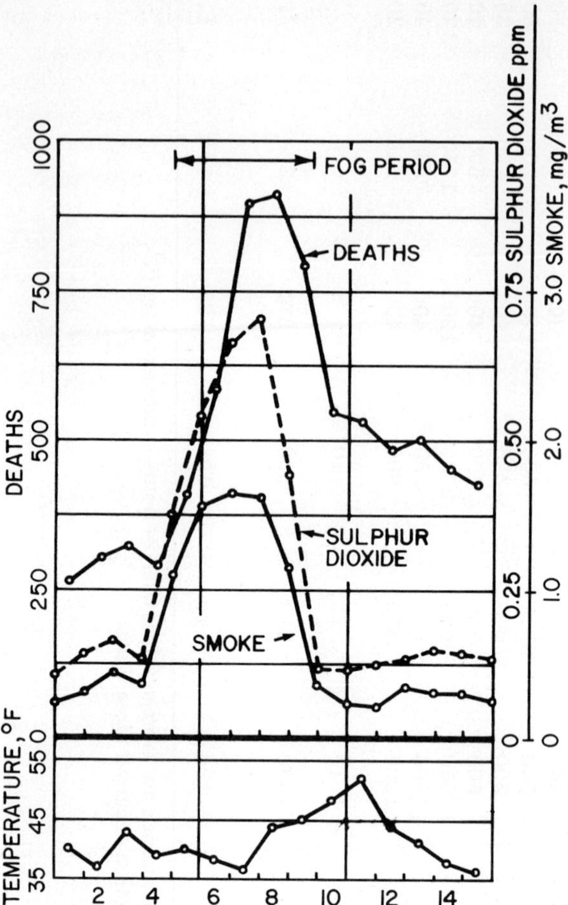

Fig. 14.6 Deaths, air pollution and temperature during December, 1952, in Greater London. (*Reprinted by permission from "Interim Report of the Committee on Air Pollution," Cmd. 9011, 1953. Crown Copyright reserved.*)

affected. The fog lifted on December 9. Figure 14.6 indicates the deaths, sulfur-dioxide level and "smoke" level during the episode. The maximum *daily* SO_2 concentration was 1.34 ppm (about 4,000 $\mu g/m^3$). Smoke levels were 4.46 mg/m^3. These figures are in agreement with those of Brasser et al. (1967) but are substantially higher than those of Fig. 14.6. This may indicate that different locations in London had substantially different concentrations. AP-49 indicates that British "smoke" and American "suspended particulate" are not exactly the same. The report notes that "limited data indicates that the American values may be higher in the same situation."

The London disaster of 1952 is charged with causing 4,000 deaths. Similar atmospheric conditions have occurred since 1952, in particular in December, 1962, with similarly high levels of SO_2. However, the smoke levels were markedly lower because of the clean up effected by the British Clean Air Act. Deaths in 1962 were much lower than in 1952. Whether this is because of the lower smoke levels or because of a population with fewer respiratory cripples left to be exposed is not clear. It is generally presumed that the lower smoke levels are responsible for the improvement. (The story of the 1952 episode is available in the book by Wise, 1971.)

New York

New York also has had several episodes of air pollution which have been responsible for excess deaths, for example, November, 1953 (Greenburg, 1962) and November, 1962 (McCarrol and Bradley, 1966). The 1966 Thanksgiving weekend episode is possibly the best known. The meteorology for this period has been described in Chap. 7. The maximum 24 hour average of hourly SO_2 values was 0.51 ppm (electroconductivity measurement) on November 23. On the 24th the value was 0.47 ppm, and on the 25th it was 0.41 ppm. The maximum *hourly* concentration was 1.02 ppm on November 25 and the smoke shade values were at or above 5 cohs on the three days (daily mean). The maximum hourly smoke shade was 8.2 cohs on the 25th. Glasser et al. (1967) have examined mortality data from this episode as compared to other control periods and conclude that there were 168 deaths. The excess was highest for older people but was observed in each of the three age groups studied: under 45, 45 to 64, and 65 and over. They also studied mortality by cause and investigated the effect of temperature, which was unusually high during this period with maximums in the 60s. Temperature effects were shown not to be the cause of the deaths.

The usefulness of studies of air pollution disasters is that they clearly indicate levels of pollution which cause immediate health effects even though the concentrations exist for only short periods of time, that is, a few days. From studies of these and other air pollution episodes we may conclude that:

1. Excess mortality is detectable in large populations if the concentrations of SO_2 rise abruptly to levels at or near 715 $\mu g/m^3$ ($\approx$ 0.25 ppm) in the presence of smoke at 750 $\mu g/m^3$ (Lawther, 1963).

2. At concentrations of about 0.19 ppm of SO_2 for 24-hour mean values, with low particulate levels, increased mortality rates may occur (Brasser, 1967).

Thus air pollution ambient standards should be placed well below these levels.

EPIDEMIOLOGICAL STUDIES

Epidemiology is the study of the factors involved in the distribution and frequency of a disease process in a given population. In terms of air pollution these studies analyze the effects from *ambient* exposures on groups of people actually living in the community. However, there are a number of difficulties involved in such studies. It is important to recognize:

1. There are a very large number of variables in large population studies, for example, smoking habits, socioeconomic levels, population density, local climate, ethnic background of the population (certain groups have high susceptibility to certain diseases). Before air pollution can be cited as the cause of health effects the interactions of these other variables must be investigated.
2. Different pollutants may cause similar effects yet it is difficult to get large populations exposed to individual pollutants. The effects of synergism and anergism of pollutants may be important. Furthermore, the concentrations of different pollutants do not necessarily follow the same function of time during the day, month, or year.

The point is not that epidemiological studies are unimportant; they are, in fact, our best source of data on long term chronic exposures. We must look with care, however, at interpreting results from such studies.

The danger in drawing conclusions between air pollutant levels and health effects, here mortality peaks, is shown in Fig. 14.7, taken from AP-50. Here data on particulate level and death rates are noted for Dublin, Ireland. From 1941–1947 particulate levels, and SO_2, were lower because during and after the war, peat was burned in place of coal. After coal was available in 1948 air pollution again rose, although not to previous levels. The war period death rates did not decrease in proportion to meet the lower levels of pollution nor did they rise after 1947. It is also true, however, that medical practice was improved in

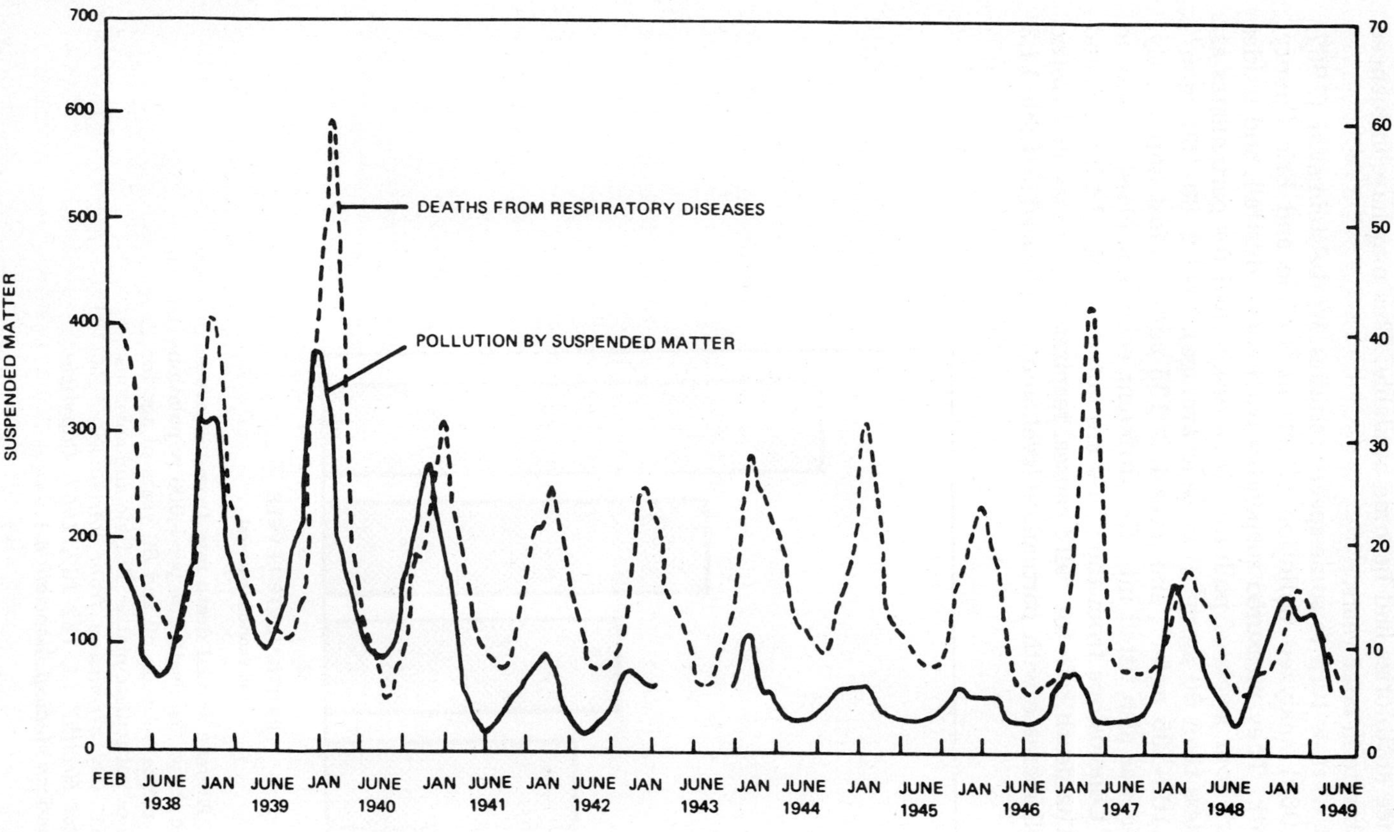

Fig. 14.7 Death rates and air pollution levels in Dublin, Ireland, for 1938–1949. (*Reprinted by permission from A. G. Leonard, et al., Royal Dublin Society Science Proceedings, 25, 166–167, 1950.*)

1948 since antibiotics had become available. This may have influenced the death rate in later years.

Let us now turn to some specific studies. Winkelstein et al. (1967, 1968, 1969) analyzed pollution effects in Buffalo and Erie County, New York. They measured suspended particulate, dustfall, and oxides of sulfur. Four levels of pollution were established for particulates as: level 1, less than 80 $\mu g/m^3$ (two year average), level 2, 80–100 $\mu g/m^3$, level 3, 100–135 $\mu g/m^3$, and level 4, > 135 $\mu g/m^3$. Most importantly, each area was also fitted into five different economic classes based on income. Conclusions from the study are noted in Figs. 14.8, 14.9, and 14.10. The death rate for "all" causes, respiratory diseases, and gastric cancer all increase with particulate level, and as indicated in Table 14.3

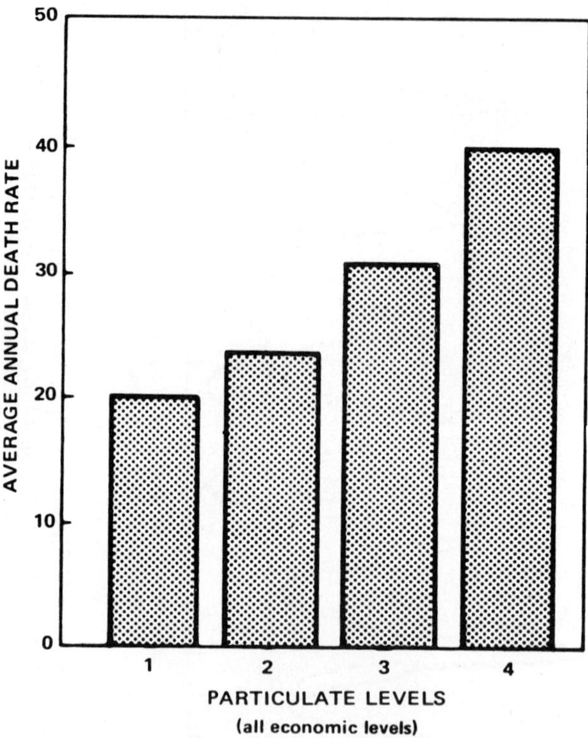

Fig. 14.8 Average annual death rate from all causes. (The graph shows the death rate per 1,000 population for white males between 50 and 69 years of age for four levels of particulate matter, Buffalo and environs, 1959–1961. (*By permission from "Archives of Environmental Health" 14, 162–169, 1967. Copyright 1967, American Medical Association.*)

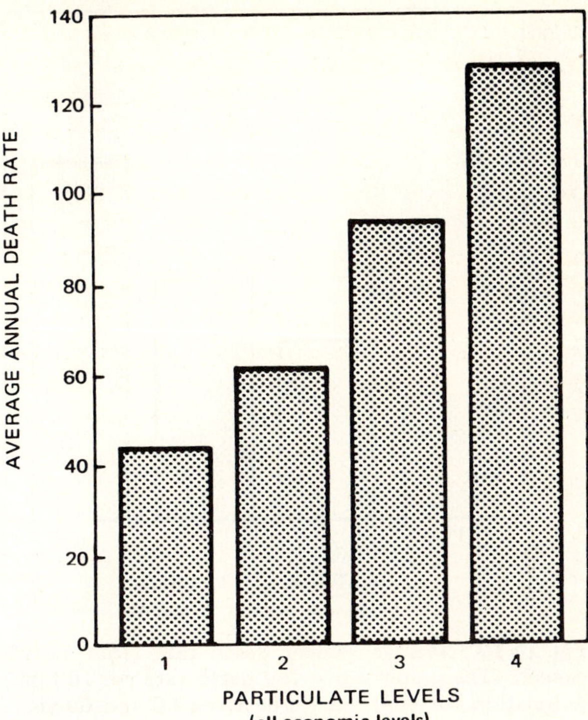

Fig. 14.9 Average annual death rate from asthma, bronchitis, or emphysema indicated on the death certificate. (The graph shows the death rate per 100,000 population for white males between 50 and 69 years of age for four levels of particulate matter, Buffalo and environs, 1959-1961. (*By permission from "Archives of Environmental Health"* **14**, *162-169, 1967. Copyright 1967, American Medical Association.*)

this conclusion is independent of economic level. The lower the economic level, in general, the higher the death rate but with only one exception (economic level 3, particulate level 3) the higher the particulate level the higher the death rate. Unfortunately, this study could not indicate smoking habits of the study group so this variable is not included.

Since it is assumed that children are nonsmokers, epidemiological studies of their health are of interest in air pollution studies. Douglas and Waller (1966) have studied 3,866 children from birth in 1946 to age 15 in 1961. Unfortunately, measured concentrations of smoke and SO_2 were not made until 1962 and 1963 so we must *assume* that similar pollution levels held throughout the study. Pollution classes

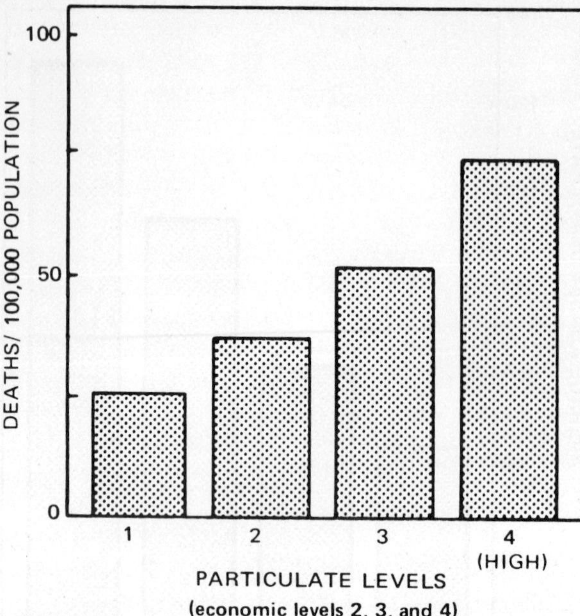

Fig. 14.10 Average annual death rate from gastric cancer. (The graph shows the death rate per 100,000 population for white males between 50 and 69 years of age for four levels of particulate matter, Buffalo and environs, 1959–1961. (*By permission from "Archives of Environmental Health"* 18, *544–547, 1969. Copyright 1969, American Medical Association.*)

TABLE 14.3 Average annual death rates per 100,000 population from chronic respiratory disease according to economic and particulate levels, and age: white males, 50–69 years of age, Buffalo and environs, 1959–1961

Economic level	Particulate level				
	1 (Low)	2	3	4 (High)	Total
1 (low)	—	0*	126	188	133
2	64	75	96	105	84
3	—	65	51	103*	64
4	35	47	114	—	52
5 (high)	42	63	0*	—	50
Total	44	62	94	129	72

*Rate based on less than five deaths.

were determined *during* the study by correlating with domestic coal consumption. The conclusions from this study were that the number and severity of upper respiratory tract infections were not related to air pollution but lower respiratory tract infections were so related. No difference was found between children in middle and working class families. The concentrations measured in 1962–63 were greater than 130 $\mu g/m^3$ for smoke and 0.046 ppm for SO_2 (also about 130 $\mu g/m^3$). Note that this study indicates the effects of very long term exposure to low levels of pollution. Lunn et al. (1967) did a similar study, also in Britain, but over a shorter time period. They indicate both upper and lower respiratory tract infections are associated with the levels of air pollution. The SO_2 level required for an "effect" is about the same as for the Douglas and Waller data, while the smoke levels were 100 $\mu g/m^3$. Social "class," number of children in the house, and sharing of bedrooms did not affect the incidence of respiratory illness. Again these pollution levels are the "ambient" levels over several years of exposure.

Studies have been made on incidence of bronchitis as a function of SO_2 and particulate air pollution levels (Wicken and Buck, 1964, and Burn and Pemberton, 1963). A positive relation was found in each study. Other pollutants have also been studied. Wayne et al. (1967) studied the effects of air pollution on athletic performance while Shy et al. (1970) studied the effects of NO_2 on human health. In Wayne's study a correlation was observed between *oxidant* levels and the percent of cross country team members whose performance decreased compared to their performance in the immediately previous home meet. There seems to be a definite causal effect but whether this is direct or one of general discomfort, for example, eye irritation, is not clear. Other pollutants were measured and did not correlate with performance.

A statistically significant correlation between oxidant levels and Los Angeles motor vehicle accidents has been found by Ury et al. (1970). They found no association with CO level and the number of accidents. While this is not a direct health effect, it certainly is an indirect one.

Shy et al. have studied the effects of community exposure to NO_2 in four areas of Chattanooga where a large TNT plant provides a plentiful source of nitrogen dioxide. Particulate and NO_2 levels were measured (SO_2 levels were less than 0.015 ppm) and used to designate regions of high NO_2—low particulate, high particulate—low NO_2, and two control areas. Socioeconomic factors were considered in the study. Home cigarette smoking was found to be similar in the areas. The conclusions were:

1. The ventilatory performance of second-grade children in the high NO_2 exposure area was significantly lower than in the control areas.
2. Illness incidence rates for each family segment—mothers, fathers, and children—in the high NO_2 areas were consistently and significantly higher than incidence in the two control areas.

The increased incidence of acute respiratory disease was observed when the 24-hour NO_2 concentration, over a six-month period, was between 117 and 205 $\mu g/m^3$ (0.062 to 0.109 ppm) and mean suspended *nitrate* level was 3.8 $\mu g/m^3$ or greater. Besides the high NO_2 area, the high particulate area also showed an excess illness in all family segments. Conclusions of this study are discussed in AP-84. Criticism of these conclusions and the methodology of the study are presented in Heuss et al. (1971). A rebuttal by EPA is given at the end of that reference. Pertinent information about NO_2 measurement techniques is in Hauser and Shy (1972).

Thus we are led to conclude that long-term exposure to presently existing ambient levels of air pollution for suspended particulate, sulfur dioxide, and nitrogen dioxide can have an adverse effect on human health. The data indicate that these pollutants can increase the incidence of respiratory disease but we cannot conclude that they cause the diseases. We should also be aware that some studies do not indicate such effects. The results of Winkelstein et al. (1968), which did show an adverse effect of particulate on health, do not show an adverse effect of sulfates at the levels that they measured.

Certainly the strongest statement to come from EPA on health effects and air pollutants is that in AP-84 which states that:[2]

> Any site that exhibits a concentration of 113 $\mu g/m^3$ (0.06 ppm) or greater exceeds the Chattanooga health effect related NO_2 value. As might be expected an analysis of the NASN data shows that the yearly NO_2 averages reflect variations according to population densities. Ten percent of the cities with populations of less than 50,000 show a yearly average equal to or exceeding 113 $\mu g/m^3$. In the population range from 50,000 to 500,000, 54 percent of the cities equal or exceed a yearly average of 113 $\mu g/m^3$. In the over 500,000 population class, 85 percent of the cities equal or exceed 113 $\mu g/m^3$ NO_2 on a yearly average.

SPECIFIC DISEASES

Lung Cancer

There is definite evidence of an "urban" effect on lung cancer. Table 14.4 from Lave and Seskin (1970) presents the results of Buell and

[2] See, however, *J. Air Pollution Control Assoc.*, vol. 23, pp. 769-772, 1973.

TABLE 14.4 A summary of lung cancer mortality studies. Number of deaths from lung cancer per 100,000 population

Standardized for age and smoking			Nonsmokers			Study
Urban	Rural	Urban/ rural	Urban	Rural	Urban/ rural	
101	80	1.26	36	11	3.27	Buell, Dunn, and Breslow[a]
52	39	1.33	15	0	∞	Hammond and Horn[b]
189	85	2.23	50	22	2.27	Stocks[c]
			38	10	3.80	Dean[d]
149	69	2.15	23	29	.79	Golledge and Wicken[e]
100	50	2.00	16	5	3.20	Haenszel et al.[f]

[a] California men; death rates by counties.
[b] American men.
[c] England and Wales.
[d] Northern Ireland.
[e] England; no adjustment for smoking.
[f] American men.
Source: Reprinted by permission from L. B. Lave and E. P. Seskin, *Science*, 169, 723–733. August 21, 1970. Copyright 1970 by the American Association for the Advancement of Science.

Dunn (1967). Among smokers we see that the likelihood of dying from lung cancer ranges from 26 percent to 123 percent greater if one lives in an urban area. For nonsmokers the range is wider and the urban effect appears to be even greater. The obvious question is: is this increased incidence of urban lung cancer caused by air pollution or is it even related to air pollution? Depending on whom you consult the answer varies. Goldsmith (Stern, 1968) answers negatively. Among his reasons are (*a*) the urban factor is not largest in the heaviest polluted countries, (*b*) rates should be highest in lifetime urban residents but instead are higher in migrants to urban areas, (*c*) women should be as much affected as men if the urban factor is due to air pollution; but they are not.

On the other hand, Stokinger and Coffin writing in the same volume of Stern but in the previous chapter state: "As pointed out previously, the inclusion of air pollution as a factor contributing to lung cancer is based on the premise that all or part of the so-called urban factor is a result of carcinogenic agents in atmospheric pollutants. While the premise cannot be definitely proved at this time, it is strongly supported by the bulk of the existing evidence." A British review article

(1970) indicates that the evidence for air pollution causing lung cancer is "inconclusive." Lave and Seskin (1970) argue that no reasonable man can object to the evidence of a substantial quantitative association between air pollution and several diseases, not just lung cancer. They argue that the possible effects for city dwellers of greater smoking, occupational exposure, less exercise, additional tension and stress are not an explanation.

There is no argument as to the urban effect in lung cancer. The argument is whether it is because, at least in part, of air pollution. It seems reasonable to conclude that while the above factors may have some effect, air pollution must also play an important role.

Bronchitis

There is agreement that exposure to air pollution in the form of suspended particulates and sulfur oxides is a causal factor in promoting and aggravating bronchitis. Reviews of this are given in Goldsmith (1968), Lave and Seskin (1970), and the British Royal College of Physicians survey (1970). Smoking habits and population density (also related to air population since fuel use is a function of the population density) are also important variables found in epidemiological studies.

Other Cancer

There is evidence of an association between stomach cancer and air pollution (Winkelstein et al., 1969). Remember that the cilia clearing action removes particulate from the bronchioles after which they are swallowed. Several chemicals which are carcinogenic (cancer-causing) have been found in the atmosphere. A review of these is given in Sawicki (1967) and AP-49, Chap. 10. The chemical of most interest as a cancer-causing agent is benzo(a)pyrene, BaP, which is one of the polynuclear aromatic hydrocarbons. This family of hydrocarbons may exist in the atmosphere as free molecules or can be adsorbed on carrier particles. As we have already seen the size of the carrier particulate affects its possibility of deposition in the lungs. In urban atmospheres there is an ample supply of particulate of about 1μ size for deposition in the lung.

SPECIFIC EFFECTS

There are an enormous number of studies of the effects of specific concentrations of single pollutants on a particular animal. The Air Quality Criteria reports list these in detail as does the article by

Stokinger and Coffin (Stern, 1968). Several of the effects noted are of particular interest.

Increased Infection Susceptibility

Air pollutants are known to increase the susceptibility to bacterial infection. For example NO_2 exposure at 0.5 ppm for several months caused increases in mortality of mice exposed to *K. pneumonial* (Ehrlich and Henry, 1968). Other studies, at higher NO_2 levels, on squirrel monkeys and hamsters showed similar effects. Ozone has also shown an ability to increase susceptibility to bacterial infection. Mice exposed to streptococcal aerosol after a 3-hour exposure to ozone from 0.08 ppm to 0.52 ppm showed an increase in mortality. Mice and hamsters exposed to higher ozone levels and then exposed to *K. pneumonial* also exhibited increased mortality. Thus it is certainly plausible that air pollutants can increase human susceptibility to respiratory disease and mortality.

Tolerance

Some pollutants seem to have the ability to initiate a tolerance response in laboratory animals. In particular, rats and mice exposed to ozone are found to develop a protection to second exposures. Tolerance is demonstrable directly at a single 1-hour exposure to levels as low as 0.3 ppm ozone (AP-63). In fact, ozone exposure can provide protection to later exposure of NO_2, H_2O_2, H_2S, and other gases. Experiments with mice exposed to NO_2 also indicate reduced effects upon subsequent exposure. Whether these results of animal studies can be extrapolated to humans has not been determined.

Cilia Effects

Since the cilia are the primary mechanism by which the bronchiole tubes are cleared of particulate any effects of gaseous pollutants in reducing their effectiveness is of interest. At SO_2 concentrations of 12 ppm (far higher than ambient levels) ciliary cessation has been shown to occur in about 4 to 6 minutes in rats. After exposure the cilia regained their mobility. Mice exposed to 0.5 ppm of NO_2 for 6 hours a day for 6 months showed a loss of cilia among other effects to the bronchioles and alveoli (AP-84). Cigarette smoking can also stop the cilia action.

SPECIFIC POLLUTANTS

We have seen in the preceding pages some evidence of effects on health of pollutants such as SO_2, NO_2, O_3, and particulates. A short description of two other pollutants of concern will be given in this section.

Lead

Lead is found in our water, food, and air. Most of the atmospheric lead comes from antiknock compounds added to *both* regular and premium gasoline. Typical premium gasoline contains between 2 to 4 grams of lead per gallon with an average of about 2.8 grams. Regular gasoline averages about 2.3 grams per gallon. Thus, in a state like California where over 13×10^6 motor vehicles are in use, the total lead consumed in gasoline is enormous. ($13 \times 10^6 \times 2$ gallons per day $\times$ 2.5 grams per gallon $\times$ 365 days per year $\times$ 1/454 lb per gram $\times$ 1/2,000 tons per lb = 26,000 tons per year.) On the average 70 to 80 percent of the lead in gasoline is exhausted out the tailpipe as particulate. Less is exhausted at low-speed driving and more is exhausted with freeway conditions. Most of the lead particulate from automobiles is less than 2μ in diameter and thus does not settle out of the air rapidly. The automobile source of atmospheric lead creates urban concentrations of inorganic lead of about 1 to 3 $\mu g/m^3$ with maximum values of 7 to 9 $\mu g/m^3$ in areas of heavy traffic (Bové and Siebenberg, 1970; Chow and Earl, 1970). Even higher concentrations are noted near freeways. Inorganic lead acts as an agent to cause a variety of effects on human health including liver and kidney damage, gastrointestinal damage, mental health effects in children, and abnormalities in fertility and pregnancy. Adults with less than 80 μg/100 grams of blood, and urine lead less than 150 μg/liter are considered to be within normal limits (some authorities argue for 40 μg/100 cc of blood as the limit).

How much lead does man take in with this background concentration? Adults ingest via food and water about 300 μg/day with an absorption of perhaps 10 percent for a total of 30 μg/day. Cigarette smoking produces about 0.5 μg per cigarette or 10 μg per pack. In contrast, atmospheric intake in *urban* areas is about 50 μg/day. Of this some 25 to 50 percent is retained in the lung for an input of 12 to 25 μg/day. An adult nonsmoker in an urban area will take up about 40 to 55 μg/day of lead of which 30 to 45 percent comes from the atmosphere. Fortunately lead is removed from the body primarily with the urine. Total excretion is about 30 to 40 μ/day which is about that taken in. People who have an additional occupational exposure can build up high concentrations of lead. With sufficient "body burdens" health effects become evident.

Figure 14.11 indicates the mean blood level of lead versus the estimated level of exposure. At present it appears that rural exposure leads to blood levels of about 15 μg/100 grams of blood. This is in agreement with the results of Kubota et al. (1968), who found an average level of 13.2 μg/100 milliliters of blood based on a sample of 249 subjects around the United States. General conclusions agreed upon are: men have greater levels of lead in their blood than women, urban residents have greater levels than rural residents, and smokers have greater levels than nonsmokers.

Carbon Monoxide

Carbon monoxide is an odorless, tasteless, and colorless gas. It is produced as a product of incomplete combustion. Upon entering the respiratory system it combines in the lung with the hemoglobin in the bloodstream to form carboxyhemoglobin, COHb. This reduces the ability of the hemoglobin to carry oxygen to the body tissues. CO has some 200+ times the affinity of O_2 for attaching itself to the hemoglobin so that low levels of CO can still result in high levels of COHb. Figure 14.12 shows the COHb level for an individual at rest as a function of exposure and CO concentration. The more strenuous the physical activity the more rapidly an equilibrium is attained although the equilibrium COHb level is not changed. This graph does not indicate

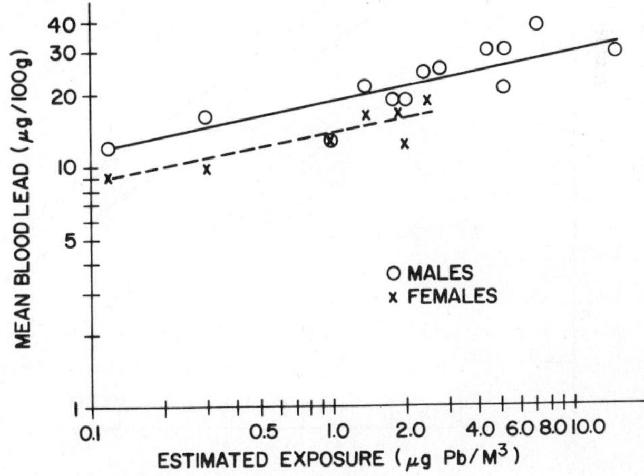

Fig. 14.11 Mean blood levels for lead. (*Lead in the Environment, Calif. ARB, 1967.*)

the equilibrium values but as an example 50 ppm leads to an equilibrium value of about 7 percent COHb and 100 ppm to about 14 percent COHb. After exposure the CO is slowly released from the blood with a clearance half-life of 3 to 4 hours.

Cigarette smokers commonly have COHb levels of about 5 percent with heavy smokers reaching values of 10 percent. This is due to the high CO concentration, about 400 ppm, in cigarette smoke. Setting ambient CO limits must take into account that many people have a "normal" high background level.

Goldsmith (Stern, 1968) comments that: "Although the mechanisms of effects are somewhat different, exposure to about 30 ppm of CO for a period long enough for equilibrium to be established (about 4 to 6 hours) is roughly comparable, in terms of oxygen saturation of the blood, to dwelling at an altitude of about 6,000 feet."

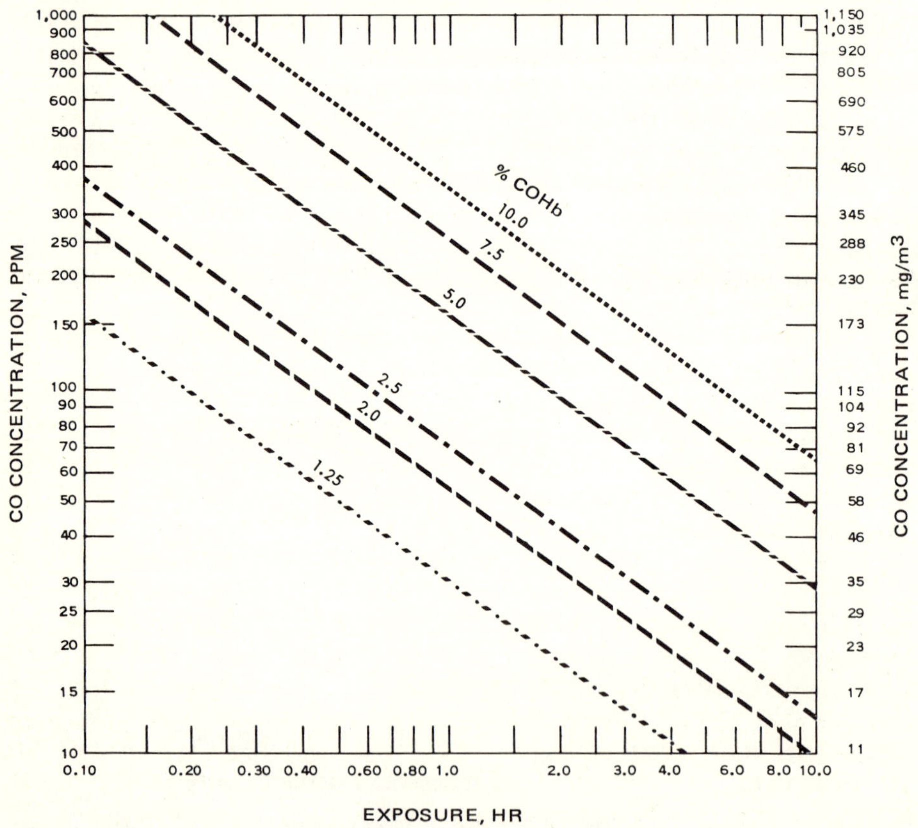

Fig. 14.12 COHb level as a function of exposure. (*AP-62.*)

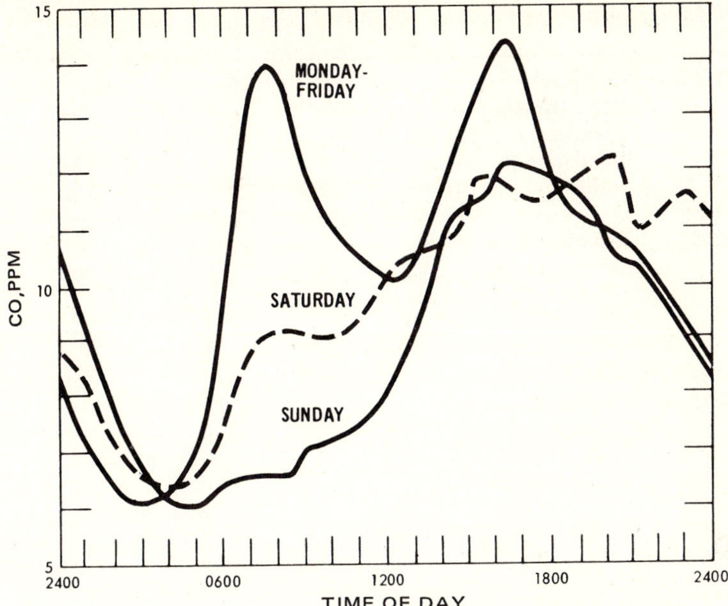

Fig. 14.13 Diurnal variation of mean carbon monoxide levels on weekdays, Saturdays, and Sundays in Chicago, 1962 through 1964. (*AP-84.*)

One must presume that heavy cigarette smokers living at high altitudes have a problem.

CO levels in the ambient air are particularly high along urban streets. Figure 14.13 notes measurements in Chicago where the very evident commuter traffic peaks are seen. Measurements taken over several weeks on a busy street in New York indicate a value of 18 ppm from 8 AM to 6 PM for a two-hour average concentration (Bové and Siebenberg, 1970). Other measurements (Jaffe, 1970) show 50 ppm for 30-minute periods with peak values over 140 ppm.

At COHb levels of 2 to 5 percent effects are found to the central nervous system such as impairment of time interval discrimination and visual acuity. At levels greater than 5 percent there are cardiac and pulmonary functional changes (National Academy of Sciences, 1969). There is evidence that CO concentrations encountered in the Los Angeles area (10 to 15 ppm ambient Los Angeles *basin* 24-hour *average* values) are associated with excess mortality (Hexter and Goldsmith, 1971). Evidence has also been presented that a relationship exists between ambient CO levels and myocardial infarction (heart attacks) by Cohen et al. (1969). Contrary to what might be expected the results of

TABLE 14.5 Properties of pollutants

Ozone	Colorless	
	Odor threshold	0.02–0.05 ppm
	Irritation of nose and throat	0.05 ppm
	Headache (in 30 minutes)	1 ppm (1962 $\mu g/m^3$)
	Industrial limit (8-hr day, WHO*)	100–200 $\mu g/m^3$
NO_2	Brown-red color	
	Odor threshold	0.12–0.22 ppm
	Increased respiratory illness over 6-month test period	0.06–0.109 ppm
	U.S. industrial limit (8-hr day)	5 ppm
SO_2	Colorless	
	Taste threshold	0.3 ppm
	Odor threshold	0.5 ppm
	Increased mortality	0.25 ppm (24 hr measurement) with 750 $\mu g/m^3$ smoke
	Possible increased mortality	0.19 ppm (24-hr measurement) with "low" particulate levels for "few" days
	Increased respiratory illness in children	0.046 ppm with smoke at 100 $\mu g/m^3$ for long term exposure
	Industrial limit (WHO, 8-hr day)	10–13 $\mu g/m^3$
CO	Colorless, tasteless, odorless (faint odor at very high concentrations)	
	Central nervous system effects	15 ppm for 10 hr
	Industrial limit (WHO, 8-hr day)	50 ppm

*World Health Organization, Technical Report Series no. 415, 1969.

Ury et al. (1970) have shown oxidant level and not CO level to be related to motor vehicle accidents in Los Angeles.

There is ample evidence that ambient air concentrations of CO in major traffic areas are high enough to produce significant effects on health. Further research will establish a clearer figure of the COHb level which produces health effects in the most susceptible individual. Present United States standards of 9 ppm for 8 hours will produce an equilibrium COHb level of about 2 percent.

Table 14.5 summarizes the properties and affects of some of the air pollutants.

REFERENCES

Air Quality Criteria for Particulate Matter, National Air Pollution Control Administration, Publication AP-49, 1969.

Air Quality Criteria for Sulfur Oxides, National Air Pollution Control Administration, Publication AP-50, 1969.

Air Quality Criteria for Carbon Monoxide, National Air Pollution Control Administration, Publication AP-62, 1970.

Air Quality Criteria for Photochemical Oxidants, National Air Pollution Control Administration, Publication AP-63, 1970.

Air Quality Criteria for Nitrogen Oxides, Air Pollution Control Office, Publication AP-84, 1971.

Amdur, M. O., and M. Corn: The Irritant Potency of Zinc Ammonium Sulfate of Different Particle Sizes, *Am. Ind. Hyg. Assoc. J.*, vol. 24, pp. 326-333, 1963.

Bové, J. L., and S. Siebenberg: Airborne Lead and Carbon Monoxide at 45th Street, New York City, *Science*, vol. 167, pp. 986-987, 1970.

Bowen, C.: Donora, Pennsylvania, *Atlantic*, pp. 27-34, November, 1970.

Brasser, L. J., P. E. Joosting, and D. Von Zuilen: Sulfur Dioxide—To What Level is it Acceptable?, Research Institute for Public Health Engineering, Delft, Netherlands, Report G-300, 1967.

Buell, P. and J. E. Dunn, Jr.: Relative Impact of Smoking and Air Pollution on Lung Cancer, *Arch. Environ. Health*, vol. 15, pp. 291-297, 1967.

Burn, J. L. and J. Pemberton: Air Pollution, Bronchitis, and Lung Cancer in Salford, *Int. J. Air and Water Pollution*, vol. 7, pp. 5-16, 1963.

Chow, T. J. and J. L. Earl: Lead Aerosols in the Atmosphere: Increasing Concentrations, *Science*, vol. 169, pp. 577-580, 1970.

Cohen, S. I., M. Deane, and J. R. Goldsmith: Carbon Monoxide and Survival from Myocardial Infarction, *Arch. Environ. Health*, vol. 19, pp. 510-520, 1969.

Douglas, J. W. B. and R. E. Waller: Air Pollution and Respiratory Infection in Children, *Brit. J. Prevent. Social Med.*, vol. 20, pp. 1-8, 1966.

Effects of Chronic Exposure to Low Levels of Carbon Monoxide on Human Health, Behavior and Performance, National Academy of Sciences and National Academy of Engineering, 1969.

Ehrlich, R. and M. C. Henry: Chronic Toxicity of Nitrogen Dioxide: 1. Effects on Resistance to Bacterial Pneumonia, *Arch. Environ. Health*, vol. 17, pp. 860-865, 1968.

Firket, J.: The Cause of the Symptoms Found in the Meuse Valley during the Fog of December, 1930, *Bull. Roy. Acad. Med.* (Belgium), vol. 11, pp. 683-739, 1931.

Firket, J.: Fog Along the Meuse Valley, *Trans. Faraday Soc.*, vol. 32, pp. 1192-1197, 1936.

Glasser, M., L. Greenburg, and F. Field: Mortality and Morbidity During a Period of High Levels of Air Pollution, *Arch. Environ. Health*, vol. 15, pp. 684-694, 1967.

Goldsmith, J. R., in Stern, "Air Pollution," vol. 1, chap. 14, Academic Press, New York, 1968.

Greenburg, L., C. Erhardt, F. Field, J. I. Reed, and N. S. Seriff: Intermittent Air Pollution Episodes in New York City, 1962, *Public Health Repts.* (U.S.), vol. 78, pp. 1061-1065, 1963.

Hauser, T. R. and C. M. Shy: Position Paper: NO_x Measurement, *Environ. Sci. Technol.*, vol. 6, pp. 890-894, 1972.

Heuss, J. M., G. J. Nebel, and J. M. Colucci: National Air Quality Standards for Automotive Pollutants—A Critical Review, *J. Air Pollution Control Assoc.*, vol. 21, pp. 535-548, 1971.

Hexter, A. C. and J. R. Goldsmith: Carbon Monoxide: Association of Community Air Pollution with Mortality, *Science*, vol. 172, pp. 265-267, 1971.

Jaffe, L. S., in Singer: "Global Effects of Environmental Pollution," Springer-Verlag, New York, 1970.

Kubota, J., V. A. Lazar, and F. Losee: Metals in Human Blood, *Arch. Environ. Health*, vol. 16, p. 788, 1968.

Lave, L. B. and E. P. Seskin: Air Pollution and Human Health, *Science*, vol. 169, pp. 723-733, 1970.

Lawther, P. J.: Compliance with the Clean Air Act: Medical Aspects, *J. Inst. Fuel* (London), vol. 36, pp. 341-344, 1963.

Lunn, J. E., J. Knowelden, and A. J. Handyside: Patterns of Respiratory Illness in Sheffield Infant School Children, *Brit. J. Prevent. Social Med.*, vol. 21, pp. 7-16, 1967.

McCarrol, J. and W. Bradley: Excess Mortality as an Indicator of Health Effects of Air Pollution, *Am. J. Public Health*, vol. 56, pp. 1933-1942, 1966.

Royal College of Physicians (British): "Air Pollution and Human Health," Pitman, London, 1970.

Sawicki, E.: Airborne Carcinogens and Allied Compounds, *Arch. Environ. Health*, vol. 14, pp. 46-53, 1967.

Schrenk, H. H., et al.: Air Pollution in Donora, Pennsylvania: Epidemiology of the Unusual Smog Episode of October 1948, *Pub. Health Bull.* 306, 1949.

Shy, C. M., J. P. Creason, M. E. Pearlman, K. E. McClain, F. B. Benson, and M. M. Young: The Chattanooga School Children Study: Effects of Community Exposure to Nitrogen Dioxide, *J. Air Pollution Control Assoc.*, vol. 20, pp. 539-545, and pp. 582-588, 1970.

Stokinger, H. E. and D. L. Coffin, in Stern: "Air Pollution," vol. 1, Chap. 13, Academic Press, New York, 1968.

Ury, H. K., J. R. Goldsmith, and N. M. Perkins: Possible Association of Motor Vehicle Accidents with Pollutant Levels in Los Angeles, Project Clean Air, Research Reports, vol. 2, University of California, 1970.

Wayne, W. S., P. F. Wehrle, and R. E. Carroll: Oxidant Air Pollution and Athletic Performance, *J. Am. Med. Assoc.*, vol. 199 (12), pp. 901-904, March 20, 1967.

Wicken, A. J. and S. F. Buck: Report on a Study of Environmental Factors Associated with Lung Cancer and Bronchitis Mortality in Areas of North East England, Tobacco Research Council, Research Paper 8, 1964.

Winkelstein, W., S. Kantor, E. W. Davis, C. S. Maneri, and W. E. Mosher: The Relationship of Air Pollution and Economic Status to Total Mortality and Selected Respiratory System Mortality in Man, *Arch. Environ. Health*, vol. 14, pp. 162-169, 1967.

Winkelstein, W., S. Kantor, E. W. Davis, C. S. Maneri, and W. E. Mosher: The Relationship of Air Pollution and Economic Status to Total Mortality and Selected Respiratory System Mortality in Man (2. Oxides of Sulfur), *Arch. Environ. Health*, vol. 16, pp. 401-405, 1968.

Winkelstein, W. and S. Kantor: The Relationship of Air Pollution to Cancer of the Stomach, *Arch. Environ. Health*, vol. 18, pp. 544-547, 1969.

Wise, W.: "Killer Smog," Audubon-Ballantine Books, New York, 1971.

Wolozin, H. (ed.): "The Economics of Air Pollution," Norton, New York, 1966.

Woolf, C.: *Arch. Environ. Health*, vol. 18, p. 715, 1969.

15

AIR QUALITY AND EMISSION STANDARDS

> Progress is achieved by being aware of the mechanisms and processes by which pollution harms us so that we steadily eliminate the worst sources and apply ever more stringent standards to the new sources.
>
> *R. S. Scorer*

UNITED STATES EFFORTS

Countries throughout the world have reacted in various ways to the task of cleaning the air; some employing serious control measures, some ignoring the problem completely. The United States in the last decade has gradually evolved from the latter position to the former. We will discuss air quality standards and the related question of emission standards, but first it is well to review briefly the history of Federal efforts in this area.

In 1955 the Air Pollution Control Act was passed. The act provided that: the Public Health Service was to do research on air pollution effects, there was to be technical assistance to the states, and there was to be training of individuals in the area of air pollution. Both intramural and extramural research was to be done.

In 1963 the Clean Air Act passed. This amended the 1955 Act and provided that:

The Congress finds—

(1) that the predominant part of the Nation's population is located in its rapidly expanding metropolitan and other urban areas, which generally

cross the boundary lines of local jurisdictions and often extend into two or more states;

(2) that the growth in the amount of complexity of air pollution brought about by urbanization, industrial development, and the increasing use of motor vehicles, has resulted in mounting dangers to the public health and welfare, including injury to agricultural crops and livestock, damage to and the deterioration of property, and hazards to air and ground transportation;

(3) that the prevention and control of air pollution at its source is the primary responsibility of states and local governments; and

(4) that Federal financial assistance and leadership is essential for the development of cooperative Federal, State, regional and local programs to prevent and control air pollution.

The act then authorized:

1. accelerated research and training,[1]
2. matching grants to state and local agencies for air pollution programs (federal government to pay two-thirds or three-fourths of the costs),
3. control of air pollution from federal facilities,
4. federal authority to abate interstate air pollution,
5. encouragement of efforts on the part of the automotive companies and fuel industries to prevent pollution.

Further laws were to be required especially in this last area. Let us look in more detail at section (4), the abatement provision. This section provided for the Secretary of the Department of Health, Education, and Welfare to call a "conference" for presentation of the issues. Subsequently, and if the Secretary believed that effective progress toward abatement was not being made, he could *recommend* to the appropriate *state* agency that the "necessary remedial action be taken." If after six months the remedial action had not occurred, the Secretary was authorized to call a public hearing before a hearing board. On the basis of the evidence presented, the board was to make recommendations to the Secretary. He was then to remit findings to the *state* agency "together with a notice specifying a reasonable time (not less than six months) to secure abatement." If action was not forthcoming with this notice, the Secretary could then request the Attorney General to bring suit.

If one wonders why the Clean Air Act of 1963 was less than a spectacular success the above description should indicate some of the

[1] Note that as early as 1963 a program was to be initiated to achieve low-cost techniques for extracting sulfur from fuels.

reasons. Before criticizing HEW, remember that it was Congress who wrote the Act, not the Air Pollution Control Administration.

In 1965 this Act was amended to provide for federal standards to control pollution from motor vehicles (Motor Vehicle Air Pollution Control Act). Additionally, the Secretary of HEW was given authority to intervene in intrastate air pollution problems "of substantial significance." This was to be done, however, through the conference technique outlined above and his "findings and recommendations shall be advisory only."

In 1967, after great debate, the Air Quality Act passed the Congress. Whether it was a victory for the forces of clean air is not clear since the Act was amended again in 1970 before ever coming close to reaching a steady state condition. National emission standards were first seriously considered, and rejected, at this time. The original Johnson bill called for these but the final Muskie bill deleted them. Either Congress or industry, or both, were not ready for national standards but preferred continued state and local control.

The 1967 bill called for the following:

1. establish areas in the United States on the basis of meteorology, topography, and climate;[2]
2. designate air quality control regions;
3. develop *criteria* of air quality reflecting identifiable effects on health and welfare;
4. give recommendation to state and local control agencies on technology to achieve the levels of air quality given in the criteria reports (this provision led to the development of the Air Pollution Control Technique Reports);
5. require on a fixed time schedule that state and local control agencies establish air quality *standards* consistent with the air quality criteria. States may set standards to achieve a higher quality than given in the criteria reports.[3]

[2] Eight such areas were designated in the United States: Appalachian, South Florida, Washington Coastal, California-Oregon Coastal, Rocky Mountain, Great Plains, Great Lakes, and mid-Atlantic Coastal. What Congress had in mind by this provision of the Act was never clear.

[3] If the state does not act then HEW can promulgate standards for each Air Quality Region. The timetable for state action is: after the Governor receives notice of criteria and control techniques for a *specific pollutant* he must file within 90 days a letter of intent to: (1) adopt within 180 days, after public hearings, ambient air control standards, (2) within an additional 180 days the state must provide a plan to implement these standards. Before HEW can set standards, if the state does not follow the above procedure, a public hearing must be held if the Governor so petitions. Note that this torturous procedure must be followed for *each* pollutant.

The complexity of setting up air quality regions, the series of hearings and the implementation plans was simply too much for the understaffed National Air Pollution Control Administration people; only a limited number of these were so designated by 1970. Politics and economics naturally played a large role. As an example, in southern Arizona, four adjacent counties (Maricopa, Pinal, Pima and Gila) were included in a proposed Air Quality Region. Santa Cruz County just south of the region was included at its Board of Supervisors' *request*. To those living in adjacent Pima County this meant that any future pollution source going into Santa Cruz County would now be required to have pollution controls. The copper industry was developing several new mines in this region so the possibility of a new, large source of sulfur dioxide from a smelter was very real. In designating this air quality control region an adjacent county, Cochise, was excluded and thus the largest copper smelter in the state escaped (only temporarily, however) being placed within a control region.

Finally in 1970 the 1967 Act was amended (Public Law 91-604, 91st Congress, 2d session, H. R. 17255, Dec. 31, 1970) to provide for:

1. additional research efforts requested and funds appropriated;
2. additional state and regional grant programs authorized; matching grants established for implementing standards;
3. *national ambient air quality standards* to be set by HEW including both primary and secondary standards;[4]
4. complete designation of air quality control regions;
5. implementation plans to meet the standards (to be set above) to fit a given timetable but to continue to be initiated by the states;
6. standards of performance for *new* stationary sources to be required.[5] Each state to implement and enforce the standards of performance;

[4] "National primary ambient air quality standard shall be ambient air quality standards the attainment and maintenance of which in the judgment of the Administrator, based on such criteria and allowing an adequate margin of safety, are requisite to protect the public health." "Any National secondary ambient air quality standard shall specify a level of air quality the attainment and maintenance of which in the judgment of the Administrator, based on such criteria, is requisite to protect the public welfare from any known or anticipated adverse effects associated with the presence of such air pollutant in the ambient air."

[5] "Section 111. The term standard of performance means a standard for emissions of air pollutants which reflects the degree of emissions limitation achievable through the application of the best system of emission reduction which (taking into account the cost of achieving such reduction) the Administrator determines has been adequately demonstrated."

7. *national emissions standards* for hazardous air pollutants to apply to existing as well as new plants (exemption possible);[6]
8. industry to monitor and maintain emissions records which are to be available to HEW officials;
9. automobile emission standards, to be generally set for a 90 percent reduction of the partially controlled 1970 levels, that is, the emissions from 1970 models;
10. development of low emissions vehicles to be encouraged with appropriations for research;
11. aircraft emissions standards to be developed by HEW;
12. citizens' suits to be allowed against "any person (including the United States) who is alleged to be in violation of an emissions standard or an order issued by the Administrator." ". . . against the Administrator where there is alleged a failure to perform any act or duty under this Act . . . ."

Federal enforcement procedures are outlined if the states fail to meet the requirements of the Act. The law calls out explicitly, however, the retention of state authority for control of air pollution emissions.

A summary of air pollution control acts is given in Table 15.1. Note that the 1970 amended Air Quality Act provides for the first national ambient air quality standards, the first national emission standards for stationary sources, and the first standards of performance. But how the ambient standards are to be implemented is still left to the

[6] As, Cl, HCl, Cu, Mn, Ni, Va, Zn, Ba, Bo, Ch, Se, pesticides, and radioactive substances are typical of the materials that may be considered.

TABLE 15.1 Summary of federal air pollution acts

1955	Air Pollution Control Act	First federal legislation
1963	Clean Air Act	Federal financial assistance but problem left to the states.
1965	Amended Clean Air Act	First federal emission standards for cars
1967	Air Quality Act	First Air Quality Criteria Reports, first requirement for ambient standards, but these to be set by the states.
1970	Amended Air Quality Act	First national ambient standards and these to be set by federal action; first national emission standards to be set for certain pollutants; first standards of performance for new plants; continued state implementation

states. Thus, while the trend in Congress has been to create more and more federal involvement, the final move to complete federal control has not been made. Section 107 of the 1970 act provides that: "(*a*) each State shall have the primary responsibility for assuring air quality within the entire geographic area comprising such State by submitting an implementation plan for such State which will specify the manner in which national primary and secondary ambient air quality standards will be achieved and maintained within each Air Quality Control Region in such State." Note that each AQCR may use its own methods but all must eventually meet these standards.

We might well ask, if conditions were as bad as indicated in the preface to the 1963 law, why it took until December 31, 1970, to get a better law? And, more important, why (as of 1970) had little or none of the pollution in the air been removed? Here we must leave the area of fact and begin to speculate.

Certainly one major cause of slow progress is industrial reluctance to clean up—it costs money. This reluctance is hardly limited to the United States or even to a capitalistic economic system. Other countries such as those in the Eastern European bloc offer ample proof of this. Another reason for slow progress is politics. Of course, politics is often intertwined with economic considerations. For example, the threat to close a mill or smelter may result in sufficient political pressure to postpone, if only temporarily, implementing emission standards. The fact that air quality inspectors only work during the day may enable a sand and gravel plant to operate during the early morning hours without the need to purchase dust control equipment. "Vanishing Air" discusses the weaknesses of the laws passed by Congress through the 1967 Act.

Certainly a third reason was the decision to allow the states to set standards (the 1967 Act) and then (the 1970 Act) to implement procedures to meet these standards. As quoted previously this decision goes back to the 1963 Act. Why should such action occur at a state level? Here are the pros and cons of the argument.

For State Action

1. Air pollutants are not the same in all geographical regions. Thus different measures may be required to deal with each separate area, that is, there is a need for regional flexibility.
2. We are better off to have many clean-up strategies rather than just one. We have an opportunity this way to find out what works.
3. An administrative agency far removed from the local scene may stipulate rules which are unfair and reflect a lack of understanding for the local scene.

4. Local problems should be solved locally—a political philosophy which is independent of the particular question of air pollution.

For Federal Action

1. Many states do not have the expertise and manpower to determine whether submitted plans will work, to carry out monitoring, and to proceed with enforcement.
2. A competitive advantage may accrue to segments of industry if the states set standards or implementation plans in each air quality region. (Note that this argument is independent of the question of national emission standards.)
3. The national regulatory agency is less influenced by local economic interests which bring pressure for the status quo.
4. In terms of global effects from pollutants, federal action is the only appropriate level at which to attack the problem.

The question, in short, is one of federal presence; and a recurring theme in our national history has been that of federal versus state control. It is clear, however, that the direction of federal legislation has been toward more national control. One can speculate that some sort of national control will prevail in the decade of the 70s unless a real show of progress materializes. The case for national emission standards was stated by Stern (1970). He noted that even national standards may vary; they can be weighted by population density or by present density of emissions. Uniform standards can also be mitigated by the condition of the plant, for example, less stringent rules applying to older plants.

SETTING THE STANDARDS

How do we differentiate among the terms "criteria," "ambient standards," and "emission standards"? Quoting from the 1967 Act: "Air quality criteria are descriptive, that is they describe the effects that can be expected to occur whenever the ambient air level of a pollutant reaches or exceeds a specific figure for a specific time period. Air quality standards are prescriptive, they prescribe pollutant levels that cannot legally be exceeded during a specific time period in a specific geographic area." From the 1970 Act, "Criteria should reflect that latest scientific knowledge useful in indicating the kind and extent of all the identifiable effects on health and welfare which may be expected in the presence of an air pollution agent in the ambient air."

However, ambient standards, which is what is usually meant by the term air quality standards, are meaningless unless one can achieve them. This, in turn, means that *one must set up emission standards*,

TABLE 15.2 Ambient air quality standards

Pollutant	Averaging time	California standards		Federal standards[d]		
		Concentration[g]	Method[a]	Primary[b,g]	Secondary[c,g]	Method[e]
Photochemical oxidants (corrected for NO_2)	1 hr	0.10 ppm (200 $\mu g/m^3$)	Neutral buffered KI	160 $\mu g/m^3$ [h] (0.08 ppm)	Same as primary std.	Chemiluminescent method
Carbon monoxide	12 hr	10 ppm (11 mg/m^3)	Nondispersive infrared spectroscopy	—	Same as primary standards	Nondispersive infrared spectroscopy
	8 hr	—		10 mg/m^3 (9 ppm)		
	1 hr	40 ppm (46 mg/m^3)		40 mg/m^3 (35 ppm)		
Nitrogen dioxide	Annual average	—	Saltzman method	100 $\mu g/m^3$ (0.05 ppm)	Same as primary standard	Colorimetric method using NaOH
	1 hr	0.25 ppm (470 $\mu g/m^3$)		—		
Sulfur dioxide	Annual average	—	Conductimetric method	80 $\mu g/m^3$ (.03 ppm)	60 $\mu g/m^3$ [i] (0.02 ppm)	Pararosaniline method
	24 hr	0.04 ppm (105 $\mu g/m^3$)		365 $\mu g/m^3$ (0.14 ppm)	260 $\mu g/m^3$ (0.10 ppm)	
	3 hr	—		—	1300 $\mu g/m^3$ (0.5 ppm)	
	1 hr	0.5 ppm (1310 $\mu g/m^3$)		—	—	
Suspended particulate matter	Annual geometric mean	60 $\mu g/m^3$	High volume sampling	75 $\mu g/m^3$	60 $\mu g/m^3$	High volume sampling
	24 hr	100 $\mu g/m^3$		260 $\mu g/m^3$	150 $\mu g/m^3$	

Lead (particulate)	30-day average	1.5 μg/m³	High volume sampling, dithizone method	—	—	—
Hydrogen sulfide	1 hr	0.03 ppm (42 μg/m³)	Cadmium hydroxide STRactan method	—	—	—
Hydrocarbons (corrected for methane)	3 hr (6–9 AM)	—	—	160 μg/m³ (0.24 ppm)	Same as primary standard	Flame ionization detection using gas chromatography
Visibility reducing particles	1 observation	In sufficient amount to reduce the prevailing visibility[f] to 10 miles when the relative humidity is less than 70%		—	—	—

Notes:

a. Any equivalent procedure which can be shown to the satisfaction of the Air Resources Board to give equivalent results at or near the level of the air quality standard may be used.

b. National Primary Standards: The levels of air quality necessary, with an adequate margin of safety, to protect the public health. Each state must attain the primary standards no later than three years after that state's implementation plan is approved by the Environmental Protection Agency (EPA).

c. National Secondary Standards: The levels of air quality necessary to protect the public welfare from any known or anticipated adverse effects of a pollutant. Each state must attain the secondary standards within a "reasonable time" after implementation plan is approved by the EPA.

d. Federal standards, other than those based on annual averages or annual geometric means, are not to be exceeded more than once per year.

e. Reference method as described by the EPA. An "equivalent method" of measurement may be used but must have a "consistent relationship to the reference method" to be approved by the EPA.

f. Prevailing visibility is defined as the greatest visibility which is attained or surpassed around at least half of the horizon circle, but not necessarily in continuous sectors.

g. Concentration expressed first in units in which it was promulgated. Equivalent units given in parentheses are based upon a reference temperature of 25°C and a reference pressure of 760 mm of mercury.

h. Corrected for SO_2 in addition to NO_2.

i. Revoked in 1973.

that is, how much each polluter may emit, such that the ambient standards can still be reached. Whenever discussing standards, it is *essential* to differentiate between ambient air quality standards and emission standards. The ambient standards are goals set on the basis of the criteria reports and are noted as concentrations not to be exceeded more than a specified number of times in a certain time period. Emission standards are amounts which may not be exceeded by any emission source. Present (1972) California and federal standards for *ambient air quality* are noted in Table 15.2. These federal standards were set after passage of the 1970 amendments and in response to requirements of that law.

It is not clear what experiments received what weight in establishing these standards. Heuss, Nebel, and Colucci (1971) discuss this question and also whether the standards are too tough. A rebuttal from EPA is included with their article.

We quoted earlier the 1970 Act regarding the criteria to be used in setting primary and secondary air quality standards. With these in mind, it is worthwhile for the reader to compare these standards to pollutant levels reported in Chaps. 13 and 14 to cause adverse effects. It is also of interest to compare the standards to ambient levels of air pollutants which have been given in previous chapters.

What still remains to be done after the goals for ambient air quality are set? Table 15.3 summarizes this situation. The ambient standards are set by federal action and it is the states which must implement plans to meet these standards. As we have described, in order to meet the federal ambient standards (at least in regions with major sources of pollution) states must set up emission standards. These standards define who can emit what and possibly even when the emissions will be allowed. In order to set emission standards and of particular interest to engineers, is the need for a diffusion model, including chemical reactions, for the air quality control region. Such a model must be able to take known sources of pollution and determine the ambient levels of each pollutant. Meteorology, topography, and chemical reaction kinetics must be incorporated into the model. The ultimate simulation will require information about the winds, the solar radiation input, the temperature as a function of altitude, and the different moving and stationary sources of pollution. This will be combined with the equations which govern motion through the atmosphere and the equations for chemical reactions and kinetics. The model will provide an output of pollution concentrations as a function of position and time. No one has yet formulated this complete a model but more and more complete computer programs are being developed.

TABLE 15.3 Setting local emission standards*

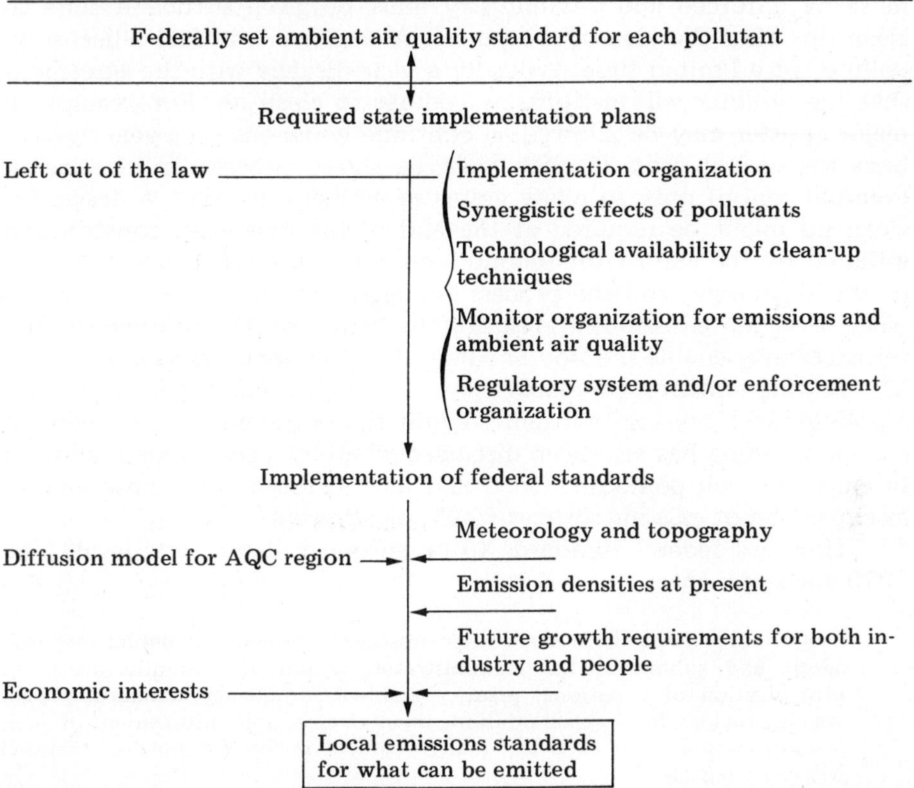

*In May, 1972, the District of Columbia Court interpreted the Clean Air Act of 1970 as requiring not only the achievement of national ambient air quality standards, but also the prevention of significant deterioration of air quality in those areas where the air is now cleaner than required by law. This decision greatly reduced options in setting local emission standards and was appealed. In June 1973 the Supreme Court upheld the decision.

One use of such a model would be to answer such questions as: if the standards allow X $\mu g/m^3$ of Y pollutant which comes from Z sources in an Air Quality Control Region and if our measurements show the concentration to be $2X$, which of the Z sources must improve its clean-up techniques or even shut down? Which is to be fined for noncompliance? We are, at present, not able to model the atmosphere this accurately when we have multiple sources of emission, particularly if there are important chemical reactions. In fact, it is difficult to predict accurately concentrations downwind from a single, well defined source.

After local emission standards are set, what happens? The limits must be enforced and yet industry must be given sufficient time to clean up. Thus the idea of a *variance*. A variance is in fact a license to pollute for a limited time, typically a year, usually with the agreement that the polluter will institute procedures to clean up. For example, a major emitter may be allowed to continue emissions on a year-by-year basis for several years if, each year, he shows progress in meeting an eventual goal of only so many grams of emission per day. A design for clean up might be required at the end of the first year, construction initiated by the end of the second, construction completed by the end of the third year, and the process debugged during the fourth year to reach a certain emission level during the fifth year. Other much simpler variances may enable the Boy Scouts to have an open fire at a picnic.

Setting emission standards may also involve emission fees whereby a polluter will pay for the right to emit the pollutant. The concept of emission zoning has also been discussed whereby a given area is allowed to emit so much pollutant. Once that level is reached, no new sources or expansion of existing sources would be allowed.

How are federal standards to be enforced at the state level? The 1970 law states that:

> Section 110. Each State shall, after reasonable notice and public hearings, adopt and submit to the Administrator, within nine months after the promulgation of a national primary ambient air quality standard, a plan which provides for implementation, maintenance, and enforcement of such primary standard in each air quality control region (or portion thereof) within such State.

Nothing else is indicated about state or local enforcement. Section 113 deals with federal enforcement and indicates that if the state does not enforce its own implementation plan, then the Administrator may step in and "issue an order to comply" or "bring a civil action." The civil action can include injunctive proceedings. A knowing violation can bring a "fine of not more than $25,000 per day of violation, or by imprisonment of not more than one year, or both." A second conviction doubles the possible penalty.

It seems likely, however, that with 50 different states and many times that number of AQCR, it will be some time before the federal government steps directly into the enforcement area. Meanwhile, it is up to the states to bring the emissions down to levels which will enable the ambient air quality standards to be met.

Table 15.4 indicates the sequence of events required by the passage of the 1970 amendments to the Air Quality Act. This sequence calls for achieving the primary standards by mid-1975. By January,

TABLE 15.4 Sequence of events following passage of the Amended Air Quality Act

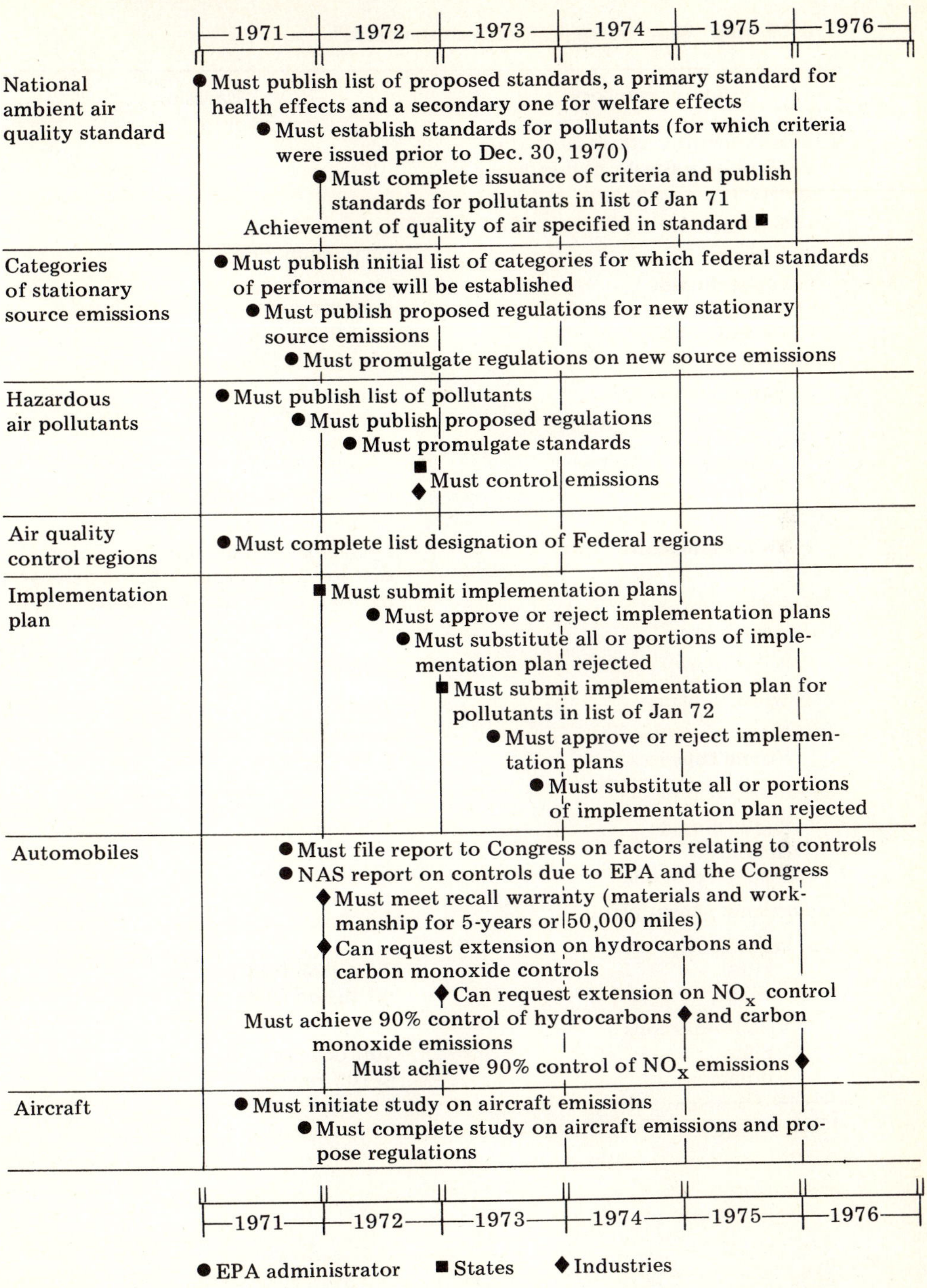

● EPA administrator ■ States ◆ Industries

TABLE 15.5 New source performance standards*

Source category	Emission
1. Fossil fuel-fired steam generators (250 million Btu/hr heat input or greater)	
Particulates	0.1 lb/10^6 Btu (maximum 2-hr average)
Sulfur dioxide	
oil-fired	0.80 lb/10^6 Btu
coal-fired	1.2 lb/10^6 Btu (maximum 2-hr average)
Nitrogen oxides	
gas-fired	0.20 lb/10^6 Btu
oil-fired	0.30 lb/10^6 Btu
coal-fired	0.70 lb/10^6 Btu (maximum 2-hr average expressed as NO_2)
Visible emissions	not to exceed 20% opacity, except that for 2 min in any 1-hr, emissions may be as great as 40% opacity
2. Incinerators	
Particulates	0.08 grains/scf
3. Nitric acid plants	
Nitrogen oxides	3 lb/ton acid
Visible emissions	$<$ 10% opacity
4. Sulfuric acid plants	
Sulfur dioxide	4 lb/ton acid
Acid mist	0.15 lb/ton acid
Visible emissions	$<$ 10% opacity
5. Portland cement plants	
Particulates	
kilns	0.3 lb/ton feed
clinker coolers	0.1 lb/ton feed
Visible emissions	
kilns	10% opacity
others	$<$ 10% opacity

*_Federal Register_, Dec. 23, 1971.

1972 the states were to have submitted implementation plans. While few states made this deadline, all have subsequently submitted plans. The Administrator must approve or reject each state's plans and this has been done.[7] Future efforts are indicated on the figure.

Some of the standards of performance called for in the Amended Air Quality Act have been set and they apply to new stationary sources of air pollution. Standards have been set for fossil fuel-fired power plants of more than 250 million Btu/hr heat input, municipal incinerators, nitric acid plants, sulfuric acid plants, and Portland cement plants. Table 15.5 summarizes the allowable emission from these sources.

One can only hope after all of this legal effort that the race to clean the air by 1975 will indeed be successful. It will surely be an interesting challenge for the engineers who have the task of cleaning up to meet the ambient standards.

[7] *Federal Register*, July 27, 1972.

REFERENCES

Esposito, J. C. (ed.): "Vanishing Air," Grossman, New York, 1970.

Heuss, J. M., G. J. Nebel, and J. M. Colucci: National Air Quality Standards, for Automotive Pollutants—A Critical Review, *J. Air Pollution Control Assoc.*, vol. 21, pp. 535-548, 1971.

Scorer, R. S.: "Air Pollution," Pergamon, London, 1968.

State Plans for Implementation of National Air Quality Standards, *Federal Register*, vol. 37, 145, pp. 15094-15113, pt. 3, July 27, 1972.

Stern, A.: National Emission Standards for Stationary Sources, *J. Air Pollution Control Assoc.*, vol. 20, pp. 524-527, 1970.

PROBLEMS

15.1 What requirements should be met in setting the ambient air quality standards?

15.2 Write a history of air pollution efforts in your state. If your state has made little or no effort, choose a state which has made an effort.

15.3 Consider the problem of reducing air pollution from used cars. Since cars last some 10 years, new car standards for 1975 will not become fully effective until after 1985. Thus there is much interest in requiring add-on devices to the present generation of cars on the road. Estimate the total costs in your state for such a program including the costs of the clean-up devices (assume 50, 100, and 250 dollars per car), the cost of inspections (how often to be required?), the costs of the inspectors, the required equipment, etc. You will be surprised at the total "bill."

15.4 Discuss which should come first, the technology to clean up or the setting of standards which will require a clean-up? Is it fair to require industry to clean up a process if the technology is not yet available? On the other hand, will industry ever find a way to clean up if it does not have to meet emission standards?

15.5 Assume you are the local pollution control officer. The ambient yearly average of particulate in your area is about 100 $\mu g/m^3$, whereas the standards allow only 75 $\mu g/m^3$. Measurements in the rural surroundings indicate background levels of 35 to 45 $\mu g/m^3$. (*a*) What do you do to reduce the levels? (*b*) If the city is a southwestern one with no burning of coal or oil, where might you look for a source of particulate? (*c*) What kind of controls might you initiate for dust?

15.6 See if your city has adopted any emission standards and, if so, what are they? Are they reasonable to meet the ambient standards? Are they reasonable for industry to meet? What ambient standards, if any, has your state adopted?

15.7 What Air Quality Control Region are you in? When and how was it established?

15.8 Standards of performance for new power plants were proposed and printed in the *Federal Register* on Aug. 17, 1971, vol. 36, 159, p. 15704. The adopted standards were printed on Dec. 23, 1971. Look up these standards and indicate what level of control they are likely to require.

15.9 Read Heuss et al. (see References) and write a summary of their arguments. Do you agree with them?

16
CONCLUSION

We have met the enemy and he is us—
Pogo © 1970 Walt Kelly

Figure 16.1 illustrates an overview of the air pollution problem we have examined in detail in the previous chapters. We see that the engineer of tomorrow will need to work with a variety of contributors from different disciplines if progress is to be made in improving the quality of air. Furthermore, the engineer will need to be more aware than in the past of both the technical and sociopolitical aspects of the issue.

We have discussed much of the information represented by the left-hand side of Fig. 16.1 but have not explored that on the right-hand side at all. Also we have not discussed instrumentation, source testing, or sampling techniques; in short, the measurement and analysis of air pollutants. Thus, while the student has had an opportunity to learn in fair detail parts of the problem, there are large areas not covered here. These include relatively narrow technical fields such as source testing for air pollutants, and broader areas such as the question of public tax policy and its effects on the purchase of control systems. The information represented on the right-hand side of the figure is equally important in improving the quality of air to that on the left-hand side.

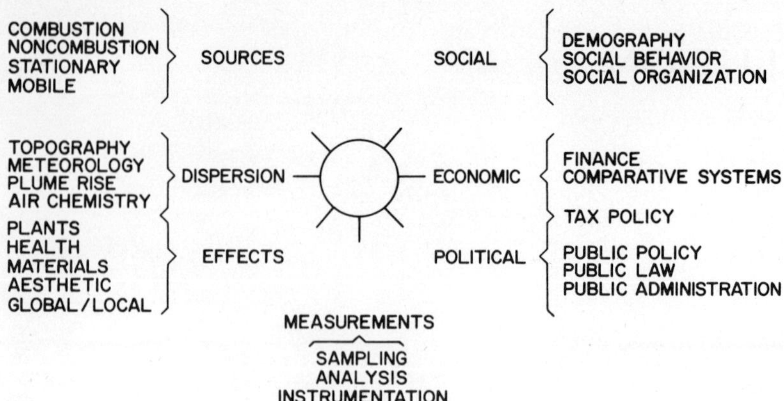

Fig. 16.1 An overview of air pollution. (*Adapted from G. S. Samuelsen, California Air Environment, vol. 2, Oct./Dec., 1971.*)

ENERGY

A problem which cuts across many fields of interest and plays a most important role in air pollution is the question of energy use. It must be clear that there are far-reaching implications to any energy-use policy. One effect of our use of energy is directly related to air pollution, namely, in the production of pollutants.

Most of our air pollution comes from combustion of fuels using the stored chemical energy of the fuel to produce heat and work. The very rapidly accelerating use of fuels, particularly for vehicles and power plants, has greatly increased air pollution. Table 16.1 indicates the use of energy by country both as kwhr/capita and as kilograms of equivalent coal per capita (Electric World, 1970). The table is for the calendar year 1967. As expected, the United States leads in energy use although not in electric power use per capita. Energy use is clearly related to standard of living and a comparison of the gross national product per capita to the energy use per capita shows a positive correlation (Cook, 1971). This indicates that the material standard of living is related to use of fuel (energy use), which in turn is related to the emission of pollutants. One of the very difficult dilemmas we in the United States must already consider is the need for increased power generation with the resulting degradation of the environment.

Table 16.2 indicates briefly the history of energy use. We see that only in the very recent past have fossil fuels been an important source of energy for industrial use. This rise in energy use and the projected continued increase have been discussed in Chap. 2. Hubbert has commented that:

TABLE 16.1 World energy use, 1967 data

Country	Electric use as kwhr / capita	Energy use as kilograms of equivalent coal used per capita	Rank
U.S.	6612	9833	1
Canada	8111	8061	2
Czechoslovakia	2813	5490	3
East Germany	3491	5338	4
United Kingdom	3796	5003	5
Sweden	6762	4805	6
Australia	3634	4790	7
Belgium and Luxembourg	2745	4669	8
Denmark	2225	4265	9
Norway	13566	3965	10
U.S.S.R.	2487	3957	11
New Zealand	4297	2591	22
Brazil	399	392	44
India	81	175	48
World average	1131	1647	

> It is difficult for people living now, who have become accustomed to the steady exponential growth in the consumption of energy fuels, to realize how transitory the fossil-fuel epoch will eventually prove to be when it is viewed over a longer span of human history. The situation can better be seen in the perspective of some 10,000 years, half before the present and half afterward. On such a scale the complete cycle of the exploitation of the world's fossil fuels will be seen to encompass perhaps 1,300 years, with the principal segment of the cycle (defined as the period during which all but the first 10 percent and the last 10 percent of the fuels are extracted and burned) covering only about 300 years. *(From "The Energy Resources of the Earth," K. Hubbert. Copyright © September 1971 by Scientific American, Inc. All rights reserved.)*

The figures on energy use are indeed staggering. For example, the United States will generate as much electricity in the next 10 years as it has generated since we began to use electricity. The world will use as much fossil fuel in the next 10 years as it has in the previous hundreds. Clearly this exponential growth cannot continue indefinitely. Some questions raised by the enormous rise in the use of energy must be considered. If we assume that sufficient energy is available throughout

TABLE 16.2 **Energy use**

Years ago	Generations ago	Use
	25,000 6,000	Control of fire
	2,000 400	Domestication of the dog, i.e., extension of man's available energy
8,000 5,000	400 250	Some use of animal power (oxen), water power (waterwheel), and wind power (sailing)
1,000 500	50 25	Waterwheel technology extended, windmill used in Europe, horse use developed with horse collar and horse shoe
500 220	25 9	Fire-water-steam use began, fossil fuel for the production of work started, Newcomen engine developed
220 70	9 3	Practical steam engine developed so fossil-fuel use became large, steam engines used to drive other machines
70 0	3 0	Exponential rise in energy use, in developments in science and in technology, and in world population. Electric and nuclear eras began

the world, whose energy resources are to be developed? Whose raw energy sources are to be used up first? Where will there be strip mines? Where will the power plants and their resulting *air pollution* be? Where is thermal energy to be rejected? Where do we put the vast number of new power plants which will have to be constructed? Where will the required capital come from to build these plants, and what other uses for this money will have to be sacrificed? In reality there is no guarantee that energy resources will be available in the future. New methods (such as solar energy, geothermal energy, and fusion energy) or at least improvements on older methods of energy transformation may well provide us with future requirements, but only if research in these directions begins now. Do we have the foresight for such long-term technological planning?

All of these questions are germane to the problem of air pollution. Alternate energy sources may reduce or eliminate air pollution. New policies for power plant locations may clean up some areas but dirty others. We can see that air pollution as an undesirable by-product of energy use is only a small part of the much larger problem of energy-resource management. The engineer who fails to have an overview of this problem may recommend technically acceptable solutions which will fail on nontechnical grounds.

SUMMARY

It is important to appreciate that the air pollution problems of one city or country may differ greatly from those of another city or country. This is true even if the sources of air pollution are similar. We have seen the important role of meteorology in determining whether pollution reaches significant levels. Thus to recommend satisfactory clean-up techniques one must take into account the local topography and meteorology. At the same time, the engineer has an obligation to design local remedies, always keeping the global problem in mind. The method of dumping SO_x into the air at higher altitudes by using taller stacks cannot be taken as a final solution when we recognize the acidifying effect of the sulfur gases.

In fact, the whole question of whether we can use science and technology as a band-aid for our present problems is one that should be carefully considered, especially by the trained technologist.

There should be little doubt by now that the problem of air pollution is indeed complex. Initiatives to clean up one situation may in fact worsen it or create another. The sources of air pollution are diverse and range from individually small mobile sources to larger stationary sources, both sending forth different pollutants. The effect of these pollutants has been difficult to document in terms of the costs of pollution versus the costs of reducing the emissions. How much air pollution can be tolerated? What are the costs of cleaning up? Should we achieve emission controls for SO_x by cleaning up the flue gas or is it better to desulfurize the fuel? Can the country afford to clean up at the cost of raising the prices of our products? How can we then compete with other countries which do not clean up? How can we assist the developing nations to avoid pollution as they industrialize? The answers to these questions must come from a cooperative effort between nations and between the technologist and the nontechnologist. Hopefully, the engineer will play an active role in this cooperation. In reading the following quotation from L. St. Pendred, consider carefully the passage as a whole, and particularly the date. Is our human inertia so great that 40 years after these hopes were expressed we are only beginning to make progress in the struggle to preserve the environment?

Beneficent Work for the Engineer[1]

> Mechanical Engineers have supplied all the peoples of the earth with marvelous inventions, but we must admit that many of those inventions are imperfect. We poison rivers with the effluents, charge the atmosphere with noxious fumes, jar the nerves with vibrations, deafen our ears with noises, and offend our eyes with ugly factories and a countryside destroyed by our activities. I can imagine no more beneficent work for the engineer and the scientist than the removal of all those things which still mar the great and wonderful work they have done. May all of you who are coming on to the field that we are leaving carry our work a step further toward a larger and greater efficiency than we have ever known, and remove the defects of applied science without diminishing the great benefits which it has conferred upon mankind.
>
> *Loughnan St. L. Pendred*

[1] From Presidential Address before the Institution of Mechanical Engineers, London, October 17, 1930.

REFERENCES

Cook, E.: The Flow of Energy in an Industrial Society, *Sci. Am.*, vol. 224, p. 134, September, 1971.

Felix, F.: Annual Growth Rate on Downward Trend, *Electric World*, July 6, 1970.

Hubbert, K.: The Energy Resources of the Earth, *Sci. Am.*, vol. 224, p. 60, September, 1971.

St. L. Pendred, L.: Beneficent Work for the Engineer, *Mech. Eng.*, vol. 52, December, 1930.

PROBLEMS

16.1 Make a list of the implications of a switchover from gasoline-powered cars to electric cars for people living in urban areas. Discuss one of these effects in detail.

16.2 The question of whether a technical solution can be found to our pollution problems is considered in L. White, The Historic Roots of our Ecologic Crisis, *Science*, vol. 155, pp. 1203–1207, March 10, 1967; G. Hardin, The Tragedy of the Commons, *Science*, vol. 162, pp. 1243–1248, Dec. 13, 1968. Read and summarize their arguments.

16.3 Show that the amount of resource used in any one doubling period is always greater than the total of that resource used in all of the previous doubling periods. What implications must this have for our present use of fossil fuels?

16.4 Plot up the United States vehicle registration figures as a function of time and the United States population figures as a function of time. Now consider the number of vehicles per person. What can you conclude?

16.5 Plot up the energy use per capita for the United States as a function of time. Is the air pollution problem solely because of our increase in population? (Note: Do the same problem, substituting water use, waste disposal, etc.)

APPENDIX A:
CONVERSION FROM ppm TO μg/m³

At a temperature of 25°C and a pressure of 760 mm Hg we have

$$\text{concentration in } \mu\text{g/m}^3 = \text{concentration in ppm} \times \frac{\hat{M}}{24{,}500} \times 10^6$$

where $\hat{M}$ is the molecular weight. For SO_2 we have for 1 ppm

$$[SO_2]\ \mu\text{g/m}^3 = (1) \times \frac{64}{24{,}500} \times 10^6 = 2{,}620\ \mu\text{g/m}^3$$

How is this conversion established?

If we are given a concentration in ppm, this is a concentration by volume and if we want μg/m³ we must convert to a concentration by weight in a given volume. We have

$$1 \text{ ppm} = \frac{1 \text{ m}^3}{10^6 \text{ m}^3} = \frac{1 \text{ cm}^3}{10^6 \text{ cm}^3}$$

The weight of species i in 1 m^3 is the density or the inverse of the specific volume:

$$\rho = \frac{1}{v} \quad \text{weight/m}^3$$

But it is also true that $v = V_{TP}/\hat{M}$ where V_{TP} is the volume occupied by 1 g-mole at a given temperature and pressure. At 25°C and 760 mm Hg this volume is 0.02450 m^3. Thus we have at this temperature and pressure

$$\frac{1 \text{ part}}{10^6 \text{ parts}} \times \frac{\hat{M}}{0.02450} \times 10^6 \frac{\mu g}{g} = \begin{array}{c}\text{concentration in}\\ \mu g/m^3\end{array}$$

Conversion factors for several air pollutants are given in Table A.1.

TABLE A.1 Conversion factors for air pollutants

Carbon monoxide (CO)	0°C, 760 mm 25°C, 760 mm	1 ppm = 1,250 $\mu g/m^3$ 1 ppm = 1,145 $\mu g/m^3$
Nitric oxide (NO)	25°C, 760 mm	1 ppm = 1,230 $\mu g/m^3$
Nitrogen dioxide (NO_2)	25°C, 760 mm	1 ppm = 1,880 $\mu g/m^3$
Ozone (O_3)	0°C, 760 mm 25°C, 760 mm	1 ppm = 2,141 $\mu g/m^3$ 1 ppm = 1,962 $\mu g/m^3$
PAN [$CH_3(CO)O_2NO_2$]	0°C, 760 mm 25°C, 760 mm	1 ppm = 5,398 $\mu g/m^3$ 1 ppm = 4,945 $\mu g/m^3$
Sulfur dioxide (SO_2)	0°C, 760 mm 25°C, 760 mm	1 ppm = 2,860 $\mu g/m^3$ 1 ppm = 2,620 $\mu g/m^3$

APPENDIX B:

PHYSICAL PROPERTIES

TABLE B.1 Nominal thermodynamic properties of gases at low pressures

Substance	$\hat{M}$, lbm/lb mole, g/gmole	c_P, Btu/lbm-°R	$\hat{c}_P$, Btu/lb mole-°R	c_v, Btu/lbm-°R	$\hat{c}_v$, Btu/lb mole-°R	R, ft-lbf/lbm-°R	R, Btu/lbm-°R	$k = c_P/c_v$
Argon, Ar	39.94	0.123	4.91	0.074	2.96	38.65	0.0496	1.67
Helium, He	4.003	1.24	5.00	0.75	3.00	386.3	0.4963	1.66
Hydrogen, H_2	2.016	3.42	6.89	2.43	4.90	767.0	0.9856	1.41
Nitrogen, N_2	28.01	0.248	6.95	0.177	4.96	55.13	0.0708	1.40
Oxygen, O_2	32.00	0.219	7.01	0.156	4.99	48.24	0.0620	1.40
Carbon monoxide, CO	28.01	0.249	6.97	0.178	4.98	55.13	0.0708	1.40
Air	28.97	0.240	6.95	0.171	4.95	53.34	0.0686	1.40
Water vapor, H_2O	18.016	0.446	8.07	0.336	6.03	85.58	0.1099	1.33
Methane, CH_4	16.04	0.532	8.53	0.403	6.46	96.4	0.1236	1.32
Carbon dioxide, CO_2	44.01	0.202	8.91	0.156	6.87	35.1	0.0451	1.30
Sulfur dioxide, SO_2	64.07	0.154	9.87	0.122	7.82	24.1	0.0310	1.26
Acetylene, C_2H_2	26.04	0.409	10.65	0.333	8.67	59.4	0.0761	1.23
Ethylene, C_2H_4	28.05	0.374	10.49	0.304	8.53	55.1	0.0707	1.23
Ethane, C_2H_6	30.07	0.422	12.69	0.357	10.73	51.3	0.0658	1.18
Propane, C_3H_8	44.09	0.404	17.81	0.360	15.87	35.0	0.0450	1.12
Isobutane, C_4H_{10}	58.12	0.420	24.41	0.387	22.49	26.6	0.0342	1.09

Source: W. C. Reynolds and H. C. Perkins, *Engineering Thermodynamics*, McGraw-Hill Book Co., New York, 1970.

TABLE B.2 Properties of air at moderate pressure

T, °F	μ, lbm/hr-ft	k, Btu/hr-ft-°F	c_P, Btu/lbm-°F	Pr
−100	0.0319	0.0104	0.239	0.739
0	0.0394	0.0131	0.240	0.718
100	0.0459	0.0157	0.240	0.706
200	0.0519	0.0181	0.241	0.693
300	0.0574	0.0203	0.243	0.686
400	0.0626	0.0225	0.245	0.681
500	0.0675	0.0246	0.248	0.680
600	0.0721	0.0265	0.250	0.680
700	0.0765	0.0284	0.254	0.682
800	0.0806	0.0303	0.257	0.684
900	0.0846	0.0320	0.260	0.687
1000	0.0884	0.0337	0.263	0.690

Source: Natl. Bur. Std. Circ. 564, 1955.

TABLE B.3 The elements

Name	Symbol	At. No.	International atomic weight 1966
Actinium	Ac	89	—
Aluminum	Al	13	26.9815
Americium	Am	95	—
Antimony, stibium	Sb	51	121.75
Argon	Ar	18	39.948
Arsenic	As	33	74.9216
Astatine	At	85	—
Barium	Ba	56	137.34
Berkelium	Bk	97	—
Beryllium	Be	4	9.0122
Bismuth	Bi	83	208.980
Boron	B	5	10.811
Bromine	Br	35	79.904
Cadmium	Cd	48	112.40
Calcium	Ca	20	40.08
Californium	Cf	98	—
Carbon	C	6	12.01115
Cerium	Ce	58	140.12
Cesium	Cs	55	132.905
Chlorine	Cl	17	35.453
Chromium	Cr	24	51.996
Cobalt	Co	27	58.9332
Columbium, see *Niobium*			
Copper	Cu	29	63.546
Curium	Cm	96	—
Dysprosium	Dy	66	162.50
Einsteinium	Es	99	—
Erbium	Er	68	167.26
Europium	Eu	63	151.96
Fermium	Fm	100	—
Fluorine	F	9	18.9984
Francium	Fr	87	—
Gadolinium	Gd	64	157.25
Gallium	Ga	31	69.72
Germanium	Ge	32	72.59
Gold, aurum	Au	79	196.967
Hafnium	Hf	72	178.49
Helium	He	2	4.0026
Holmium	Ho	67	164.930
Hydrogen	H	1	1.00797
Indium	In	49	114.82
Iodine	I	53	126.9044
Iridium	Ir	77	192.2
Iron, ferrum	Fe	26	55.847
Krypton	Kr	36	83.80
Lanthanum	La	57	138.91
Lawrencium	Lr	103	(257)
Lead, plumbum	Pb	82	207.19
Lithium	Li	3	6.939
Lutetium	Lu	71	174.97
Magnesium	Mg	12	24.312
Manganese	Mn	25	54.9380
Mendelevium	Md	101	—
Mercury, hydrargyrum	Hg	80	200.59
Molybdenum	Mo	42	95.94

TABLE B.3 The elements (*continued*)

Name	Symbol	At. No.	International atomic weight 1966	Name	Symbol	At. No.	International atomic weight 1966
Neodymium	Nd	60	144.24	Samarium	Sm	62	150.35
Neon	Ne	10	20.183	Scandium	Sc	21	44.956
Neptunium	Np	93	—	Selenium	Se	34	78.96
Nickel	Ni	28	58.71	Silicon	Si	14	28.086
Niobium (columbium)	Nb	41	92.906	Silver, argentum	Ag	47	107.868
Nitrogen	N	7	14.0067	Sodium, natrium	Na	11	22.9898
Nobelium	No	102	—	Strontium	Sr	38	87.62
Osmium	Os	76	190.2	Sulfur	S	16	32.064
Oxygen	O	8	15.9994	Tantalum	Ta	73	180.948
Palladium	Pd	46	106.4	Technetium	Tc	43	—
Phosphorus	P	15	30.9738	Tellurium	Te	52	127.60
Platinum	Pt	78	195.09	Terbium	Tb	65	158.924
Plutonium	Pu	94	—	Thallium	Tl	81	204.37
Polonium	Po	84	—	Thorium	Th	90	232.038
Potassium, kalium	K	19	39.102	Thulium	Tm	69	168.934
Praseodymium	Pr	59	140.907	Tin, stannum	Sn	50	118.69
Promethium	Pm	61	—	Titanium	Ti	22	47.90
Protactinium	Pa	91	—	Tungsten (wolfram)	W	74	183.85
Radium	Ra	88	—	Uranium	U	92	238.03
Radon	Rn	86	—	Vanadium	V	23	50.942
Rhenium	Re	75	186.2	Xenon	Xe	54	131.30
Rhodium	Rh	45	102.905	Ytterbium	Yb	70	173.04
Rubidium	Rb	37	85.47	Yttrium	Y	39	88.905
Ruthenium	Ru	44	101.07	Zinc	Zn	30	65.37
				Zirconium	Zr	40	91.22

APPENDIX C:

CONVERSION FACTORS AND PHYSICAL CONSTANTS

TABLE C.1 Physical constants

Avogadro's number	$N_0 = 6.0221 \times 10^{23}$ /g mole
Boltzmann's constant	$\mathbf{k} = 1.380 \times 10^{-23}$ joule/°K
Gas constant	$\mathcal{R} = 1545.33$ ft/lbf/lb mole-°R = 8.314 joule/g mole-°K = 1.987 Btu/lb mole-°R = 1.987 cal/g mole-°K
Planck's constant	$\mathbf{h} = 6.626 \times 10^{-34}$ joule-sec
Electronic charge	$\mathbf{e} = -1.6021 \times 10^{-19}$ coul
Speed of light	$\mathbf{c} = 2.998 \times 10^{8}$ m/sec
Newton constant	$g_c = 32.174$ ft-lbm/lb f-sec^2
Gravitational constant	$k_G = 6.67 \times 10^{-11}$ m^3/kg-sec^2
Stefan-Boltzmann constant	$\sigma = 5.669 \times 10^{-8}$ watt/m^2 $-$ ^{0}K^4

TABLE C.2 Selected dimensional equivalents

Length	1 m = 3.280 ft = 39.37 in. 1 cm $\equiv 10^{-2}$ m = 0.394 in. = 0.0328 ft 1 mm $\equiv 10^{-3}$ m 1 micron (μ) $\equiv 10^{-6}$ m 1 angstrom (Å) $\equiv 10^{-10}$ m
Time	1 hr $\equiv$ 3600 sec = 60 min 1 millisec $\equiv 10^{-3}$ sec 1 microsec (μsec) $\equiv 10^{-6}$ sec 1 nanosec (nsec) = 10^{-9} sec
Mass	1 kg $\equiv$ 1,000 g = 2.2046 lbm = 6.8521 $\times 10^{-2}$ slugs 1 slug $\equiv$ 1 lbf-sec^2/ft = 32.174 lb m
Force	1 newton $\equiv$ 1 kg-m/sec^2 1 dyne $\equiv$ 1 g-cm/sec^2 1 lbf = 4.448 $\times 10^5$ dynes = 4.448 newtons
Energy	1 joule $\equiv$ 1 kg-m^2/sec^2 1 Btu $\equiv$ 778.16 ft-lbf = 1.055 $\times 10^{10}$ ergs = 252 cal 1 cal $\equiv$ 4.186 joules 1 kcal $\equiv$ 4,186 joules = 1,000 cal 1 erg $\equiv$ 1 g-cm^2/sec^2 1 ev $\equiv$ 1.602 $\times 10^{-19}$ joules
Power	1 watt $\equiv$ 1 kg-m^2/sec^3 = 1 joule/sec 1 hp $\equiv$ 550 ft-lbf/sec 1 hp = 2545 Btu/hr = 746 watts 1 kw $\equiv$ 1000 watts = 3413 Btu/hr
Pressure	1 atm $\equiv$ 14.696 lbf/in.^{2} = 760 torr 1 mmHg = 0.01934 lbf/in.^{2} $\equiv$ 1 torr 1 dyne/cm^2 = 145.04 $\times 10^{-7}$ lbf/in.^{2} 1 bar = 14.504 lbf/in.^{2} $\equiv 10^6$ dynes/cm^2 1 micron (μ) $\equiv 10^{-6}$ mHg = 10^{-3} mmHg 1 atm = 101,325 newton/m^2
Volume	1 gal $\equiv$ 0.13368 ft^3 1 liter $\equiv$ 1,000.028 cm^3 1 gal = 3.785 liters
Temperature	1 K° = 1 C° = 1.8 F° = 1.8 R° 0°C corresponds to 32°F, 273.16°K, and 491.69°R

TABLE C.3 Conversion factors: Mass

Given units \ Desired units	Gram	Microgram	Kilogram	Metric ton	Short ton	Long ton	Grain	Ounce (avdp)	Lb (avdp)
Gram	1.0000 E 00	1.0000 E 06	1.0000 E–03	1.0000 E–06	1.1023 E–06	9.8421 E–07	1.5432 E 01	3.5274 E–02	2.2046 E–03
Microgram	1.0000 E–06	1.0000 E 00	1.0000 E–09	1.0000 E–12	1.1023 E–12	9.8421 E–13	1.5432 E–05	3.5274 E–08	2.2046 E–09
Kilogram	1.0000 E 03	1.0000 E 09	1.0000 E 00	1.0000 E–03	1.1023 E–03	9.8421 E–04	1.5432 E 04	3.5274 E 01	2.2046 E 00
Metric ton	1.0000 E 06	1.0000 E 12	1.0000 E 03	1.0000 E 00	1.1023 E 00	9.8421 E–01	1.5432 E 07	3.5274 E 04	2.2046 E 03
Short ton	9.0718 E 05	9.0718 E 11	9.0718 E 02	9.0718 E–01	1.0000 E 00	8.9286 E–01	1.4000 E 07	3.2000 E 04	2.0000 E 03
Long ton	1.0160 E 06	1.0160 E 12	1.0160 E 03	1.0160 E 00	1.1200 E 00	1.0000 E 00	1.5680 E 07	3.5840 E 04	2.2400 E 03
Grain	6.4799 E–02	6.4799 E 04	6.4799 E–05	6.4799 E–08	7.1428 E–08	6.3775 E–08	1.0000 E 00	2.2857 E–03	1.4286 E–04
Ounce (avdp)	2.8349 E 01	2.8349 E 07	2.8349 E–02	2.8349 E–05	3.1250 E–05	2.7902 E–05	4.3750 E 02	1.0000 E 00	6.2500 E–02
Lb (avdp)	4.5359 E 02	4.5359 E 08	4.5359 E–01	4.5359 E–04	5.0000 E–04	4.4643 E–04	7.0000 E 03	1.6000 E 01	1.0000 E 00

Note: To convert a value from a given unit to a desired unit, multiply the given value by the factor opposite the given units and beneath the desired unit. Note that E–XX means 10 to the –XX power.
Source: Workbook of Atmospheric Dispersion Coefficients, AP-26, 1970.

TABLE C.4 Conversion factors: Concentration, density

Given units \ Desired units	Gram per cu meter	Mg per cu meter	Microgram per cu m	Microgram per liter	Grain per cu ft	Ounce per cu ft	Lb per cu ft	Gram per cu ft	Lb per cu meter
Gram per cu meter	1.0000 E 00	1.0000 E 03	1.0000 E 06	1.0000 E 03	4.3700 E−01	9.9885 E−04	6.2428 E−05	2.8317 E−02	2.2046 E−03
Mg per cu meter	1.0000 E−03	1.0000 E 00	1.0000 E 03	1.0000 E 00	4.3700 E−04	9.9885 E−07	6.2428 E−08	2.8317 E−05	2.2046 E−06
Microgram per cu m	1.0000 E−06	1.0000 E−03	1.0000 E 00	1.0000 E−03	4.3700 E−07	9.9885 E−10	6.2428 E−11	2.8317 E−08	2.2046 E−09
Microgram per liter	9.9997 E−04	9.9997 E−01	9.9997 E 02	1.0000 E 00	4.3699 E−04	9.9883 E−07	6.2427 E−08	2.8316 E−05	2.2046 E−06
Grain per cu ft	2.2883 E 00	2.2883 E 03	2.2883 E 06	2.2884 E 03	1.0000 E 00	2.2857 E−03	1.4286 E−04	6.4799 E−02	5.0449 E−03
Ounce per cu ft	1.0011 E 03	1.0011 E 06	1.0011 E 09	1.0012 E 06	4.3750 E 02	1.0000 E 00	6.2500 E−02	2.8349 E 01	2.2072 E 00
Lb per cu ft	1.6018 E 04	1.6018 E 07	1.6018 E 10	1.6019 E 07	7.0000 E 03	1.6000 E 01	1.0000 E 00	4.5359 E 02	3.5314 E 01
Gram per cu ft	3.5314 E 01	3.5314 E 04	3.5314 E 07	3.5315 E 04	1.5432 E 01	3.5274 E−02	2.2046 E−03	1.0000 E 00	7.7855 E−02
Lb per cu meter	4.5359 E 02	4.5359 E 05	4.5359 E 08	4.5360 E 05	1.9822 E 02	4.5307 E−01	2.8317 E−02	1.2844 E 01	1.0000 E 00

Note: To convert a value from a given unit to a desired unit, multiply the given value by the factor opposite the given units and beneath the desired unit. Note that E−XX means 10 to the −XX power.
Source: Workbook of Atmospheric Dispersion Coefficients, AP-26, 1970.

TABLE C.5 Conversion factors: Power

Given units \ Desired units	Watt (int)	Kilowatt (int)	Megawatt (int)	Cal (int) per sec	Btu per min	Btu per hr	Joules abs per sec	Watt (abs)	Elect. horsepower
Watt (int)	1.0000 E 00	1.0000 E−03	1.0000 E−06	2.3880 E−01	5.6857 E−02	3.4114 E 00	1.0002 E 00	1.0002 E 00	1.3407 E−03
Kilowatt (int)	1.0000 E 03	1.0000 E 00	1.0000 E−03	2.3880 E 02	5.6857 E 01	3.4114 E 03	1.0002 E 03	1.0002 E 03	1.3407 E 00
Megawatt (int)	1.0000 E 06	1.0000 E 03	1.0000 E 00	2.3880 E 05	5.6857 E 04	3.4114 E 06	1.0002 E 06	1.0002 E 06	1.3407 E 03
Cal (int) per sec	4.1876 E 00	4.1876 E−03	4.1876 E−06	1.0000 E 00	2.3810 E−01	1.4286 E 01	4.1884 E 00	4.1884 E 00	5.6145 E−03
Btu per min	1.7588 E 01	1.7588 E−02	1.7588 E−05	4.2000 E 00	1.0000 E 00	6.0000 E 01	1.7591 E 01	1.7591 E 01	2.3581 E−02
Btu per hr	2.9313 E−01	2.9313 E−04	2.9313 E−07	7.0000 E−02	1.6667 E−02	1.0000 E 00	2.9319 E−01	2.9319 E−01	3.9301 E−04
Joules abs per sec	9.9981 E−01	9.9981 E−04	9.9981 E−07	2.3875 E−01	5.6846 E−02	3.4108 E 00	1.0000 E 00	1.0000 E 00	1.3405 E−03
Watt (abs)	9.9981 E−01	9.9981 E−04	9.9981 E−07	2.3875 E−01	5.6846 E−02	3.4108 E 00	1.0000 E 00	1.0000 E 00	1.3405 E−03
Elect. horsepower	7.4586 E 02	7.4586 E−01	7.4586 E−04	1.7811 E 02	4.2407 E 01	2.5444 E 03	7.4600 E 02	7.4600 E 02	1.0000 E 00

Note: To convert a value from a given unit to a desired unit, multiply the given value by the factor opposite the given units and beneath the desired unit. Note that E−XX means 10 to the −XX power.

Source: Workbook of Atmospheric Dispersion Coefficients, AP-26, 1970.

TABLE C.6 Conversion factors: Emission rates

Given units \ Desired units	g/sec	g/min	kg/hr	kg/day	lb/min	lb/hr	lb/day	tons/hr	tons/day	tons/yr
g/sec	1.0	60.0	3.6	8.640×10	1.3228×10^{-1}	7.9367	1.9048×10^{2}	3.9683×10^{-3}	9.5240×10^{-2}	3.4763×10
g/min	1.6667×10^{-2}	1.0	6.0×10^{-2}	1.4400	2.2046×10^{-3}	1.3228×10^{-1}	3.1747	6.6139×10^{-5}	1.5873×10^{-3}	5.7938×10^{-1}
kg/hr	2.7778×10^{-1}	16.667	1.0	2.4000×10	3.6744×10^{-2}	2.2046	5.2911×10	1.1023×10^{-3}	2.6456×10^{-2}	9.6563
kg/day	1.1574×10^{-2}	6.9444×10^{-1}	4.1667×10^{-2}	1.0	1.5310×10^{-3}	9.1860×10^{-2}	2.2046	4.5930×10^{-5}	1.1023×10^{-3}	4.0235×10^{-1}
lb/min	7.5598	4.5359×10^{2}	2.7215×10	6.5317×10^{2}	1.0	60.0	1.44×10^{3}	3.000×10^{-2}	7.2000×10^{-1}	2.6280×10^{2}
lb/hr	1.2600×10^{-1}	7.5598	4.5359×10^{-1}	1.0886×10	1.6667×10^{-2}	1.0	24.0	5.0000×10^{-4}	1.2000×10^{-2}	4.3800
lb/day	5.2499×10^{-3}	3.1499×10^{-1}	1.8900×10^{-2}	4.5359×10^{-1}	6.9444×10^{-4}	4.1667×10^{-2}	1.0	2.0833×10^{-5}	5.0000×10^{-4}	1.8250×10^{-1}
tons/hr	2.5199×10^{2}	1.5120×10^{4}	9.0718×10^{2}	2.1772×10^{4}	3.3333×10	2.0×10^{3}	4.8000×10^{4}	1.0	24.0	8.7600×10^{3}
tons/day	1.0500×10	6.2999×10^{2}	3.7799×10	9.0718×10^{2}	1.3889	8.3333×10	2.0×10^{3}	4.1667×10^{-2}	1.0	365.0
tons/yr	2.8766×10^{-2}	1.7260	1.0356×10^{-1}	2.4854	3.8052×10^{-3}	2.2831×10^{-1}	5.4795	1.1416×10^{-4}	2.7397×10^{-3}	1.0

Note: To convert a value from a given unit to a desired unit, multiply the given value by the factor opposite the given units and beneath the desired units.
Source: Handbook of Air Pollution, AP-44.

TABLE C.7 Conversion factors: Pressure

Given units \ Desired units	$\frac{g_m}{cm\text{-}sec^2}$	$\frac{dynes}{cm^2}$	$\frac{lbm}{ft\text{-}sec^2}$	$\frac{poundals}{ft^2}$	$\frac{g_f}{cm^2}$	$\frac{lbf}{ft^2}$	$\frac{lbf}{in.^2}$	"Atmospheres"
$\frac{g_m}{cm\text{-}sec^2}$	1	1	6.7197×10^{-2}	6.7197×10^{-2}	1.0197×10^{-3}	2.0885×10^{-3}	1.4504×10^{-5}	9.8692×10^{-7}
$\frac{dynes}{cm^2}$	1	1	6.7197×10^{-2}	6.7197×10^{-2}	1.0197×10^{-3}	2.0885×10^{-3}	1.4504×10^{-5}	9.8692×10^{-7}
$\frac{lbm}{ft\text{-}sec^2}$	14.882	14.882	1	1	1.5175×10^{-2}	3.1081×10^{-2}	2.1584×10^{-4}	1.4687×10^{-5}
$\frac{poundals}{ft^2}$	14.882	14.882	1	1	1.5175×10^{-2}	3.1081×10^{-2}	2.1584×10^{-4}	1.4687×10^{-5}
$\frac{g_f}{cm^2}$	980.665	980.665	65.898	65.898	1	2.0482	1.4223×10^{-2}	9.6784×10^{-4}
$\frac{lbf}{ft^2}$	478.80	478.80	32.174	32.174	4.8824×10^{-1}	1	6.9444×10^{-3}	4.7254×10^{-4}
$\frac{lbf}{in.^2}$	6.8948×10^{4}	6.8948×10^{4}	4.6331×10^{3}	4.6331×10^{3}	70.307	144.00	1	6.8046×10^{-2}
"Atmospheres"	1.0133×10^{6}	1.0133×10^{6}	5.8087×10^{4}	6.8087×10^{4}	1.0332×10^{3}	2.1162×10^{3}	14.696	1

Note: To convert a value from a given unit to a desired unit, multiply the given value by the factor opposite the given units and beneath the desired units.

Source: Handbook of Air Pollution, AP-44.

TABLE C.8 Conversion factors: Pressure head

Given units \ Desired units	Bar	Millibar	In Hg	Cm Hg	Mm Hg	In. H_2O	"Atmospheres"
Bar	1	1000	29.530	75.006	750.06	401.47	9.8692×10^{-1}
Millibar	1×10^{-3}	1	2.9530×10^{-2}	7.5006×10^{-2}	7.5006×10^{-1}	4.0147×10^{-1}	9.8692×10^{-4}
In Hg	3.3864×10^{-2}	33.864	1	2.5400	25.400	13.546	3.3421×10^{-2}
Cm Hg	1.3332×10^{-2}	13.332	3.9370×10^{-1}	1	10	5.3525	1.3158×10^{-2}
Mm Hg	1.3332×10^{-3}	1.3332	3.9370×10^{-2}	1×10^{-1}	1	5.3525×10^{-1}	1.3158×10^{-3}
In H_2O	2.4909×10^{-3}	2.4909	7.3823×10^{-2}	1.8683×10^{-1}	1.8683	1	2.4583×10^{-3}
"Atmospheres"	1.0133	1.0133×10^{3}	29.291	76	760	406.79	1

Note: To convert a value from a given unit to a desired unit, multiply the given value by the factor opposite the given units and beneath the desired units.
Source: Handbook of Air Pollution, AP-44.

TABLE C.9 Conversion factors: Volume

Given units \ Desired units	Cubic yard	Cubic foot	Cubic inch	Cubic meter	Cubic decimeter	Cubic centimeter	Liter
Cubic yard	1	27	4.6656×10^{4}	0.764559	764.559	7.64559×10^{5}	764.54
Cubic foot	3.7037×10^{-2}	1	1728	2.8317×10^{-2}	28.317	2.8317×10^{4}	28.316
Cubic inch	2.143347×10^{-5}	5.78704×10^{-4}	1	1.63872×10^{-5}	1.63872×10^{-2}	16.3872	1.63868×10^{-2}
Cubic meter	1.30794	35.314445	6.1023×10^{4}	1	1000	1×10^{6}	999.973
Cubic decimeter	1.3079×10^{-3}	3.5314×10^{-2}	61.023	0.001	1	1000	.99997
Cubic centimeter	1.3079×10^{-6}	3.5314×10^{-5}	6.1023×10^{-2}	1×10^{-6}	1×10^{-3}	1	9.99973×10^{-4}
Liter	1.3080×10^{-3}	3.5316×10^{-2}	61.025	1.000027×10^{-3}	1.000027	1000.027	1

Note: To convert a value from a given unit to a desired unit, multiply the given value by the factor opposite the given units and beneath the desired units.
Source: Handbook of Air Pollution, AP-44.

TABLE C.10 Conversion factors: Velocity

Given units \ Desired units	m/sec	ft/sec	ft/min	km/hr	mi/hr	knots	mi/day
m/sec	1.0	3.2808	1.9685×10^2	3.6	2.2369	1.9425	5.3687×10
ft/sec	3.0480×10^{-1}	1.0	60	1.0973	6.8182×10^{-1}	5.9209×10^{-1}	1.6364×10
ft/min	5.0080×10^{-3}	1.6667×10^{-2}	1.0	1.8288×10^{-2}	1.1364×10^{-2}	9.8681×10^{-3}	2.7273×10^{-1}
km/hr	2.7778×10^{-1}	9.1134×10^{-1}	5.4681×10	1.0	6.2137×10^{-1}	5.3959×10^{-1}	1.4913×10
mi/hr	4.4707×10^{-1}	1.4667	88.0	1.6093	1.0	8.6839×10^{-1}	24
knots	5.1479×10^{-1}	1.6890	1.0134×10^2	1.8533	1.1516	1.0	2.7637×10
mi/day	1.8627×10^{-2}	6.1111×10^{-2}	3.6667	6.7056×10^{-2}	4.1667×10^{-2}	3.6183×10^{-2}	1.0

Note: To convert a value from a given unit to a desired unit, multiply the given value by the factor opposite the given units and beneath the desired units.
Source: Handbook of Air Pollution, AP-44.

INDEX